**Netherlands Organization for
Applied Scientific Research**

Thermal Storage of Solar Energy

PROCEEDINGS OF AN INTERNATIONAL TNO-SYMPOSIUM
HELD IN AMSTERDAM, THE NETHERLANDS, 5-6 NOVEMBER 1980
Co-sponsored by:
Commission of the European Communities
Dutch Section of the International Solar Energy Society
The Royal Institution of Engineers in The Netherlands

C. DEN OUDEN
(editor)

SPRINGER-SCIENCE+BUSINESS MEDIA, B.V.

Distributors

for the United States and Canada

Kluwer Boston, Inc.
190 Old Derby Street
Hingham, MA 02043
USA

for all other countries

Kluwer Academic Publishers Group
Distribution Center
P.O.Box 322
3300 AH Dordrecht
The Netherlands

This volume is listed in the Library of Congress Cataloging in Publication Data

ISBN 978-94-009-8304-5   ISBN 978-94-009-8302-1(eBook)
DOI 10.1007/ 978-94-009-8302-1

LEGAL NOTICE

Neither TNO nor any of the co-sponsors of this Symposium or any person acting on behalf of them
is responsible for the use which might be made of the following information.

THERMAL STORAGE OF SOLAR ENERGY

INTERNATIONAL TNO-SYMPOSIUM

THERMAL STORAGE OF SOLAR ENERGY

SCIENTIFIC COMMITTEE

DR. E. ARANOVITCH (E.C.)

PROF. J. BOUGARD (BELGIUM)

PROF. B.J. BRINKWORTH (UK)

PROF. F. FITTIPALDI (ITALY)

PROF. J.C. FRANCKEN (THE NETHERLANDS)

PROF. E. HAHNE (FRG)

PROF. C.J. HOOGENDOORN (THE NETHERLANDS)

DR. A. STRUB (E.C.)

DR. R. TORRENTI (FRANCE)

ORGANIZING COMMITTEE

PROF. C.W.J. VAN KOPPEN, CHAIRMAN

IR. C. DEN OUDEN, TECHNICAL PROGRAMME

IR. E. VAN GALEN, SECRETARY

MR T.C. STEEMERS, ADVISOR

MS W.J.M. VAN GIERSBERGEN, ADMINISTRATIVE MATTERS

MRS E.L.S. JANSSEN, ADMINISTRATIVE MATTERS

PREFACE

The International TNO-Symposium 'Thermal Storage of Solar Energy' was held on November 5 and 6 in Amsterdam, The Netherlands. The symposium was organized by the Netherlands Organization for Applied Scientific Research (TNO) and co-sponsored by:
- the Commission of the European Communities;
- the Dutch Section of the International Solar Energy Society;
- the Royal Institution of Engineers in the Netherlands.

The objectives of the TNO-Symposium were:
- to provide and overall picture and a state of the art of the various possibilities for thermal storage of solar energy,
- to present and discuss the latest progress with respect to research on thermal storage devices for solar applications.

In order to guarantee excellent quality the Scientific Committee has selected and invited more than 20 speakers to present their work on four different subjects: sensible heat storage; heat storage in phase change materials; long term or seasonal heat storage and chemical storage of thermal energy.

More than 200 experts from 16 different countries participated in this conference. There were four Technical Sessions: I Sensible Heat Storage, II Latent heat Storage, III Long Term Storage, IV Chemical Storage of Thermal Energy, and a closing session in which the chairmen of the four technical sessions summarized the presented work and the status of the work.

The full text of all papers presented has been included in this book and each chapter is preceeded by an overview and summary prepared by the session chairmen. Together with the indtroductory papers of some of the lecturers the book reflects the objectives of the symposium.

The Organizing Committee seizes this opportunity to thank the co-sponsors, the chairmen, the speakers and all participants whose active and constructive contribution made the success of the Symposium possible.

C. den Ouden
Technical Programme

CONTENTS

OPENING ADDRESSES
AND ADDRESSES OF WELCOME

OPENING SPEECH

Prof. Ir. W.A. de Jong
Chairman Netherlands Organization
for Applied Scientific Research

Mr Chairman, Ladies and Gentlemen,

It is a pleasure to welcome so many of you to this International
Symposium on the Thermal Storage of Solar Energy. A special word of welcome
is extended to the representatives of the three co-sponsors viz.,
- the Commission of the European Communities,
- the Dutch Section of the International Solar Energy Society, and
- the Royal Institution of Engineers in The Netherlands,
as well as to the speakers from various countries, especially those who
have come from very far to join us here in Amsterdam. I think that it is
only fitting that the organizers of a symposium on Solar Energy Storage,
a field that is in a very rapid and perhaps turbulent development, have
chosen this ancient city as a meeting place, a city where developments
are equally turbulent - if not hectic.

I think I should place the role of Solar Energy Storage into the
wider perspective of our energy problems. Since the oil embargo of the
first half of the past decade we have become aware more and more how much
we depend on adequate supplies of energy. If we, in the Netherlands, do
not find means to reduce our dependence on imported fossil fuels, we will
have to spend around 10% of our national revenue on imported fuels in
1990. An obvious and necessary remedy is to conserve energy as much as
possible. To stimulate this, the International Energy Agency has chosen
the month of October 1980 to start a campaign on energy conservation.
In this connection, TNO participated in organizing a national meeting on
energy conservation last week; judging from the attendance, around 1000
representatives of every branch of industry and of many other institutions,
the need for conservation is obvious to many.

However, conservation must be regarded merely as a means to give us
more time for new developments, particularly in the field of diversification
of our energy sources.

It is in this connection that solar energy is believed to be a very
promising substitute, which is not only renewable but also non-polluting.
Of the various conversion methods of solar energy, viz. by biosynthesis,
photovoltaic conversion and direct conversion into heat, expecially the
latter has developed rapidly during the last 6 years or so. Since almost
all solar heating systems are equiped with thermal storage systems, the
same applies to them. The need for developing efficient storage systems
for solar installations is evident: not only are the demand for domestic
heat and the supply of solar energy poorly matched over short periods,
but there is even a greater mismatch over the year - especially under the
climatic conditions of North-Western Europe. This is why I welcome the
decision to devote the entire symposium to thermal storage of solar energy.

If solar heating is to become economically viable, it will be
necessary to improve the present thermal storage systems, i.e. the interface between the collector unit and the heating unit.
Improvement of the energy density of storage systems and, more generally
speaking, cost reduction will probably lead to faster penetration of
solar installations for various applications.
This is why TNO, the Netherlands Organization for Applied Scientific
Research, is actively participating in the Dutch National Solar Energy
Programme. However, our activities in energy research cover a much
wider range: we are also involved in
- the National Energy Conservation R & D Programme,
- a similar Programme on Wind energy
- various Energy Research Programmes of the Commission of the European
  Communities, and
- similar Programmes of the International Energy Agency.
Thus, the fraction of our employees involved in research on energy and
related subjects has increased steadily over the past 6 or 7 years.
In all, about 200 researchers are actively working on R & D projects on
energy, whereas about 600, i.e. 1 out of every 8 persons in TNO, are
either full-time or part-time involved in projects related to energy.
I am convinced that the results of their efforts will contribute to a more
rational use of energy in domestic and office applications, industry and
the transportation sector.

This is the first time that TNO organizes an international symposium dealing with thermal storage of solar energy. I sincerely hope that the formal meeting as well as the informal discussions will provide you with new ideas and information for your own research, and that the initiative to hold this symposium will result in closer cooperation between the many organizations working on the subject in the various countries represented here.

Ladies and Gentlemen,

I wish it were possible for me to attend some of the sessions of this symposium, which look very attractive to me, but unfortunately this is not possible.

Therefore, I can only express the hope that you will have a stimulating and profitable meeting which will help you in your efforts to utilize solar energy for the benefit of us all.

Wishing you an enjoyable symposium and a pleasant stay in Amsterdam, I open this symposium on the Thermal Storage of Solar Energy.

OPENING ADDRESS

Prof. dr. ir. J.C. Francken
Chairman of the Dutch Section of the
International Solar Energy Society

Mr Chairman, Ladies and Gentlemen,

On behalf of the Dutch Section of the International Solar Energy
Society, first of all I should like to thank the Board of TNO and the
Organizing Committee for inviting our Society to co-sponsor this
Symposium. As a matter of fact, the theme of this conference exactly
suits the purpose of ISES, i.e. the advancement of basic and applied
research and development, as well as the utilization of solar energy.
Therefore, it is not by chance that the storage of solar energy happened
to be the theme of the meeting of our Section in the Spring of 1978.
In the extensive field of the science and the technology concerning the
application of solar energy, the *thermal* storage of solar energy is but
one of the many areas to be investigated. However, the restriction of
this symposium to that specific area seems to have some very good reasons.

One is the vital part the storage of energy plays in all cases where
the supply cannot be controlled so as to meet the demand. Obviously, in
the case of solar energy, with its great daily and seasonal fluctuations
in supply and its varying demand, some kind of storage is needed in the
majority of its applications. Solar radiation cannot be stored as such,
so first of all an energy conversion has to be brought about and, depending
on this conversion, an electrical, mechanical or thermal storage device
is needed.

Conversion of solar energy into thermal energy is the easiest and
the most used method. Furthermore, the greater part of the use of energy
in countries with moderate and cold climates is in the form of heat
at low and intermediate temperatures.

The choice of the topic of this symposium, therefore, is fully
justified. The more so, because it appears that, even the restricted
subject "thermal storage", has opened up wide fields of research,

development and engineering.

This symposium covers large areas of these fields, but leaves also other parts bare. Obviously, in view of the limited time available, a selection of topics had to be made.

From an inspection of the presented program it appears that the emphasis lays on storage at low temperatures, being of the greater importance in climates like our own.

Storage of thermal energy is possible in different physical or chemical ways. The choice to be made is entirely governed by the economics of the system, a fact rather disappointing for scientists and inventors of advanced systems. An economic fact to keep in mind here is, that the more often a certain storage element is loaded and unloaded in a year, the lower the price of the stored energy per MJ (or, if you like, kWhr) becomes. This means that the longer the average storage time is, the heavier the invested costs weigh upon the price of the stored energy. This may be one of the main reasons that seasonal storage of solar energy has made little progress so far. However, research continues in this field, as may appear from the program of this symposium. From the abstracts presented it seems that, at least some of the studies show hopeful openings and we are looking forward towards more information.

The other sessions of this conference are grouped according to the physics or chemistry of the method of storing, that is as sensible heat, latent heat or chemical reaction energy.

It may seem rather amazing that such a straightforward and, physically speaking, elementary method as the *storing of sensible heat* should still be a topic at a scientific symposium on the thermal storage of solar energy.

However, economic engineering solutions of elementary physical problems are sometimes difficult to find and a breakthrough may occur as a result of one original idea or invention. The development of stratified storage, using the floating inlet, as developed by Prof. Van Koppen and his group, is a striking example. Undoubtedly Mr Veltkamp will enlarge upon this subject later this morning.

An area not treated in this symposium, but interesting for cooling and industrial use of solar energy, is storing sensible heat in the *intermediate temperature range* (that is from $100^0 - 500^0C$). For this application, fluids are needed (oils, molten salts or liquid metals), that meet the requirements of large heat capacity, thermal stability and compatibility with container materials, at reasonable costs. To reduce these costs, a combination of a solid storage medium and a transfer fluid is conceivable. Both material and system studies are still necessary to arrive at optimum designs.

More sophisticated than storage of sensible heat, is the *latent heat storage*. In this case, use is made of reversible phase-changes where, in a narrow temperature range, high energy densities can be stored. In practice, only solid-liquid transitions involving heat of fusion, meet the requirement of a high volumetric density.

In the *low temperature range* paraffins and salt hydrates are possible candidates and this afternoon several papers will deal with these. Again, systems operating in the intermediate temperature range seem to be missing. For this range, salts and salt-mixtures have been proposed and investigated. Although the costs of some of these competitive with the costs of heat liquids, problems in containment, heat exchange and corrosion still have to be solved.

The last session of this symposium deals with *chemical storage* of thermal energy. Generally speaking, energy is stored in this case as the bond energy of a chemcial compound, and energy can be repeatedly stored and released in the same materials by reversible chemical reactions. The use of the strong chemical bonding forces, rather than the much weaker forces acting in phase changes, results in higher energy storage densities. An added advantage is, that considerable storage densities are even possible at ambient conditions, so that thermal insulation of the system would not be needed.

However, the technology seems to be, either in the research stage or in the early development stage and practical systems have not yet been demonstrated.

Furthermore, no experience is available about the cycling properties in the long term.

In view of the promising possibilities of chemical storage, large and intensive efforts in research and development in this field would be very much justified.

As a matter of fact, in my opinion, and without detracting anything from the great merits of physicists and engineers, the future of solar energy lies very much in the hands - or rather the heads - of chemists and material technologists. With their great resources with respect to material design, real break-throughs may be expected from their research efforts.

This applies to the whole field of energy-oriented research. It is, therefore, gratifying that the Royal Dutch Chemical Society (Koninklijke Nederlandse Chemische Vereniging) has recently recommended ten research aims in the energy field, including solar energy, in particular with respect to the production of hydrogen. Not being a chemist myself, I might suggest to the Dutch chemists and technologists, present at this meeting, to add research on materials for chemical storage of thermal energy to that list. Here, I like to repeat the remark I made before:

the problem of storing thermal energy has opened up a wide field of research, development and engineering, and this applies to some other areas of the application of solar energy as well.
But in order to explore these fields, means have to be provided for the necessary research, both applied and fundamental. In our country, a rather modest national solar energy program was started three years ago. In this program, the emphasis lies on short term practical projects, and a relatively small part is devoted to fundamental, solar energy oriented research.
Incidentally, some funding from other sources has taken place in that field.

Our society, convinced that in future solar energy has to be a major resource for the energy needs of the world, is concerned about the limited scale of fundamental, solar energy oriented research in our country. Therefore, we have recently made an urgent appeal to the responsible Ministers, to mark out a part of their research funds to that aim.

A fair-sized program for fundamental research is, of course, not the only condition for the introduction of solar energy. However, I have meant to mention such research in particular since, like all fundamental research, its benefits can only be expected in the long run and, therefore, it is apt to receive a low priority.

Mr Chairman, ladies and gentlemen, on behalf of the Dutch Section of ISES, I wish you an interesting symposium, with fruitful discussions and, last but not least, a useful spin-off for your future work. Thank you for your attention.

OPENING ADDRESS

Prof. Ir. C.W.J. van Koppen
Chairman of the Organizing Committee

Prof. De Jong, Ladies and Gentlemen,

Before entering upon some practical subjects concerning this meeting I first want to thank Prof. Jan Francken for his clear and careful survey of the field of solar heat storage at low temperatures. Even this restricted field appears to encompass so many questions and problems that it is real achievement to indicate in such a lucid way the main features of each of the four directions of research that can be distinguished in this field.

I am the more grateful because his comprehensive address enables me to apologize - on behalf of the Organizing Committee - for two shortcomings in the programme which the Organizing Committee did not succeed in avoiding:
Firstly it proved impossible to include in the programme all the contributions that were brought to our attention. As a consequence some interesting information may be missing in the programme we have prepared for you. However, we hope that this shortcoming may, at least partially, be compensated for by direct contacts and discussions during the coffee and lunch breaks and in the evenings.
Secondly, we are aware of the fact that we do not offer an easy meeting to you as participants but just some days of hard work, not unlike most meetings on solar energy. More than ten papers per day on often very intricate subjects and presented by highly qualified experts certainly demand a great effort on the part of the audience.

In this context the question seems justified when and which final results are to be expected from these efforts, in terms of practically feasible methods and systems for the storage of solar heat. Considering the storage devices that are currently being applied I think we all must agree with the memorable remark made by Dr. Fred Morse last year, when

presenting the U.S. R. D & D programme at the ISES Congress in Atlanta:
"As regards heat storage in practice we are still living in the water and
stone age". May we, to put the question in a different way, expect that
this symposium will change this state of affairs, or will the situation
remain as it is for many years to come?

For two - I admit not conclusive - reasons I do not believe so. The
first being that, like most if not all of you, I share the confidence
just expressed by Prof. Jan Francken that the research and development
work presented at this symposium will lead to viable means for even
the long term storage of solar heat in not too distant future.
My second reason is connected to the remark just quoted, but takes it in
a much broader sense by relating it not only to heat storage but to
energy supply in general. I believe that the observation is correct that
up to now man has behaved as an energy-nomad as regards his supply of fuel.
Searching in his environment for any combustible material, collecting and
utilizing his booty, thus depleting the resources found and then, by
sheer necessity, migrating to new hunting and collecting grounds. As you
know a similar behaviour has characterized man in relation to his food
supply until ten thousand years ago, when the increased population
density forced him into the Neolithic revolution from which agriculture,
sedentary societies and the ancient civilisations have emerged. It is
my firm belief that the thrill, suspense and confidence which are so
striking for solar research and solar meetings originate from the deep
awareness of being involved in a revolution of a similar nature and
significance, but now concerning man's energy supply. The common and
final objective of all efforts being to pave the way for an environmentally
acceptable energy supply system capable of supporting the future world
population increased by a hundred fold since the Neolithic revolution
took place.

Certainly this task is not an easy one and certainly it will not be
quickly accomplished and take many generations. In this framework our
symposium may represent no more than a single brick in the pavement of
the road we still have to clear. But laying this brick correctly is
definitely worth-while, considering the impact it will have on future
work and the important role storage will have to play in any solar energy
supply system. From our experience during the preparation of the symposium
I draw the full confidence that your support and co-operation will meet
all chalenges this symposium may put to them.

SENSIBLE HEAT STORAGE

SUMMARY AND OVERVIEW

Dr. E. Aranovitch
Joint Research Centre, ISPRA, ITALY

In this chapter four papers are presented, one on solar ponds, one on rock-bed storage and two on stratification. They give a good idea about the scientific problems which are yet to be solved with respect to the classical problem of sensible heat storage.

In his paper on solar ponds - a combination of a solar collector and a storage device - Tabor shows the feasibility of solar ponds, even for the production of electricity, which is rather a new application for these low temperature devices. Solar ponds prove to be attractive because of their large size capability (up to MW's) and their low investment costs. These factors may lead to a significant energy impact in regions where these solar ponds can be applied. In Europe, compared to Israel and the United States, too few comparative experiments between solar ponds and other large scale storage devices are going on and therefore there is a lack of comparative data and study results concerning the performance versus cost effectiveness. Because of the great need for heat in Europe, the use of solar ponds for heat production for various applications might be considered instead of electricity production at this stage. A further investigation of the technological, climatological and geographical conditions which are to be met for the optimum utilization of solar ponds has to be carried out.

With respect to solar air heating systems with rock-bed storage it seems that they have had too little attention in Europe so far. This is probably due to the tradition with liquid heating systems in Europe, although in the United States there is a long tradition with air heating systems and even there solar liquid heating systems are used more frequently.

In his paper Dutré presents the potential of rock-bed storage in a solar heating system. As such systems generally operate under dynamic and transient conditions there should be an interest to back up his

theoretical analysis with more experiments, because of possible
uncertainties in the heat transfer correlations.
The general opinion is that air collectors are less effective than liquid
collectors, but because they can be used with inlet-temperatures near
ambient temperature the complete solar air heating system may often
compete advantageously with the liquid systems and this point should be
investigated further. It could stimulate work on rock-bed storage as well.

Thermal stratification in heat storage devices in solar energy
systems leads to a considerable reduction of the pumping energy and to
an increase of the useful heat delivered. In his paper Veltkamp shows
that stratification may bring an increase in performance in the range
of 10 - 20%. He demonstrates that a floating inlet provides an interesting
technological solution to exploit the potential of stratification.

In his paper Rademaker shows various strategies of operation for
solar systems using stratified storages. In the near future these
strategies should be validated further by comparative experimentation.
Moreover, simple control systems are available, which, with a few
adaptations, can be used to exploit the stratified nature of the heat
storage so as to collect more solar energy at the expense of less pumping
energy than if the system is operated in the customary fashion.

Conclusions on sensible heat storage devices:
- Sensible heat storage systems are the most widely used systems to
  store solar energy.
- There is a good understanding about their operating principle,
  although there are still some scientific problems - especially on the
  optimum use of stratification - which need to be studied and
  experimentally verified.
- A further investigation of the technological, climatological and
  geographical conditions which should be satisfied for the optimum
  utilization of solar ponds has to be carried out.

STORAGE CAPABILITY OF SOLAR PONDS

H. TABOR

THE SCIENTIFIC RESEARCH FOUNDATION, JERUSALEM, ISRAEL

ABSTRACT

The non-convecting solar pond is a large-area solar collector of 1-2m deep, having built-in storage. This storage is in part due to the mass of water in the pond and in part to the ground under the pond (where ground insulation is not used). Diurnal variations of insolation or of energy withdrawal rate have negligible effect on temperature over periods of a week or so. Long-term (seasonal) storage is obtained by having additional depth of a few metres.

INTRODUCTION

Energy storage, particularly long-term (seasonal), is a key requirement for viable exploitation of solar energy for all applications where the time pattern of demand is not to be dictated by the solar supply. It is true that, by using an auxiliary conventional energy source, it is possible to manage without storage, but this would involve jettisoning excess solar energy received at times of minimal demand; this loss can be reduced by using a smaller collector, but in such a case the share of the load carried by solar would be small. The heating of buildings during the winter season is a classical case of mis-match between the solar source and the load, and constitutes one of the major inhibiting factors in advancing solar heating of buildings.

In solar thermal systems, the functions of collection and storage are separated: indeed it is a teaching in classical solar collector theory that the thermal capacity of the *collector* must be kept to a minimum since any heat 'stored' in the collector will be lost to the environment during periods of zero or very low insolation: the storage should be in a separate system that can be adequately insulated against heat loss.

Short term storage (for a day or so) is reasonably developed, particularly for temperatures below 100°C, where water can be used: water has the advantage of being convenient as a heat transport medium and is low in cost, though the insulated containers can be quite expensive. (Above 100°C – if expensive pressurised vessels are to be avoided – other fluids have to be used, leading to higher costs and possible chemical decomposition of the fluid.) For storage periods of more than a few days, not only must the store be very large, but the integrated heat losses can be considerable.

In the present paper we discuss a very interesting solar collector where the storage is "built-in".

2) The NON-CONVECTING SOLAR POND

The theory and technical aspects of solar ponds – as large area solar collectors – have been described in a number of reports (Refs 5, 7, 8, 10). If we consider the usual solar collector – as seen on roof-tops in Israel and many other countries – this is a small device of a few square metres in area: to harness solar energy by the square *kilometre* requires a different approach.

The oceans have been recognised as large-area solar collectors – with very large storage capacity – and attempts have been made to exploit the small temperature differences that occur between the upper water – where the solar radiation is absorbed and which is therefore hotter – and the deep water that is colder.

But the temperature difference – at best 20°C – is very small so that conversion to power by a heat engine is very difficult, and the absolute levels, 5°C to 25°C, are of little practical use for heating purposes.

The non-convecting solar pond has a similar philosophy of using a mass of water as the solar collector, but it is relatively shallow – a metre or two deep in its basic form – so that the solar radiation is absorbed at the *bottom*. In a normal pond, the heated water would rise to the top and the heat would be dissipated to the atmosphere.

By imposing a density gradient – by the dissolution of salt in increasing concentration with depth – convection can be suppressed* and water

---

* An alternative method of suppressing convection is the 'saturated' pond described in Appendix II. This method has not been developed to the same degree as the "salt gradient" pond discussed here.

heated at the bottom of the pond stays there and is insulated by the non-convecting mass of water above.

Because of absorption of some solar radiation on the way down, the temperature gradient is not linear but curved (see Fig.1) and may show dT/dz - where z is depth measured positively downwards - of near zero at the bottom. This means that there can be negligible heat-flow upwards; the water behaving as if it were almost a perfect insulator. As a result, temperatures of over 100°C have been recorded at the bottom of such ponds, though 90°C is a practical design point temperature.

At the bottom of the pond there must be a mixed layer (induced by the heat-extraction process) as otherwise only a very small fraction of the heat accumulating there could be exploited. Heat withdrawal can be carried out by laying a heat exchanger - in the form of an array of parallel pipes - in the bottom of the pond (Hipsher & Boehm, 1976), or by the method of decanting the hot layer which is possible because of the density gradient. (Hydrodynamics teaches that, where a vertical density gradient exists in a mass of fluid, it is possible to remove a horizontal slice without disturbing the fluid above or below*.) The decanted hot layer is passed through a heat exchanger external to the pond and then returned to the other end of the pond for reheating.

As a source of calories, the non-convecting pond is almost an order of magnitude cheaper than other solar collectors (Tabor, 1979): indeed, the calories are so cheap that conversion to power, even at the low Carnot efficiency resulting from a 90°C source temperature, is possible and viable. A 150 kW solar pond station (SPPS) was inaugurated December 1979 at Ein Bokek, on the shores of the Dead Sea.

Apart from low cost (current estimates are about \$13 per m^2 for large ponds, with collection efficiencies in the order of 20% cf. \$130/m^2 for flat-plate collectors of efficiency 50%) and some other advantages - such as no windows to clean and ease of collecting solar energy over large areas with little plumbing - an extremely important feature of solar ponds is the built-in storage resulting from the mass of water involved and to the ground under the pond. As an example of the storage

---

* See the original paper by Elata and Levin (1962) or the summary in Tabor (1980)

capacity of solar ponds, the turbo-generator at the Ein Bokek solar pond
power plant delivered 150 kW over prolonged periods whilst the pond itself
is undersized by a factor of about 7* i.e. the pond delivers, on request,
7 times its rated capacity. (Power engineers will recognise this as a
remarkable technological capability.)

3)   HEAT STORAGE CAPABILITY OF SOLAR PONDS - GENERAL

Weinberger (1964) in a classical paper on the physics of solar ponds,
has determined the theoretical rise of temperature at the bottom of a
pond of given depth subject to a given annual insolation pattern. In
order to reduce the very large number of parameters,he assumed that the
thermal diffusivity of the pond and of the ground under the pond were
equal (a reasonable first assumption if the ground is saturated with
water).

Fig.2 is a reproduction of Weinberger's Fig.8 which shows the computed rise of temperature at the bottom of a pond one metre deep, due to
the solar energy absorbed at the bottom**:

(a) shows the case of the pond first exposed to the solar radiation
in the spring, and (b) shows the case for the pond first exposed to solar
radiation in the autumn. In crude terms, we see a sine-wave of one year
period (resulting from the annual solar cycle) superimposed on an expo-
nental-type rise of the mean temperature as the mass of water and the
ground under the pond heats up.

(If the bottom of the pond were insulated - a step which may be
needed if the pond overlies a heat sink such as an aquifer - the exponen-
tal term would have a shorter time constant: this would result in the
pond reaching its design capacity output more quickly.)

Fig.3 (taken from Ref 6) presents the same data as Fig.1 with the
actual measured results from a small experimental pond built in 1959.
The general similarity in temperature rise for the experimental pond
with that computed is quite striking.

---

*   The pond was built originally to provide heating and cooling for a
hotel; it was 'borrowed' to test out the coupling of a solar pond heat
source to a low-temperature turbogenerator for the production of power.

**   No energy is extracted and it is assumed that, by some means, the
pond does not boil.

For short-time heat storage, the pond is very effective. For the parameter variables he used, Weinberger's Fig.12 - reproduced here as Fig.4 -
shows an expected daily swing of temperature at the bottom of a 1m deep
pond of about 15°C with no convective zone at the bottom. Since such a
zone will invariably exist due to the heat-extraction process, Fig.4
shows the effect of a 20 cm convecting zone: the 24 hour variation in
temperature is reduced to just over 5°C i.e. ±2½° about the mean. In
the various solar ponds built in the field to date in Israel (five in
number), all somewhat deeper than 1m, the output temperature has shown
little variation over a week or so whatever the insolation conditions.

Of special interest is the possibility of long-term storage - by
providing a thick mixed zone at the bottom of the pond. Not only can
this option be of great help when a solar pond is used as a source of
heat in the winter period - by exploiting the stored summer insolation -
but also for solar pond power stations: the storage of electricity on an
intermediate to large scale is technically very difficult, yet there is,
in general, little or no correlation between the electrical load and the
solar insolation. In the solar pond, we store the *heat*, so that, provided
the turbogenerators and heat-exchangers are designed to handle peak power,
the energy can be withdrawn as the load demands. The Israel Electric
Corporation is considering solar ponds to provide *peak* power to replace
expensive gas-turbine power now used (Bronicki et al, 1980).

4)   LONG-TERM STORAGE

The possibilities of long-term storage have been indicated by Tabor
& Weinberger (1980) for large ponds where only the heat-capacity of a
thick mixed zone at the bottom of the pond was considered. In the treatment below, the effect of thermal diffusion into the static non-convecting
zone above, and into the ground below, is indicated. A paper by Rabl &
Nielsen (1974) - where the interest was primarily in small ponds - offers
a similar treatment except that they start with an assumed sinusoidal
incoming solar radiation wave (of one-year period) and assumed optical
properties of the pond solution: in the present treatment we assume a
sinusoidal energy wave reaching the bottom of the pond.

We first note that, for large ponds, the extra cost of making the
pond deeper is small - except for the increased inventory of salt. This
is because large ponds are not made by digging out the ground but by

flattening the ground and building a wall or embankment round: the length
and hence cost of the wall per m^2 of pond area decreases as $1/A^{\frac{1}{2}}$ where A
is the area. However, the cost per metre run of wall increases approximately as the square of the height because its average thickness also
increases with height. The cost of the lining (used to prevent leakage)
on the inner wall face will also increase linearly with height. As a
rough guide, a pond of about 1 hectare $(10,000m^2)$ area and $1\frac{1}{2}m$ total
depth and costing $12/m^2$ with free brine, would increase in cost by
$2-3/m^2$ for each additional metre of depth added for storage. If salt
has to be paid for at $S per ton, it is necessary to add $0.35S per m^2
for each additional metre of depth.

The simple first-order approach for large and deep ponds is to take
the mass of water in the convecting storage zone, assume this to be well
mixed and to be perfectly insulated at the bottom, top and walls; the
annual insolation reaching and absorbed at the bottom is approximated as
a sine-wave of one year period. (We can ignore daily variations that
are, for all practical purposes, completely damped out by the large heat
capacity.)

Then the rate of change of temperature Θ of a volume V of liquid
for a uniform extraction rate $A\overline{I}$ is:

$$V\rho c \frac{d\Theta}{dt} = A(\overline{I} + \overset{\curlyvee}{I} \sin \omega t) - A\overline{I} = A\rho cd \frac{d\Theta}{dt} \qquad (1)$$

where ρ = density of liquid, kg/m^3

c = specific heat, $J/kg, °K$

A = area of pond, m^2

t = time in seconds; $t = 0$ at spring equinox

ω = angular velocity = 0.1988×10^{-6} rads/sec for an annual cycle

$\overline{I}$ = annual mean insolation reaching the bottom of the pond, W/m^2

$\overset{\curlyvee}{I}$ = amplitude of sinusoidal component of insolation reaching the
pond bottom, W/m^2

d = depth of the mixed volume V, metres

This at once yields:

$$\Theta = \frac{\overset{\curlyvee}{I}}{\omega\rho cd} \sin(\omega t - \pi/2) + \Theta \text{ mean} \qquad (2)$$

and the annual swing of Θ about the mean is:

$$\pm\Theta = \frac{\overset{\curlyvee}{\mathrm{I}}}{.1988 \times 10^{-6}(\rho c)} = 1.62\mathrm{I}/d \text{ for } (\rho c) = 3.1 \times 10^{6}\mathrm{J.m^{-3}K^{-1}} \tag{3}$$

The minimum is at the spring equinox (t = 0).

As an example, under Israel conditions (latitude 32°N) for a pond with a non-convecting zone 1m deep, and a saline solution having trans-missivity similar to sea-water No.3 (Ref. 12) the computed energy yield is approx 35 W/m^2 in mid-winter and 107 W/m^2 in mid-summer i.e. $\bar{\mathrm{I}}$ = 71, $\overset{\curlyvee}{\mathrm{I}}$ = 36. This gives $\pm\Theta$ = 58/d $^{\circ}$K. With d, say, 5m the temperature swing is $\pm11.6^{\circ}$ about the annual mean level.

In the above treatment, we have ignored the flow of heat into and out of the ground, treating the ground as an insulator.

In Appendix I we consider the diffusion of heat into a semi-infinite body subject to a sinucoidal temperature at its surface. It is shown that, if the temperature variation is:

$$\Theta = \overset{\curlyvee}{\Theta} \sin(\omega t - \epsilon) \tag{4}$$

(where ϵ is an arbitrary phase angle) then the flux into the ground (after attenuation of the transient component) is:

$$F_g = \overset{\curlyvee}{F}_g \overset{\curlyvee}{\Theta} \sin(\omega t - \epsilon + \pi/4) \tag{5}$$

where $\overset{\curlyvee}{F}_g = (\omega\rho c K)^{\frac{1}{2}}$ $\tag{6}$

where K, ρ, c are the conductivity; density and specific heat respectively of the ground. Assuming the ground to have thermal properties similar to water gives:

$$\overset{\curlyvee}{F}_g = 0.718 \text{ W/m}^2 \text{ }^{\circ}\text{C} \tag{7}$$

and the flux $F_g = 0.718\overset{\curlyvee}{\Theta} \sin(\omega t - \epsilon + \pi/4)$ $\tag{8}$

There is also a heat flux upwards into the non-convecting layer above the mixed storage zone. If this were infinitely thick, the sinucoidal flux F_u would be similar to F_g if we assume similar thermal properties. As, however, the non-convecting layer is of finite thickness, the theory of a slab of thickness ℓ subject to a sinusoidal temperature wave at the surface is treated in Appendix I and gives a flux

$$F_u = \overset{\curlyvee}{F}_u \overset{\curlyvee}{\Theta} \sin(\omega t - \epsilon + \psi) \tag{9}$$

$\overset{\curlyvee}{F}_u$ depends upon the thickness ℓ and in practical cases is close to the value of $\overset{\curlyvee}{F}_g$ calculated for water properties. The phase angle ψ approaches the value $\pi/4$ for large ℓ.

Before considering the effect of the thermal diffusion terms F_g and

F_u, we will include the case where the withdrawal rate is not constant during the year, for example the energy is needed primarily for heating in winter. Treating the extracted energy as a sine-wave, this component may be written as:

$$F_L = -\overset{\smallfrown}{L} \sin(\omega t - \phi_L) \qquad (10)$$

where $2\overset{\smallfrown}{L}$ is the range of energy extracted about the mean $\overline{L}$ - which mean must be equal to $\overline{I}$ on a long-term basis - and ϕ_L is the phase lag of the load.

Thus the net or effective energy $\overset{\smallfrown}{I}_e$ going into the mass of storage fluid is:

$$\overset{\smallfrown}{I}_e \sin(\omega t - \delta) = \overset{\smallfrown}{I} \sin \omega t - \overset{\smallfrown}{F}_g \overset{\smallfrown}{\theta} \sin(\omega t - \epsilon + \pi/4) - \overset{\smallfrown}{F}_u \overset{\smallfrown}{\theta} \sin(\omega t - \epsilon + \psi) - \overset{\smallfrown}{L} \sin(\omega t - \phi_L) \qquad (11)$$

where δ is the phase angle (presently unknown) for the net energy wave.

From equation (2) the temperature wave $\overset{\smallfrown}{\theta}$ lags the net energy input by $\pi/2$ i.e. the phase angle ϵ in (4) is $(\delta + \pi/2)$, since the phase of the input is $-\delta$. Equation (11) becomes:

$$\overset{\smallfrown}{I}_e \sin(\omega t - \delta) = (\omega \rho c d)\theta \sin(\omega t - \delta - \pi/2) \quad \text{using (2)}$$
$$= \overset{\smallfrown}{I} \sin \omega t - \overset{\smallfrown}{F}_g \overset{\smallfrown}{\theta} \sin(\omega t - \delta - \pi/4) -$$
$$F_u \overset{\smallfrown}{\theta} \sin(\omega t - \delta - \pi/2 + \psi) - \overset{\smallfrown}{L} \sin(\omega t - \phi_L)$$

i.e. $\overset{\smallfrown}{\theta}\left[\omega \rho c d \sin(\omega t - \delta - \pi/2) + \overset{\smallfrown}{F}_g \sin(\omega t - \delta - \pi/4) + F_u \sin(\omega t - \delta - \pi/2 + \psi)\right] = \overset{\smallfrown}{I} \sin \omega t - \overset{\smallfrown}{L} \sin(\omega t - \phi_L) \qquad (12)$

Summing the sine-waves on each side gives:

$$\overset{\smallfrown}{\theta} R \sin(\omega t - \delta + \beta) = (\overset{\smallfrown}{I}^2 - 2\overset{\smallfrown}{I}\overset{\smallfrown}{L} \cos \phi_L + \overset{\smallfrown}{L}^2)^{\frac{1}{2}} \sin(\omega t + \Omega) \qquad (13)$$

where $\tan \Omega = \dfrac{\overset{\smallfrown}{L} \sin \phi_L}{\overset{\smallfrown}{I} - \overset{\smallfrown}{L} \cos \phi_L} \qquad (14)$

$$\tan \beta = \frac{\omega \rho c d + .707 \overset{\smallfrown}{F}_g + \overset{\smallfrown}{F}_u \cos \psi}{.707 \overset{\smallfrown}{F}_g + \overset{\smallfrown}{F}_u \cos \psi} \qquad (15)$$

and $R^2 = (\omega \rho c d)^2 + 2\omega \rho c d(.707F_g + \overset{\smallfrown}{F}_u \cos \psi) + \overset{\smallfrown}{F}_g^2 + \overset{\smallfrown}{F}_u^2 + \sqrt{2} \overset{\smallfrown}{F}_g \overset{\smallfrown}{F}_u (\sin \psi + \cos \psi) \quad (16)$

For equation (12) to be an identity i.e. true for all values of t, the phases must be equal i.e.

$$\delta = (\beta - \Omega) \qquad (17)$$

and (12) becomes: $\overset{\smallfrown}{\theta} = \dfrac{(\overset{\smallfrown}{I}^2 - 2\overset{\smallfrown}{I}\overset{\smallfrown}{L} \cos \phi_L + \overset{\smallfrown}{L}^2)^{\frac{1}{2}}}{R} \qquad (18)$

Writing $p = 1/\omega \rho c$ (=1.62 in the present example) we write

$$R = \frac{1}{p} f(d)$$

$$f(d) = \left[d^2 + 2p(.707\overset{\curvearrowright}{F}_g + \overset{\curvearrowright}{F}_u \cos \psi) + p^2\{\overset{\curvearrowright}{F}_g{}^2 + F_u{}^2 + \sqrt{2}\overset{\curvearrowright}{F}_g\overset{\curvearrowright}{F}_u(\sin \psi + \cos \psi)\}\right]^{\frac{1}{2}} \quad (19)$$

and (17) becomes $\overset{\curvearrowright}{\theta} = \dfrac{p(\overset{\curvearrowright}{I}^2 - 2\overset{\curvearrowright}{I}\overset{\curvearrowright}{L} \cos \phi_L + \overset{\curvearrowright}{L}^2)^{\frac{1}{2}}}{f(d)}$ (18a)

We note that the bracketted term of the numerator is a function of the incoming radiation and the load: the denominator is a function of the depth of the storage layer and the properties of the ground and the non-convecting layer.

For $p = 1.62$ and using a value of $\overset{\curvearrowright}{F}_g = 0.718$ (for ground with thermal properties similar to water) Appendix I calculates the values of $\overset{\curvearrowright}{F}_u$ and ψ for three depths of the non-convecting layer and applying these to equation (19) gives $f(d)$.

For $\ell = 0.8$m $f(d) = (d^2 + 4.20d + 5.80)^{\frac{1}{2}}$

   $= 1.0$m $f(d) = (d^2 + 3.73d + 5.09)^{\frac{1}{2}}$

   $= 1.2$m $f(d) = (d^2 + 3.44d + 4.76)^{\frac{1}{2}}$

These yield Table I for $f(d)$ in terms of d.

TABLE I

Depth of mixed layer, d = 0	1	2	3	4	5m
$\ell = 0.8$m $f(d) =$	2.41	3.32	4.27	5.23	6.21	7.20
$d_o = f(d)-d=$	2.41	2.32	2.27	2.23	2.21	2.20
$\ell = 1.0$m $f(d) =$	2.26	3.13	4.07	5.03	6.00	6.98
$d_o = f(d)-d=$	2.26	2.13	2.07	2.03	2.00	1.98
$\ell = 1.2$m $f(d) =$	2.18	3.03	3.95	4.91	5.88	6.86
$d_o = f(d)-d=$	2.18	2.03	1.95	1.91	1.88	1.86

We can express $f(d)$ as $d + d_o$ where d_o varies from 1.9 to 2.4 i.e. is of the order of 2m for ℓ in the expected range of $0.8 - 1.2$ m

(18a) becomes $\overset{\curvearrowright}{\theta} = \dfrac{1.62(\overset{\curvearrowright}{I}^2 - 2\overset{\curvearrowright}{I}\overset{\curvearrowright}{L} \cos \phi_L + \overset{\curvearrowright}{L}^2)^{\frac{1}{2}}}{d + d_o}$ (18b)

For constant load (or a load with small random variations during the year)

$$\overset{\curvearrowright}{\theta} = \frac{1.62 \overset{\curvearrowright}{I}}{d + d_o} \quad (20)$$

For the extreme case where the load is a maximum in winter and zero in summer i.e. $\phi_L = \pi$ and $\tilde{L} = \bar{I}$ (the load varies from zero to $2\bar{I}$ during the year), (18b) gives:

$$\tilde{\theta} = \frac{1.62(\tilde{I} + \bar{I})}{d + d_o} \tag{21}$$

Thus for the example quoted of $I = 36$, $\bar{I} = 71$

$$\tilde{\theta} = \frac{173}{d + d_o} \sim \frac{173}{d + 2}$$

d	=	1	2	3	3.76	m
$\tilde{\theta}$	=	58	43	34.6	30	$^{\circ}$C

If we assume that a safe maximum temperature is 100°C (somewhat below the boiling point of the solution) and 40°C is the lowest useful temperature, the maximum value of $\tilde{\theta}$ is 30° i.e. the pond should have a storage zone of at least 3.8m - for the extreme case considered.

If the bottom of the pond is well insulated so that $\tilde{F}_g \rightarrow 0$, Appendix I shows f(d) to be:

For $\ell = 0.8$m $\quad f(d) = (d^2 + 2.55d + 1.76)^{\frac{1}{2}}$

$\quad\quad 1.0$m $\quad f(d) = (d^2 + 2.09d + 1.29)^{\frac{1}{2}}$

$\quad\quad 1.2$m $\quad f(d) = (d^2 + 1.795d + 1.075)^{\frac{1}{2}}$

which yields Table II.

TABLE II

Depth of mixed layer, d =	0	1	2	3	4	5m
For ℓ= 0.8 f(d) =	1.325	2.30	3.30	4.29	5.29	6.29
d_o =	1.325	1.30	1.30	1.29	1.29	1.29
= 1.0m f(d) =	1.13	2.09	3.08	4.07	5.06	6.06
d_o =	1.13	1.09	1.08	1.07	1.06	1.06
= 1.2m f(d) =	1.04	1.97	2.94	3.93	4.92	5.92
d_o =	1.04	0.97	0.94	0.93	0.92	0.92

i.e. $f(d) \sim d + d_o$ where $d_o \sim 1$m i.e. for practical cases $d_o \sim 2$m for the case of wet ground, 1m for an insulated pond and an intermediate value for dry ground without insulation.

Phase of the temperature wave

Note that, from (17) and (13) the phase of the flux wave into storage is Ω, as given by equation (14) i.e. for constant load $\Omega = 0$ and the temperature wave is in phase with the incident radiation. In the extreme case of heat extraction in winter ($\theta_L = \pi$) Ω is again zero. The temperature wave lags the flux wave by $\pi/2$ i.e. the temperature wave will lag the incident wave by $\pi/2$ for the two cases considered i.e. will show a maximum in the autumn and a minimum in spring.

REFERENCES

1. Bronicki L., Lev-Er J., Porat Y., 1980. "Large Solar Electric Power Plant Based on Solar Ponds", World Power Conf., Munich.
2. Elata C. & Levin O., 1962. "Selective Flow in a Pond with Density Gradient", Hydraulic Report, Technion, Haifa, Israel.
3. Hipsher M.S. & Boehm R.F., 1976. "Heat Transfer Considerations of a Non-Convecting Solar Pond Exchanger", Am. Soc. Mech. Engrs. 76-WA/Sol 4.
4. Nielsen C.E., 1975. "Salt Gradients for Solar Ponds for Solar Energy Utilisation", ENVIRONMENTAL CONSERVATION, 2, pp.289-292.
5. Nielsen C.E., 1979. Chapter on "Non-convective Salt Gradient Solar Ponds" in SOLAR ENERGY HANDBOOK, Eds. Dickenson & Cheremishoff, Marcel Decker.
6. Rabl A. & Nielsen C.E., 1975. "Solar Ponds for Space Heating", SOLAR ENERGY, 17, No.1, April 1975, pp.1 - 12.
7. Tabor H., 1961. "Large-area Solar Collectors (Solar Ponds) for Power Production", UN Conf. New Sources of Energy, Rome - reprinted SOLAR ENERGY VII, No.4, pp.189-194 (Oct. 1963).
8. Tabor H., 1966. "Solar Ponds", SCIENCE JOURNAL, pp. 66-71 (1968).
9. Tabor H., 1979. "Solar Ponds (Non-convecting)", UNITAR Conf. on Long-Term Energy Sources, Montreal.
10. Tabor H., 1980. "Non-Convecting Solar Ponds", Phil. Trans. R. Soc. Lond., A295, pp.423-433. Reprinted in book "Solar Energy" published by Royal Society of London (1980).
11. Tabor H. & Weinberger H.Z., 1980. Chapter on "Non-Convecting Solar Ponds", SOLAR ENERGY HANDBOOK, Ed. Kreider, McGraw-Hill, N.Y.
12. Weinberger H., 1964. "The Physics of the Solar Pond", SOLAR ENERGY VIII, No.2, pp.45-56 (April 1964).
13. Carslaw H.S., Jaeger J.C., 1959. "Conduction of Heat in Solids" Second Edition, Oxford University Press.

ACKNOWLEDGEMENTS.

The author wishes to thank his long-time colleague Mr. H.Z. Weinberger for helpful discussion in the preparation of this paper.

APPENDIX I

CYCLIC HEAT FLOW INTO THE GROUND

Consider the mixed convecting zone at the bottom of the pond to have an oscillating temperature with an annual cycle

$$\Theta = \Theta_m + \overset{\curlyvee}{\Theta} \sin (\omega t - \varepsilon) \qquad (\omega = 0.1988 \times 10^{-6} \text{ rads/sec})$$

(A) We wish to determine the oscillating flux that enters and leaves the ground. From Carslaw and Jaeger (1959) p.65, the temperature at any depth x from the surface of a semi-infinite body subject to a harmonic surface temperature $\Theta = \overset{\curlyvee}{\Theta} \sin (\omega t - \pi/2)$ is given by:

$$\Theta = \overset{\curlyvee}{\Theta} e^{-kx} \sin(\omega t - \pi/2 - kx) \tag{I1}$$

where $k \equiv (\omega/2D)^{\frac{1}{2}}$ $D =$ diffusivity $K/\rho c$

$$= (\omega \rho c/2K)^{\frac{1}{2}} \tag{I2}$$

The heat flux F_g at the surface is:

$$F_g = -K(\partial\Theta/\partial x)_x = 0$$

$$= \sqrt{2} k K \overset{\curlyvee}{\Theta} \sin(\omega t - \varepsilon + \pi/4) \tag{I3}$$

$$= (\omega \rho c K)^{\frac{1}{2}} \overset{\curlyvee}{\Theta}(\sin \omega t - \varepsilon + \pi/4) \tag{I4}$$

$$= \overset{\curlyvee}{F}_g \overset{\curlyvee}{\Theta} \sin(\omega t - \varepsilon + \pi/4) \tag{I5}$$

Consider the ground to have thermal properties similar to water i.e.

K (conductivity)	$= 0.62$ W/m$^\circ$C
ρ (density)	$= 1000$ kg/m^3
c (specific heat)	$= 4.18 \times 10^3$ J/kg$^\circ$C
ρcK	$= 2.592 \times 10^6$ J^{2}sec^{-1}m^{-4} $^\circ$C^{-2}
$\omega \rho cK$	$= 0.515$ for $\omega = 0.1988 \times 10^{-6}$ rads/sec
Amplitude $\overset{\curlyvee}{F}_g \overset{\curlyvee}{\Theta}$	$= (\omega \rho cK)^{\frac{1}{2}} \overset{\curlyvee}{\Theta} = 0.718 \overset{\curlyvee}{\Theta}$ W/m^2 $^\circ$C $\tag{I6}$

(B) We wish to determine the flux upwards into the non-convecting layer, having a depth ℓ. For a sinusoidal temperature $\Theta = \overset{\curlyvee}{\Theta} \sin(\omega t - \pi/2)$ at $x = \ell$ and zero fluctuation at the other face $x = 0$ (which is in contact with a well-mixed layer of high surface conductance) Carslaw & Jaeger (p.105-6) give the solution for Θ at any depth

$$\Theta = A' \overset{\curlyvee}{\Theta} \sin(\omega t - \pi/2 + \phi) \tag{I7}$$

where $A' = \left| \dfrac{\cosh 2kx - \cos 2kx}{\cosh 2k\ell - \cos 2k\ell} \right|^{\frac{1}{2}}$

$$\tag{I8}$$

$$\phi = \arg. \frac{\sin kx \, (1+i)}{\sin k\ell \, (1+i)}$$

The flux F_u at $x = \ell$ is:

$$F_u = K \left(\frac{\partial \Theta}{\partial x}\right)_{x = \ell}$$

$$= \frac{K\bar{\Theta} \left[\sin \omega t (\sin 2k\ell + \sinh 2k\ell) + \cos \omega t (\sin 2k\ell - \sinh 2k\ell) \right]}{\cosh 2k\ell - \cos 2k\ell} \quad (I9)$$

$$= \left[\frac{\omega \rho c K}{2}\right]^{\frac{1}{2}} . \; \tilde{\Theta} r \sin(\omega t - \frac{\pi}{2} + \psi) \quad (I10)$$

where $r^2 = 2(\sinh^2 2k\ell + \sin^2 2k\ell)/(\cosh 2k\ell - \cos 2k\ell)^2$ \quad (I11)

$$\tan \psi = \frac{\sinh 2k\ell - \sin 2k\ell}{\sinh 2k\ell + \sin 2k\ell} \quad (\to 1 \text{ for } k\ell \text{ large}) \quad (I12)$$

$$\tilde{F}_u = (\frac{\omega c \rho K}{2})^{\frac{1}{2}} r \tilde{\Theta}$$

We can calculate r and ψ for a range of values of $k\ell$: hence such quantities as F_u, F_u^2, $F_u \cos\psi$. For water, $k = (\frac{\omega c}{2K})^{\frac{1}{2}} = 0.819$. For the saline solution used for the non-convecting layer, the density will be up about 15%, the specific heat down by about the same amount and K will be down about 5%. We will thus assume k for the solution to be substantially the same as for water. The quantity $(\omega \rho c k)^{\frac{1}{2}}$ for water is 0.718 and will not be very different for the saline solution.

We obtain:

ℓ	=	0.8	1.0	1.2	m.
r	=	1.6108	1.3791	1.2610	
$\tilde{F}_u = (\frac{\omega \rho c K}{2})^{\frac{1}{2}} r$	=	0.818	0.700	0.640	
ψ	=	0.2733	0.4022	0.5242	radians

Inserting $p = 1.62$ and $\tilde{F}_g = 0.718$ in equation (19) gives:

for $\ell = 0.8$ m $\quad f(d) = (d^2 + 4.197d + 5.80)^{\frac{1}{2}}$

$\quad\quad = 1.0$ m $\quad\quad\quad = (d^2 + 3.732d + 5.086)^{\frac{1}{2}}$

$\quad\quad = 1.2$ m $\quad\quad\quad = (d^2 + 3.440d + 4.758)^{\frac{1}{2}}$

For the case of a well insulated pond i.e. $\tilde{F}_g \to 0$, equation (19) gives:

for $\ell = 0.8$ m $\quad f(d) = (d^2 + 2.55d + 1.76)^{\frac{1}{2}}$

$\quad\quad = 1.0$ m $\quad\quad\quad = (d^2 + 2.09d + 1.29)^{\frac{1}{2}}$

$\quad\quad = 1.2$ m $\quad\quad\quad = (d^2 + 1.795d + 1.075)^{\frac{1}{2}}$

(It is interesting to note that the storage effect $\overset{\curlyvee}{F}_u$ of the non-convecting layer *increases* as its thickness *decreases*. This perhaps unexpected result is explained by the fact that an oscillating flux passes through a thin layer more readily than through a thick one. As the outer face is 'washed' by a high-conductance, mixed-liquid layer, the flux that gets through is removed with negligible temperature variation: in effect, heat would appear to be stored in the 'infinite sink' of the upper mixed layer and returned to the system during the negative part of the cycle. Physically this appears impossible as the infinite sink is at the lowest temperature. The explanation is that we have been discussing the sinusoidal diffusion component: there is also a constant conduction flux so that the total net flux at the upper surface need never go negative.)

APPENDIX II
SATURATED SALT SOLAR PONDS

In the salt gradient ponds discussed in the main text, a density gradient increasing from the surface downwards is obtained by having a concentration of salt increasing with depth. If the gradient is adequate, convection does not occur even when the bottom is hotter than the top. Such a pond requires salt "management" i.e. some measures are needed to deal with the slow natural diffusion of salt that occurs.

An alternative procedure is to choose a salt (borax and potassium nitrate are two examples) which exhibits a high increase in solubility with increase in temperature and to fill a pond with a *saturated* solution of the salt. (Additional solid salt may be required at the bottom to ensure saturation.) If the pond heats at the bottom, more salt dissolves thereby increasing the density at the bottom sufficiently to overcome expansion due to heating, and convection does not occur. Such ponds have been demonstrated on a small scale but have not progressed to the same level as the salt gradient ponds. Amongst the difficulties reported is possible precipitation at the bottom of white salt crystals that reflect instead of absorb the penetrating solar radiation.

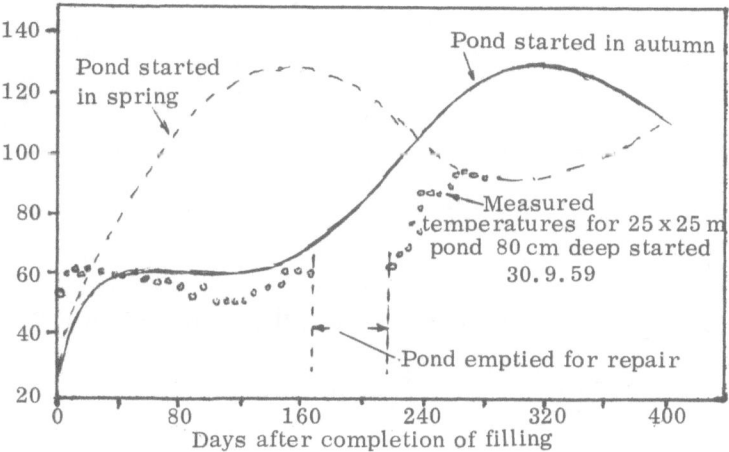

FIG. 3   TEMPERATURE RISE ABOVE AMBIENT IN   SMALL EXPERIMENTAL
POND COMPARED WITH COMPUTED TEMPERATURE RISE

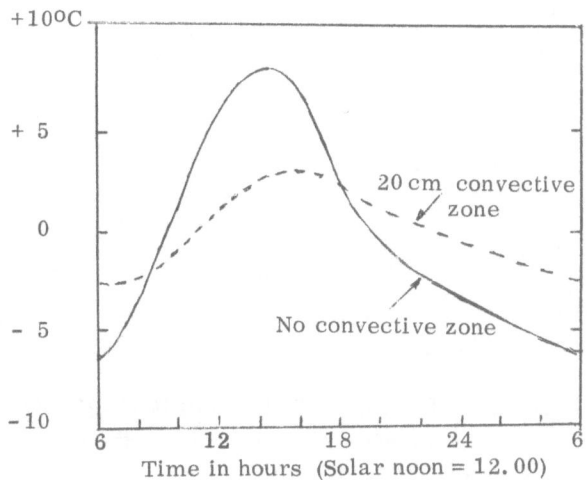

FIG. 4   HOURLY VARIATION ABOUT MEAN TEMPERATURE AT BOTTOM OF A
POND 1 M DEEP WITH AND WITHOUT A CONVECTIVE ZONE AT
THE BOTTOM

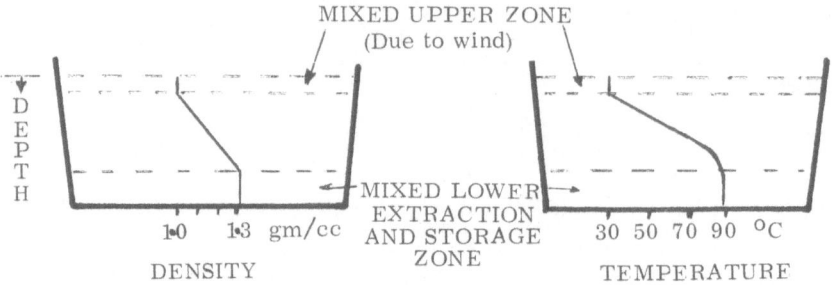

MIXED UPPER ZONE
(Due to wind)

MIXED LOWER
EXTRACTION
AND STORAGE
ZONE

DEPTH

1·0 1·3 gm/cc

DENSITY

30 50 70 90 °C

TEMPERATURE

IG. 1   DENSITY AND TEMPERATURE GRADIENTS IN SOLAR POND

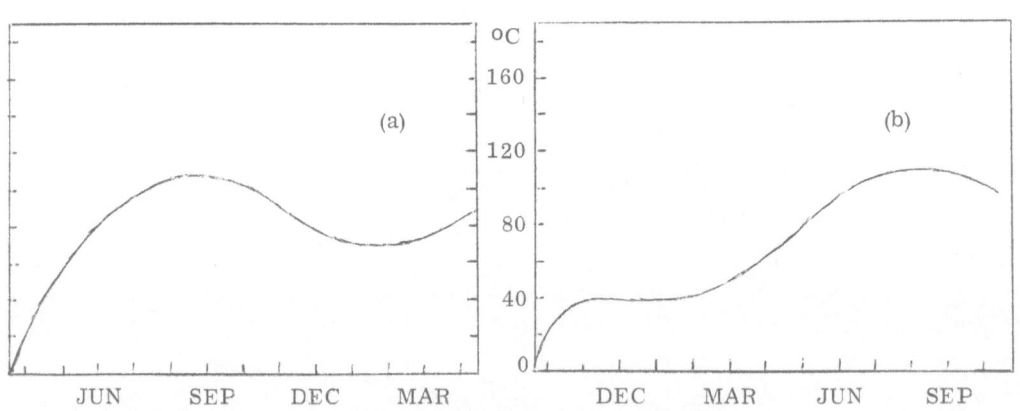

FIG. 2   INITIAL TEMPERATURE RISE ABOVE AMBIENT AT BOTTOM OF 1 M DEEP
POND   (a) Pond first exposed to insolation in the spring   (b) Pond first
exposed to solar radiation in the fall

PARAMETRIC INVESTIGATION OF THE PERFORMANCE OF SOLAR HEATING SYSTEMS
WITH ROCK BED STORAGE.

W.L.DUTRE and J.VANHEELEN

*The work reported in this paper was supported by the Belgian government
as part of the Belgian R&D-Programme on solar energy applications in
dwellings.*
*All rights to this work are reserved to the "Prime Ministers Office –
Science Policy Programme – Belgium".*

ABSTRACT

The performance of solar air collector domestic heating systems with
rock bed storage, as obtained from theoretical analysis, is discussed.
In the simulation model used for the annual performance calculations,
a segmented rock store model, ignoring the heat conduction terms, has
been used. The results of the parametric study show the relative importance of various measures which can be introduced to improve the system
performance.

I. INTRODUCTION

Solar systems for space heating are frequently named after the kind
of heat transfer fluid being used to withdraw the heat from the solar
collectors. Two main categories can be distinguished in present solar
system design : water (or water - glycol mixture) systems and air systems.
The specific advantages and disadvantages of using air collectors instead of liquid collectors have been discussed in many textbooks and
papers and are sufficiently well known. The fact that air collector
systems are still much less important on the solar market is probably
more related to the history of solar collector development and the
choice of using water storage systems, rather than to basic objections
against air collectors.

C. den Ouden (ed.), Thermal Storage of Solar Energy. All rights reserved.
Copyright © by TNO and Martinus Nijhoff Publishers, The Hague/Boston/London.

Although, air collectors can also be combined with a water storage and water heat distribution systems when an intermediate air-water heat exchanger is used, it is obvious that they can be used much more effectively when air is the only heat transporting fluid throughout the whole system. The use of only one heat transfer fluid can eliminate the need for the intermediate heat exchanger when the solar storage is of a regenerative type in which heat is exchanged between the storage material and the heat transporting fluid. For air systems, rock bed storages have since long been recognised as an attractive solution to this problem. From a system point of view, an interesting feature of air systems then consists of the possibility to heat or preheat the air of the dwelling in the collectors directly whenever the operating conditions are in favour of such a direct loop. This operational mode avoids the intermediate temperature drop in a heat exchanger as well as storage losses. The collector inlet temperature then equals the house air temperature and thus increases the collector efficiency. Incorporating this operational mode in an air system may result in a sufficient compensation for the lower efficiency which air collectors usually have because of the poor heat transfer from the absorber to the air flowing through the collectors.

The effective collector internal heat transfer can however be improved considerably when porous absorbers with large heat transfer area per unit outer collector area are used. One of the major obstacles to the use of rock bed storage systems in domestic applications is however the low specific thermal capacity which approximately triples the occupied volume as compared to an equivalent water storage tank.

For air systems as well as for water systems, a large number of different systems can be designed, mainly depending on the number of operational modes being desired, according to the possible energy transfers between the system components and the corresponding regulation criteria. In this paper, only air systems with rock storage and an air heat distribution system are considered, for space heating as well as for domestic hot water production.

In the simulation programmes used to calculate the system performance, various simplifying assumptions are introduced as a compromise between accuracy and the computer time required to calculate the annual performance.

In the air systems discussed in this paper, the direct energy transfer from the collectors to the dwelling is always included, such that, according to the instantaneous value of various system variables and the assumed regulation criteria, the system can be working in one of the following modes :
1. collector-storage loop, to store energy in the rock bed storage system,
2. storage-dwelling loop, to extract heat from the storage,
3. collector-dwelling loop, to deliver solar heat to the dwelling directly,
4. collector-storage-dwelling loop, with preheating of the dwelling air in the collectors and additional heating in the storage,
5. auxiliary heating system-dwelling loop, when the solar system cannot contribute to the house heat demand,
6. storage-domestic hot water preheating loop.

Depending on the kind of system, some of these loops may be operational simultaneously. Changing from one operating mode to another is accomplished by servomotor actuated valves in the air ducts or in a combined air handler, according to the setting requested by the electronic control devices.

The system performance calculations are based on a Belgian meteorological reference year such that the sensitivity of the system performance to some of the parameters may not be applicable to other types of climate.

II. MATHEMATICAL SIMULATION MODEL

In the simulation programme, the system differential equations are solved with a half-hour time step, using half-hourly averaged meteorological data. Since the operating mode to be selected at the beginning of each time step also depends on the thermal conditions of the dwelling, and some operating modes may be excluded while heat demand occurs, the thermal behaviour of the house must also be simulated dynamically and simultaneously in every run of the system simulation programme. Besides the thermal capacities of the building structure, the incident solar radiation on the building, internal heat dissipation from electrical appliances and occupancy and outdoor temperature as a function of time, the house model should also account for the hysteresis around the

thermostat setting point. Otherwise the time sequence of selected modes
may be very unrealistic, which of course affects the calculated performance and yields erroneous results on the relative importance of
different operating modes.

For the purpose of general parametric studies of the solar systems, a
simplified three point model of the house has been used.

The collector array is modeled dynamically with a variable heat loss
coefficient according the Klein-formula, based on the absorber temperature. The heat transfer from the absorber to the air flowing through
the collectors is described separately by an overall heat transfer coefficient per unit outer collector area.

Because the value of the thermal capacity flow in air systems can be
considerably lower than in water collectors, resulting in larger temperature differences across the collector, a segmented collector model
has been used.

The pebble bed of the storage is characterised by its heat capacity
per unit volume, the heat loss coefficient to the ambient, the total
volume and the heat transfer coefficient per unit storage volume between
the storage material and the air flow. Because the volumetric heat
transfer coefficient can be quite large, a significant temperature
gradient may arise in the flow direction in the pebble bed.

Therefore, the storage has also been subdivided in several consecutive
segments in the flow direction, each characterised by a single temperature, governed by a differential equation in which the heat conduction
term is ignored. The flow direction through the store is reversed when
the storage is being discharged as compared to the charging operation.
The correlation of Handley and Heggs [1] can be used to determine the
heat transfer coefficient in the rock bed storage. This correlation
has been used by several authors and the agreement between experimental
and theoretical results seems to be quite satisfactory [2]. The simplified segmented storage model, based on an average overall heat transfer
coefficient has been the subject of a limited validation, based on
short term measurements on a small experimental system including the
air collectors.

Reasonable agreement was obtained for the overall energy balance of the
storage system [3].

III. AIR COLLECTOR SYSTEMS WITH ROCK BED STORAGE

Different systems have been considered for a comparative study and
to evaluate the influence of some of the basic design parameters on the
annual system performance. Details on the working principles and regulation criteria will be omitted in this text because these aspects may
be clear enough from schematic representations as given in the figures.
In these figures, valves having the same identification number are
simultaneously actuated and take only the fully opened or closed positions except for the valves identified with M, which is a modulated
valve for the regulation of the air distribution temperature according
to the requested value which is assumed to depend linearly on the outdoor ambient temperature.
The domestic water preheating is connected to the rock bed storage instead of using an air-water heat exchanger between the collector outlet
and the rock store as it is in common practice in air collector systems.
Although this measure reduces the efficiency of the system during the
summer when the rock store thermal capacity together with the water preheating tank capacity largely exceeds the optimal storage capacity for
a domestic hot water system, it limits the degree of priority given to
the domestic water system during the heating season and also reduces the
head losses in the collector loop.

A first system, hereafter referred to as system 1 and represented
schematically in figure 1, is a single rock store - two ventilator system, in which the operational mode 4 (i.e. the collector-storage-dwelling loop as defined in the introduction) has been excluded in view of
simplifying the control criteria. System 2, shown in figure 2, is a
one-ventilator-version of system 1 and may, depending on the spatial
lay-out of the system in a house, be more convenient. Also the control
actions are somewhat simpler than in system 1 and the electricity consumption of the system is somewhat lower because only one ventilator
is used in the direct collector-dwelling mode. System 3, represented
in figure 3, is a double-rock store system in which the direct collector-
dwelling mode has been excluded. This is however compensated because
the separation of the rock store in two distinct compartments of equal
size, allows the collector-storage and the storage-dwelling loops to be
operational simultaneously.

38

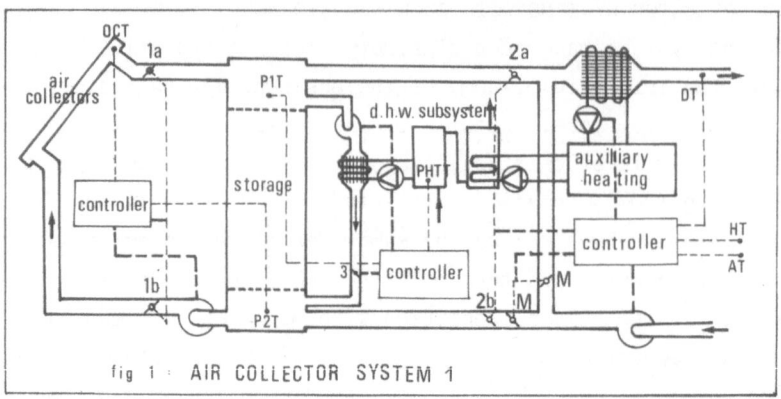

fig 1 : AIR COLLECTOR SYSTEM 1

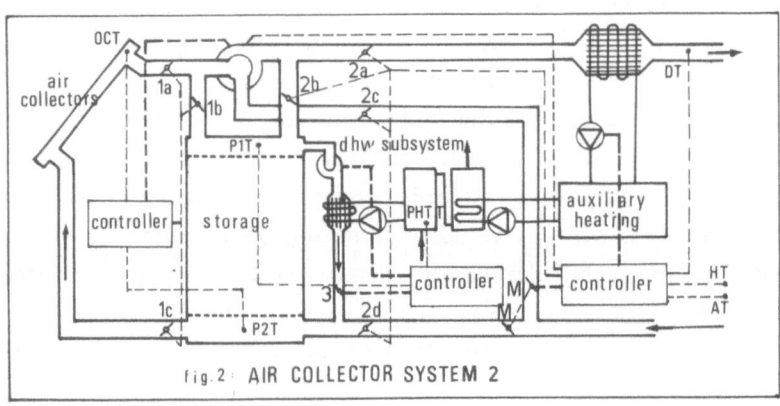

fig. 2 : AIR COLLECTOR SYSTEM 2

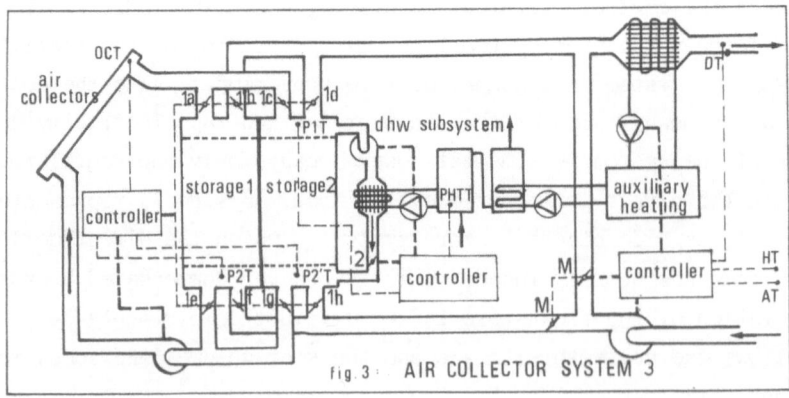

fig. 3 : AIR COLLECTOR SYSTEM 3

IV. ANALYSIS OF THE ANNUAL SYSTEM PERFORMANCE

For the systems briefly described above, annual performance calculations have been performed with the model described in paragraph II, for Belgian climatological conditions, with south facing collectors at 60° inclination angle.

Unless specified otherwise, non-selective single glass collectors have been assumed, with an average n_o = 76 % at an absorber temperature equal to the ambient temperature, a back side heat loss coefficient of 0.4 W/K m^2 and an internal absorber to air flow heat transfer coefficient of 50 W/K m^2, such that the average usual n_o (with coolant temperature at ambient temperature) approximately equals 69 %. The absorption coefficient of the absorber plate is assumed to be 95 %.

The preheating tank for domestic hot water production is in all cases a 200 liter tank.

The simulated house has an annual useful energy demand of 15,250 kWh for space heating and 2,555 kWh for domestic hot water, corresponding to a daily hot water consumption of 150 liters at 50°C. The storage heat losses are not considered as contributing to the heat demand.

For the system 1 and 2, assuming a very heavily insulated storage, equivalent to the thermal resistance of 50 cm of mineral wool, the solar contribution to the total annual useful energy demand (i.e. 17,805 kWh) is represented in figure 4 as a function of the thermal capacity of the storage, for some values of the collector area. The percentage of solar contribution to the space heating demand and to the domestic hot water production are separately represented in the figures 5 and 6 respectively. For a few cases, the numerical values are given in table 1 for space heating and domestic hot water separately as well as the overall percentage of the solar contribution, indicated in the table as SH, DHW and T respectively.

40

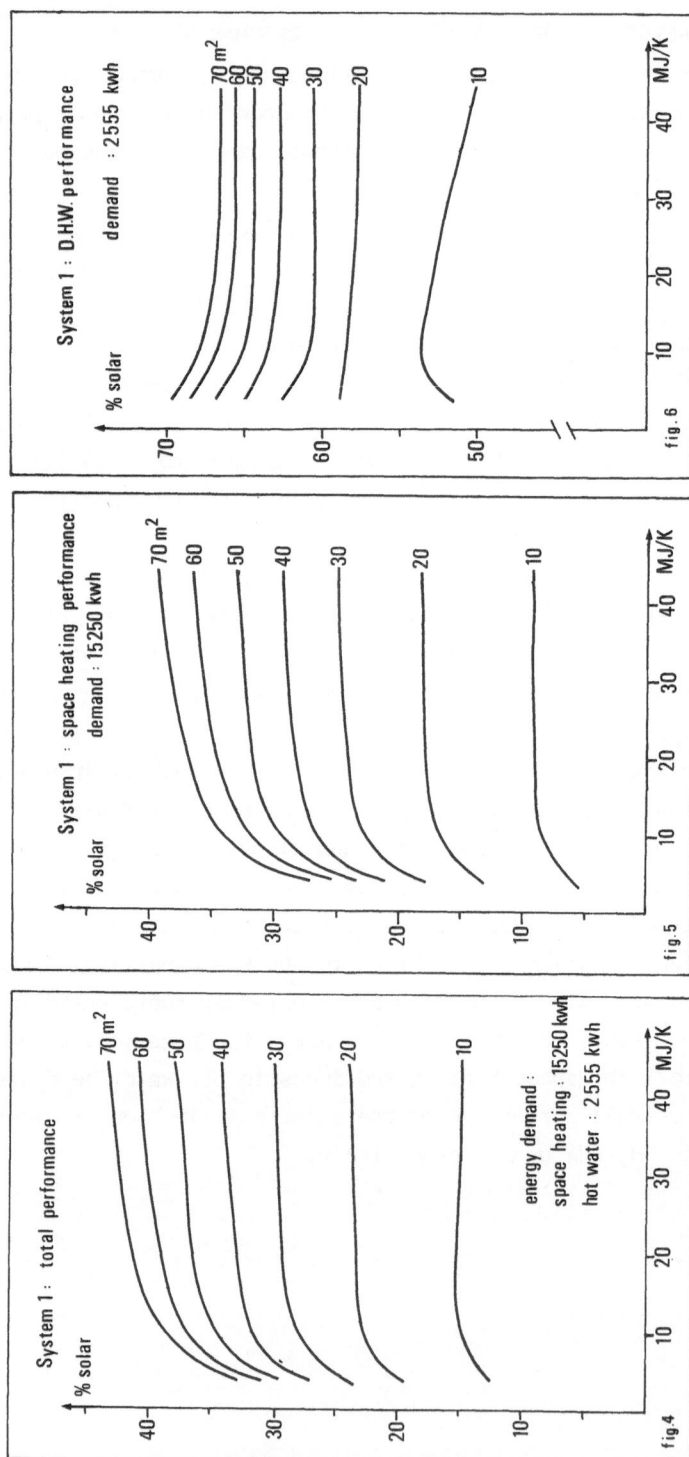

System 1 : D.H.W. performance
demand : 2555 kwh

% solar

70 m²
60
50
40
30
20
10

MJ/K

fig. 6

System 1 : space heating performance
demand : 15250 kwh

% solar

70 m²
60
50
40
30
20
10

MJ/K

fig.5

System 1 : total performance

% solar

70 m²
60
50
40
30
20
10

energy demand :
space heating : 15250 kwh
hot water : 2555 kwh

MJ/K

fig.4

Table 1.   Performance of System 1 (%)

Collector area (m²)	Thermal Capacity of the rock bed storage (MJ/K)			
	4	12	20	
10	5.9 (8.9)	8.7 (12.1)	8.9 (12.2)	SH
	51.5 (29.9)	53.5 (31.8)	52.8 (31.4)	DHW
	12.5 (11.9)	15.1 (15.0)	15.2 (15.0)	T
30	18.2 (21.3)	23.3 (26.2)	24.1 (27.2)	SH
	62.5 (33.2)	60.7 (33.4)	60.6 (33.4)	DHW
	24.6 (23.1)	28.7 (27.3)	29.4 (28.3)	T
50	23.7 (26.6)	30.3 (33.3)	31.8 (35.3)	SH
	66.8 (33.4)	64.9 (33.4)	64.4 (33.4)	DHW
	29.8 (27.7)	35.3 (33.4)	36.4 (35.1)	T

It appears from these results that above a storage capacity of about 10 MJ/K, the system performance increases only very slowly with increasing capacity. The solar contribution to the domestic hot water production is rather insensitive to the rock store thermal capacity, but decreases slightly at increasing thermal capacity of the rock store. This is a direct consequence of the connection of the water preheating system to the storage as it was mentioned in paragraph III. The same systems, when used for space heating only during the heating season of course give a higher performance for space heating but the contribution to the hot water production decreases considerably. The results for this option are given in table 1 between brackets.

For system 3, with two separated compartments in the rock store, nearly the same performance is obtained, as it can be seen from table 2, for the same cases as given in table 1. For the largest systems the performance increases slightly as compared to system 1.

Table 2. Performance of system 3 (%)

Collector area (m^2)	Thermal capacity of the rock bed storage (MJ/K)			
	4	12	20	
10	5.4	8.0 -	9.0	SH
	49.0	51.1	51.5	DHW
	11.6	14.2	15.1	T
30	17.4	23.2	24.7	SH
	61.8	61.1	60.9	DHW
	23.7	28.6	29.9	T
50	23.1	30.5	32.6	SH
	66.3	65.6	65.2	DHW
	29.3	35.6	37.3	T

Some consideration has been given to the influence of the volumetric heat transfer coefficient in the rock bed storage. For small pebble sizes, heat transfer coefficients as large as 6000 W/m^{3}K can be obtained, however at the expense of large head losses across the storage system. The results given in the previous tables are related to a value of 1000 W/m^{3}K, corresponding to a rock size of about 30 mm. For some cases, the overall performance of system 1 is given in table 3 for rock bed heat transfer coefficients of 500 W/m^{3}K, 1000 W/m^{3}K and 2000 W/m^{3}K. The corresponding values for system 3 are given between brackets.

It follows that the rock store heat transfer coefficient does not very much influence the annual system performance, although the short term performance can be very sensitive to this heat transfer coefficient. For long term performance calculations it is therefore not required to look for very accurate heat transfer correlations in pebble beds and the pebble size can be increased to 40 or 50 mm in order to reduce the pressure drop across the pebble bed.

System 3 appears to be more sensitive to this parameter for small storage capacities, but this sensitivity can be reduced significantly when the system is not allowed to switch over from one storage half to the other during the summer. This measure which can be taken manually just by switching off the corresponding control device also makes system 3 equivalent to system 1.

Table 3.   Performance of systems 1 and 3 (%)
Influence of the storage heat transfer coefficient

Collector area (m^2)	Thermal capacity of the rock bed storage (MJ/K)			heat transfer coefficient (W/m^3K)
	12	20	44	
10	14.1 (12.5)	15.1 (13.6)	15.3 (15.4)	500
	15.1 (14.2)	15.2 (15.1)	15.0 (14.9)	1000
	14.9 (16.7)	15.2 (15.1)	15.0 (14.6)	2000
30	28.2 (26.4)	29.4 (28.5)	30.2 (30.6)	500
	28.7 (28.6)	29.4 (29.9)	30.1 (30.7)	1000
	29.2 (32.9)	29.4 (31.1)	30.1 (30.7)	2000
50	34.8 (33.3)	36.4 (35.9)	38.0 (38.8)	500
	35.3 (35.6)	36.4 (37.3)	38.0 (39.2)	1000
	35.9 (39.9)	36.6 (38.9)	38.0 (39.4)	2000

The influence of the storage system insulation has been evaluated for system 1. This influence appears from the comparison of table 1 with table 4. The latter gives results obtained with a rock bed storage insulation equivalent to the thermal resistance of only 10 cm of mineral wool, i.e. a five times larger heat loss coefficient than in the table 1 results.

The 50 W/m^2K collector internal heat transfer coefficient as assumed in all the calculations discussed above, cannot be obtained with simple flat plate absorbers. With porous absorbers, this value can be reached quite easily. A lower value for this heat transfer coefficient increases the absorber temperature at a given value of the air collector inlet temperature and therefore decreases the collector efficiency. With a moderate value of 20 W/m^2K, some calculations have been performed; the results are given in table 5.

Table 4. Performance of system 1 (%)
         (5X larger storage heat loss coefficient)

Collector area (m^2)	Thermal capacity of the rock bed storage (MJ/K)			
	4	12	20	
10	4.7	6.5	6.2	SH
	44.7	44.4	41.5	DHW
	10.5	11.9	11.3	T
30	16.4	20.7	20.9	SH
	60.2	58.0	57.1	DHW
	22.7	26.1	26.1	T
50	22.0	28.3	28.9	SH
	65.4	62.7	61.9	DHW
	28.2	33.3	33.6	T

Table 5. Performance of system 1 (%)
         (Collector internal heat transfer coefficient = 20 W/m^2K)

Collector area (m^2)	Thermal capacity of the rock bed storage (MJ/K)			
	4	12	20	
10	4.7	7.1	7.4	SH
	48.5	51.3	50.7	DHW
	11.0	13.4	13.6	T
30	16.4	20.9	21.8	SH
	61.0	59.7	59.4	DHW
	22.8	26.5	27.2	T
50	22.0	28.1	29.6	SH
	65.8	63.7	63.1	DHW
	28.3	33.2	34.4	T

Comparing these results to table 1, it follows that the system performance is only about 2 % lower, such that a collector heat transfer coefficient of about 30 W/m^{2}K can be considered to be sufficient.

The performance of the systems can be improved considerably when selective collectors are used. For system 1, this influence has been evaluated, assuming air collectors with an absorption coefficient of 0.9 and an emissivity of 0.1, all other system characteristics being the same as in the cases considered in table 1. The results are given in table 6.

Table 6. Performance of system 1 (%)
          Selective collectors

Collector area (m^2)	Thermal capacity of the rock bed storage (MJ/K)			
	4	12	20	
10	8.3	11.7	11.9	SH
	56.8	56.9	56.4	DHW
	15.2	18.2	18.3	T
30	23.8	28.7	29.7	SH
	66.7	64.8	64.4	DHW
	30.0	33.8	34.7	T
50	30.7	37.7	39.6	SH
	72.5	69.9	68.9	DHW
	36.7	42.3	43.8	T

The improvement of the system performance increases with increasing collector area and ranges from 3 to 7 %. For large collector areas, this effect also increases with increasing rock store capacity.

Systems including the combined operational mode (collectors-storage-dwelling) and used for space heating only as well as such systems in combination with a heat pump are discussed in detail in |3|, |4| and |5|, mainly for systems with a very large rock bed storage capacity.

V. CONCLUSIONS

Simplified equations to describe the rock bed storage in air collector systems can be used for long term system performance calculations.

The accuracy of the rock bed heat transfer correlation is not a very critical issue for long term system simulations. The performance of solar heating systems depends only slightly on this heat transfer coefficient within a wide range, such that the choice of the pebble size should depend more on head loss considerations than on the heat transfer aspect. For the collector internal heat transfer coefficient in air collectors, a value of 30 W/m^{2}K is a good compromise between system efficiency and collector head losses. Beyond this value, the system performance increases only very little.

The system performance is most sensitive to storage insulation and collector selectivity. Increasing the storage insulation improves the system efficiency however much less than improving the collector performance. The future development of air collector systems should therefore emphasise the development of high performance air collectors rather than improvements on the rock store characteristics.

REFERENCES

[1] Handley D, Heggs PJ, 1969. Int.J.Heat and Mass Transfer, 12, 549.
[2] Gupta CP, Mehrotra RK, 1975. Intern.seminar for Heat and Mass Transfer, Dubrovnik, Yugoslavia, August 1975.
[3] Dutré WL et al, National R&D programme Energy, Belgium, Technical reports and final report E1.
[4] Dutré WL et al, Proceedings of the International Symposium-Workshop on Solar Energy, 697-730, Pergamon Press, Cairo, Egypt, 1978.
[5] Dutré WL et al, EUR 6667.

THERMAL STRATIFICATION IN HEAT STORAGES.

W.B. VELTKAMP

1. INTRODUCTION

The term "thermal stratification" suggests a discontinuous temperaturefield. Though such is impossible the term corresponds well with our image of the phenomenon. In general we understand by thermal stratification a non scalar temperature-field, - more particular in a fluid medium, - which is an increasing function of only the elevation. In a solar energy system, thermally stratified storing leads to a considerable increase in solar heat and a reduction of pumping energy. In some multipurpose installations stratification may also have the additional advantage of making heat available at different temperatures. Although the advantages of stratified storing have been noticed in the early days of solar development, they are still grossly underestimated in almost all literature and hand-books. The reasons for this state of affairs have been indicated by van Koppen e.a. |1|. Following their lines of thought further progress in establishing a better design method for solar systems and storing has been made at the Eindhoven University of Technology. This paper is mainly concerned with some of the conclusions that have been reached in this research. The main objective of the paper is to demonstrate that these conclusions hold for actual systems.

2. ANALYSIS

A perfect solar system would encompass such unattainable features as infinitely large heat exchangers operating in counter-flow and incurring no temperature drop, a stratified storage without any mixing or heatconduction, fully absorbing collector

plate coatings and so on. It may seem futile to enter upon the
analysis of such perfect systems as economics will certainly
make their realisation impossible. Yet, assuming all features
in a system but a few as ideal sometimes leads to a simpler
analysis of the system, and consequently to a better insight
in the factors that determine the output of a solar system and
the objectives that are to be selected for research and
development. This is, according to the experience at the EUT,
certainly true for the analysis of the role which stratified
storing can play in improving the output of solar systems.
However in this paper I shall only follow this approach with
regard to the options to establish thermal stratification in the
storage. Instead of a perfect system I shall present some of
the results we have obtained by means of careful measurements
and a well validated model of the solar heating and hot water
system of the laboratories of the Food Inspection Department
at Enschede, the Netherlands. In this way I hope to demonstrate:
. that thermally stratified storing can enhance the output of
  a real system by 20% or more,
. that to this end the mass flow rate through the collector has
  to be reduced to about a third of the commonly used value,
. that in order to reach this result the average heat capacity
  flow through the collector has to be related to, and in general
  slightly higher, than the heat capacity flow through the
  distribution circuits, and
. that a thermally well stratified storage can practically be
  realised in various ways.
The third aspect is one of the features that has been analysed
in more detail in the EUT laboratories. A relation between this
"average equality" and the heat capacity of the storage has also
been established in our work, and will be the basis of the next
paper by Professor Rademaker. A further account of other topics
of our work will be given in some papers that are currently
prepared for professional periodicals.

3. MODELLING A STRATIFIED STORAGE
   The model of a stratified storage that we have developed is

of a fairly general nature and can be used for most of the system
configurations commonly used, both for liquids and solid storage
materials. The most important simplification in the model is
the one-dimensionality, i.e. the temperature depends on only one
space direction. If for any reason a temperature inversion is
introduced in a fluid storage, the behaviour is described as an
instantly mixing of the adjacent layers, with complete temperature
equalisation. The discretisation of the storage is in a number
of segments with fixed positions. Contrary to the usual statements
in the literature, the convergence of such a model proved to
be slow, namely linear with the place discretisation, i.e. the
error is inversely proportional to the number of segments.
Further the error in the final result appeared to be proportional
to the heat loss factor of the collector. E.g. a loss factor of
6 $Wm^{-2}K^{-1}$, a discretisation of the storage in 16 segments and
a sequentialisation in steps of 5 minutes leads to an error in
the yearly energy output of about 1.5% and an error in the
collector throughput of about 5%. The error is an underestimation
compared to reality. In this context it should be noted that
the investigations reported in the open literature on the
benefits of stratification are mostly carried out with a
discretisation in only 3 segments. Our systematic investigations
have shown that as a consequence the output of the solar system
is underestimated by as much as 10%.

4. MEASUREMENTS AND VALIDATION OF THE STRATIFIED STORAGE MODEL
    In the Netherlands the solar heating and hot water projects
in Eindhoven (Solar House of the EUT) and in Enschede (Laboratory
of the Food Inspection Department) have, from the very start,
been designed to take full advantage of the stratification.
Therefore floating inlets are applied in the water storages in
both systems. The system configuration in Enschede (see Fig. 1)
is notable by the two floating inlets, one for the return of
the collector and one for the return of the heating system. By
means of the last inlet (left in Fig. 1) it was possible to
introduce a temperature inversion in the storage and so to
observe and analyse the behaviour with locally disturbed

50

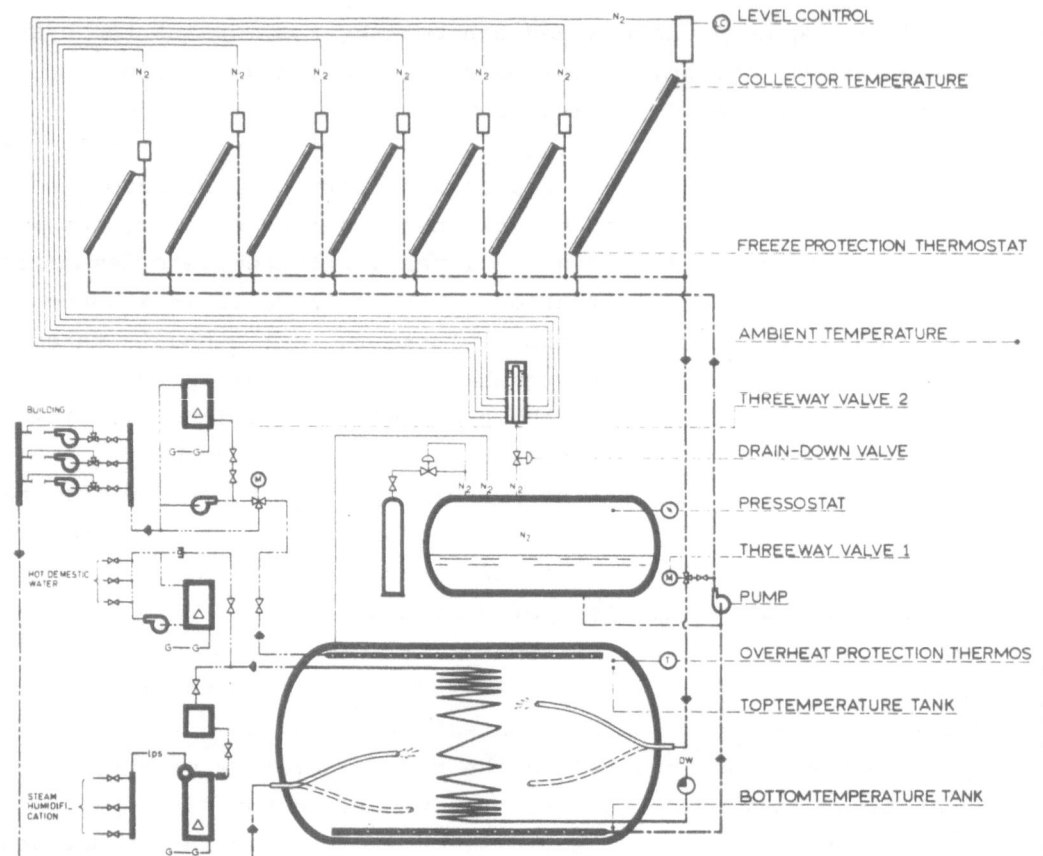

FIGURE 1. Schematic of the installation at Enschede (collector area 344 m^2, storage volume 31 m^3, working fluid demineralised water).

stratification. In order to introduce the disturbance the flow through the (air) heating system was set at maximum, thereby bringing the hot water from the top of the storage to the ultimate reach of the floating inlet, which is about halfway up the tank. Figure 2 permits a judgement of the quality of the stratified storage and system model that has been developed. The curves cover a three day period (29-31 aug. 1979) during which the storage was heated via the collectors during the day and its upper half was "inverted" via the heating circuit during the night. Moreover hot water was used during daytimes, and also some heating was applied, resulting in a really complex

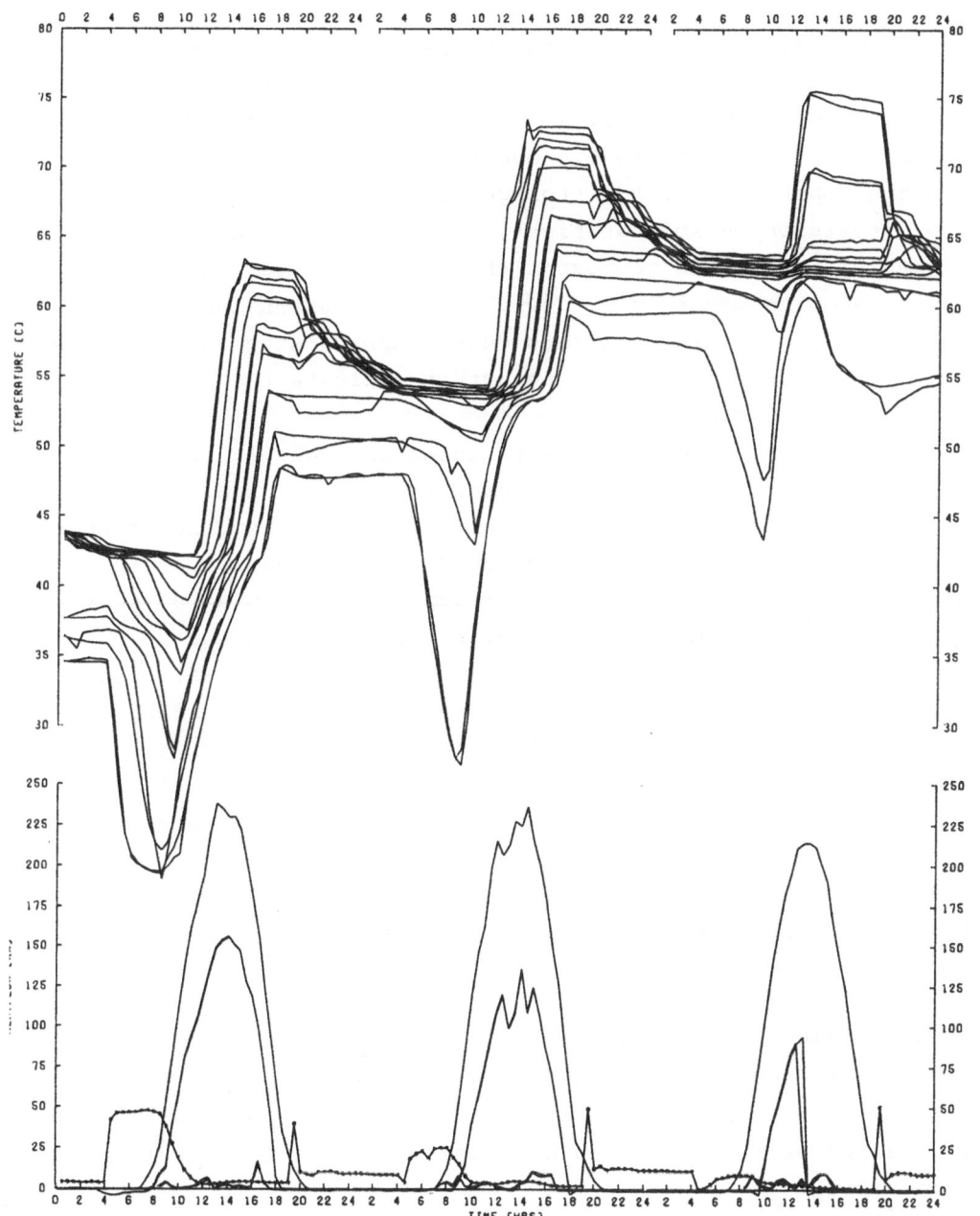

FIGURE 2. Measurements and simulation of the Enschede system during 3 days (29-31 Aug., 1979). Light lines represent measurement heavy lines simulation. Upper part: 8 storage temperatures for subsequent segments of equal volume. Lower part: higher curve = incident irradiation, second curve = heat flow from collector, dotted curve = load of heating system, lower curve = hot service water load.

supply and demand pattern. The curves in Fig. 2 - both the
experimental and calculated - show how the stratification is
established during the day (only in the upper half of the tank
on the third day because the overheating protection acts) and
how the upper part of the storage is gradually mixed during
the "inversion period" at night. It should be noted that the
slopes of the curves represent the speed at which the water
in the storage moves up or down. Parallel curves indicate
the absence of mixing, whereas curves approaching each other
demonstrate mixing and parting curves indicate the presence
of a floating inlet.

The agreement between the experimental and calculated curves
is as good as can be expected under the complex load and supply
pattern investigated. Important however is the experience that
this agreement could only be reached with a division of the
storage in 64 segments. A smaller number inevitable led to
deviations, particularly as regards the slopes of the
temperature curves, the cause being an artificial heat conduction
connected with the transport of water across the boundaries
of the segments. The penalty for this good agreement in terms
of computing time is considerable, For instance a year's run
with only 16 segments, hourly weather data and 5 minutes time
steps already takes a procestime of about 200 sec on a fast
Burroughs 7600. Even then the performance is still underestimated
by 1.5% for a collector heat loss factor of 6 $Wm^{-2}K^{-1}$, as
mentioned before.

5. THE OPTIMUM COLLECTORFLOW

As Rademaker |2| and van Koppen e.a.|1| have shown an optimum
non infinite collectorflow exists for a solar system with a
stratified storage (here we consider only on-off control and
fixed flow).

With the validated model, the dependence of the energy gain
on the collectorflow is now analysed for the Enschede system
for the actual configuration and some modifications. These
modifications were selected with the objective to disclose:
- the benefits of stratification as compared to uniform storage,

- the performance difference between a floating inlet and three
  other simple means of obtaining stratification, namely an amply
  dimensioned heat exchanging coil, supplying the water from
  the collectors in the top of the tank and storing the heat
  in some porous medium (e.g. a pebble bed or soil),
- the influence of the quality of the collector, by substituting
  $U_L = 1.8$ Wm^{-2}K^{-1} and $\alpha\tau = .67$ in place of the actually measured
  $U_L = 6.2$ Wm^{-2}K^{-1} and $\alpha\tau = .76$ and
- the usefulness of the second floating inlet for the heating
  system.

The calculations were made with the Dutch reference year developed
by the EUT |3|, as hourly-climatic input. The heating load of
the laboratory was determined by means of the repeatedly
validated computerprogram KLI |4|, with timesteps of 5 minutes
to preserve any dynamic effects, the annual heating load
amounted to 124 MWh. The hot water load was assumed to be
evenly distributed over the working hours and to total up to
an annual 70 MWh. In this way a close approximation to reality
was attained, the most important deviation probably being that
the discretisation of the storage was restricted by 16 segments
in order to save computing time. As mentioned before this leads
to an underestimation of the heat gain of 1.5% for the conventional
spectral selective, collector and of .5% for the vacuum collector.
In fig. 3 the useful heat gains calculated are depicted versus
the collector flow rate per unit of collector area. The upper
group of curves concerns the vacuum collector, the lower group
the actual, conventional collector. From the results the
following conclusions can be drawn:

- in both groups all the systems with a stratified storage
  exhibit a maximum heat gain at the low collectorflow rate of
  about .012 m^{3}m^{-2}hr^{-1}. In all cases this maximum is higher
  than the best output obtained with a uniform storage at an
  infinite flow rate. The best stratified system (with floating
  inlets) shows an extra heat gain of 7% for the vacuum collectors
  and of as much as 19% for the conventional collectors,
- replacing the collector's floating inlet by a heat exchanging
  coil or an inlet in the top of the tank reduces the heat gain

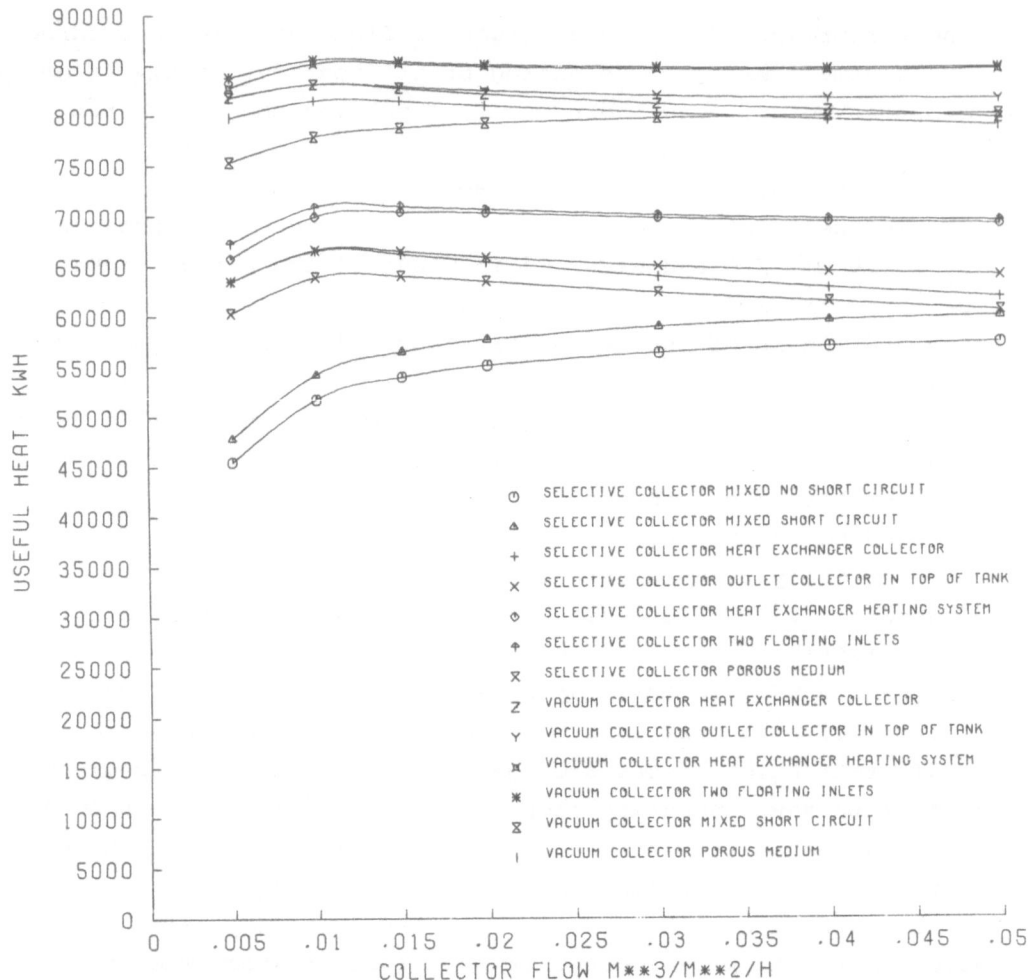

FIGURE 3. The useful heat versus the collectorflow, for several system configurations and two collectortypes.

by 3% for the vacuum- and by 7% for the conventional collectors,
- storing in a porous (say solid) medium leads to a 5% lower
  heat gain for the vacuum collectors and a 11% lower heat gain
  for the conventional collectors, again compared with the
  floating inlet system,
- if no floating inlet for the collector is applied, the
  stratified and uniform storages converge to the same value
  as the flow goes to infinity. Contrary this convergency does
  not excist for a system with a floating inlet for the collector.

- in the best stratified system the vacuum collectors perform
  20% better than the conventional ones; for the uniform
  storage the difference amounts to as much as 32%. As expected
  the vacuum collector systems are less sensitive to the quality
  of the other components of the installation than the systems
  with the conventional collectors. The sensitivity seems to be
  roughly proportional to U_L, which is not surprising as the
  heat losses of the collectors are proportional to U_L,
- finally, replacing the heating system's floating inlet by an
  ample dimensioned heat exchanging coil leads to minor
  reductions in performance, accounting to .4% for the vacuum
  collector and 1% for the conventional collectors.

Although calculated for the Enschede system these results are
in the opinion of the author typical for the majority of the
heating and/or hot water systems in a humid, meso thermal marine
climate, like the Dutch. There is also little reason why they
would be essentially different for other climates or for systems
with a seasonal storage as long as the stratification is
maintained.

From an energy conservation point of view the comparison just
presented is incomplete as the pumping energy has to be taken
into account when determining the energy displaced. However the
pumping energy strongly depends on the layout of the installation
and no general conclusions can be drawn from results determined
for a particular system! Therefore we have chosen to take into
account a fairly low pumping energy of .4 Wm^{-2} at a flow rate
of .015 $m^3 m^{-2} hr^{-1}$. Further the pumping energy was assumed to be
proportional to the flow rate to the power 7/4, and the ratio
between primary and electric energy to be equal to 3.

Figure 4 shows the so calculated displaced energies. As expected
the optimum collectorflow rate shifts to a somewhat lower value,
the optima are more pronounced and the curves for a uniform
storage now show a maximum at a finite collector flow rate. In
table 1 some absolute and relative numerical results for the
displaced energy are presented; they permit a comparison between
the various system configurations. Obviously the pumping energy
does hardly change the optimum value of the flow rate.

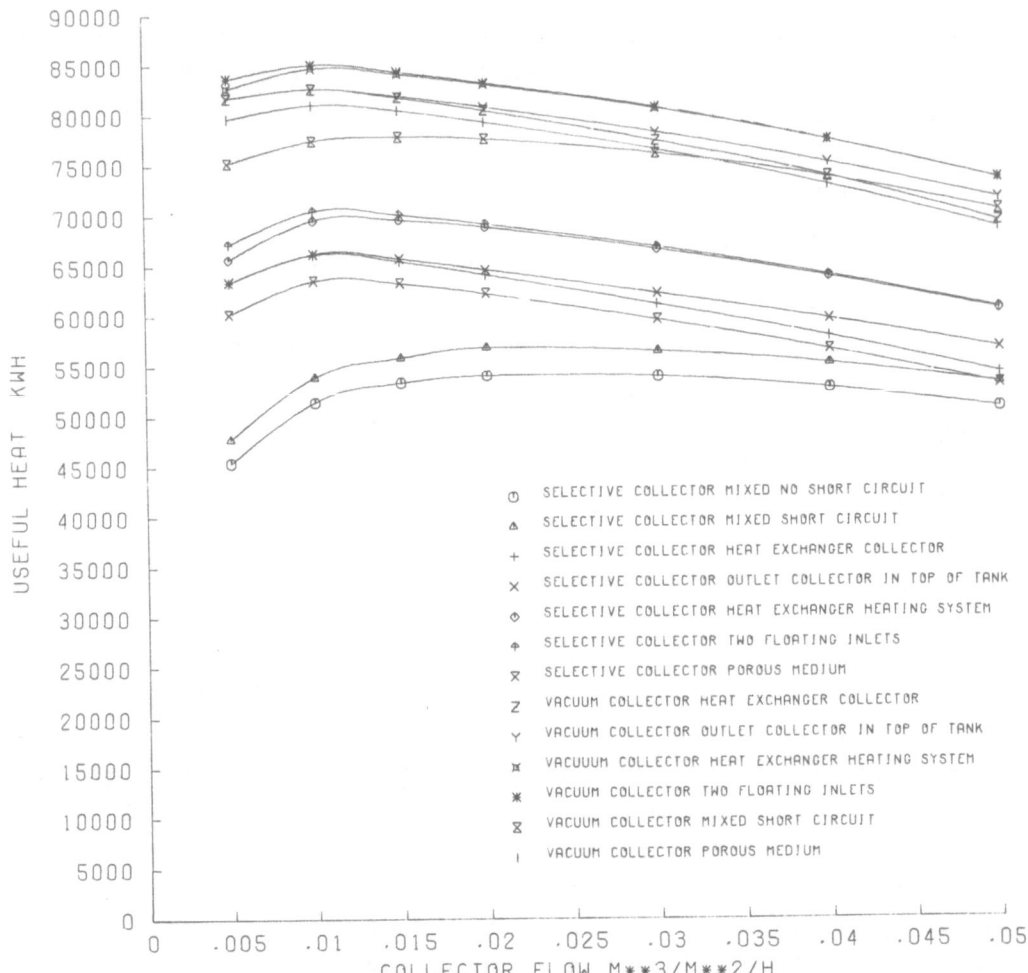

FIGURE 4. The displaced primary energy versus the collectorflow for a pumping energy of .4 Wm^{-2} at a flow rate of .015 $m^3m^{-2}hr^{-1}$ for several system configurations and two collectortypes.

	System	Conv.coll.U_L=6.2$Wm^{-2}K^{-1}$			Vac.coll:U_L=1.8Wm^{-2}		
		kWh/yr	%	%	kWh/yr	%	%
S	Two floating inlets	70500	126	100	85100	110	100
S	Heatexchanging coil or inlet top tank	66200	119	94	82700	107	97
S	Porous medium	63800	114	91	81100	105	95
S	Coil for heating system	69800	125	99	84700	109	100
U	Uniform storage	56600	100	79	77900	100	91

Table 1.

Table 1 shows absolute and relative energies displaced for various system configurations at a pumping energy of .4 Wm^{-2} for a collector flow rate of .015 $m^3m^{-2}hr^{-1}$. S = stratified storage U = uniform storage. All values refer to the particular optimum operating point of the individual system.

6. COLLECTOR- AND DISTRIBUTION CIRCUIT FLOW RATE

One of the most striking results of the optimisation studies at the EUT was that in order to obtain a maximum heat gain the average collector flow rate in systems with a stratified storage should be roughly equal to the average flow rate in the distribution system |5|. From systems with an on-off control and a fixed collector flow rate the best ratio between the flow rates was found to lie between 1 and 1.5 depending on secondary circumstances (for different working fluids in the collector and distribution circuit heat capacity flow rates should be read instead of (mass) flow rates). The objective of this paper being to demonstrate that these theoretical results hold for actual systems we also transformed the results presented so far to that flow ratio. The resulting curves, depicted in Figure 5 clearly show the optima to lie in the region mentioned. The optimum being rather flat a rule of thumb for the pre-design stage might be that this average flow ratio should be chosen at 1.2.

For the sake of completeness I want to add that not only the collector/distribution flow ratio, but also the storage volume (heat capacity) and the effectiveness of the heating system play a role in the position of the optimum operating point. Finally I want to point out that the very essence of the optimising of such systems can be reduced to the rule: Avoid as far as possible any temperature equalisation anywhere in the system |6|.

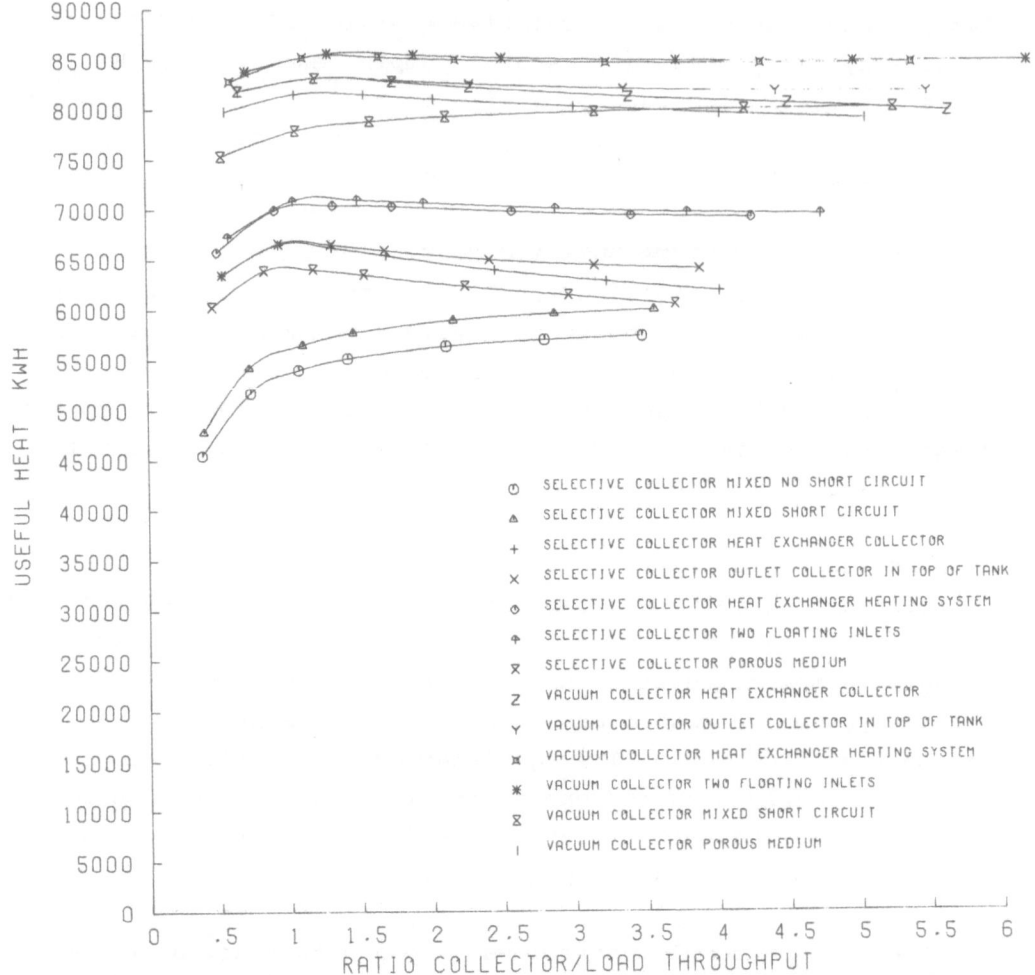

FIGURE 5. The useful heat versus the ratio between collector and distribution throughput, for several system configurations and two collectortypes.

6. CONCLUSIONS

Starting from measurements and a well validated model for the Enschede solar heating and hot water system it has been shown:
1) that the proper utilisation of thermally stratified storage enhances the output of low temperature solar systems by 10% and more,
2) that to this end the mass flow rate of the collector has to be low and about a third of the commonly used value,

3) that moreover, in order to reach this result, the average heat capacity flow through the collector has to be roughly equal and in general slightly higher than the average heat capacity flow through the distribution circuit, and

4) that a thermally well stratified storage can practically be realised in various ways, the floating inlet giving the best results.

Further important by-products of the investigations are:

5) that vacuum collectors ($U_L \approx 1.8 \ Wm^{-2}K^{-1}$) perform up to 20% better than conventional flat plate collectors ($U_L \approx 6 \ Wm^{-2}K^{-1}$) in a humid, mesothermal marine climate like the Dutch, and

6) that the accurate modelling of a stratified storage requires a discretisation of the storage in at least some 20 segments.

Further research is necessary to broaden the conclusions to a wider field of climates, systems and loads.

A promosing application can be found in the seasonal storage of solar heat in soil or water.

REFERENCES

1. van Koppen CWJ, Simon Thomas JP, Veltkamp WB, The actual benefits of thermally stratified storage in a small and a medium size solar system, Proceedings of the International Solar Energy Society, Atlanta, May 1979.
2. Rademaker O, On the dynamics and control of solar systems using stratified storage, Proceedings of the International TNO-Symposium "Thermal Storage of Solar Energy", Nov. 1980.
3. van der Hoogen AAJ, Referentiejaar voor Nederland (A reference meteorological year for the Netherlands), Verwarming en Ventilatie, 33 nr. 2 (febr. 1976), pp 81-83.
4. van der Bruggen RJA, Energy consumption for heating and cooling in relation to building design, Thesis, Eindhoven University of Technology, Sept. 1978.
5. van Koppen, CWJ, Eindhoven University of Technology, private communication, July 1979.
6. Veltkamp WB, A closed drain-down system in a medium size solar heating system, ISES International Congres, Atlanta May 1979, Poster Session P-93.

ON THE DYNAMICS AND CONTROL OF (THERMAL SOLAR) SYSTEMS
USING STRATIFIED STORAGE

Prof.ir. O. Rademaker

1. Introduction

Stratified storage of sensible heat may be regarded as an example of
distributed storage of energy in which the governing intensity variable
(in most cases: temperature) has an essentially non-uniform spatial
distribution.
The potential benefits of stratified storage accrue from the fact that
at least one - and preferably most if not all - of the streams supplying
or withdrawing energy are associated with subsystem(s) (e.g. collector,
space heating) of which the performance depends on the level of the
intensity variable (e.g. temperature) at which supplied energy is
stored or withdrawn energy is taken out.

Translated into more practical terms for sensible heat storage,
these benefits may be formulated as follows:
(1a)   *higher* storage outlet temperature to *demand* subsystem,
(1b)   *lower* storage outlet temperature to *supply* subsystem,
(2a,b) the outlet temperatures are *adjustable* (within certain limits),
(3)   unique *dynamic* advantages.
While benefits (1) and (2) are usually fairly obvious and easily
understood, those related to the *dynamic* advantages and their exploitation through appropriate control are less widely appreciated,
probably mainly because of some conceptual difficulties. In case
energy supply and demand vary according to a regular periodic and preferably
predetermined pattern - as in *heat recuperators or regenerators* such
as used with blast furnaces in the iron industry - then most dynamical
aspects are easy to grasp, but in systems where both energy supply and
demand vary highly irregularly and unpredictably, the consequences
of dynamic storage are more difficult to see through.

2. Current practice

As is well known, a collector captures more heat if more cooling medium
is pumped through it (all other conditions being equal), and hence it
has become common practice to use as high a flow rate as can be re-
conciled with the power consumption of the circulation pump. If water
or a similar liquid is used, flow rates in the order of 10^2 kg/h per m^2
collector are customary. The total amount of water passing through a
typical solar boiler collector during a single day may amount to some
$3 - 10$ m^3; for space heating the volume is an order of magnitude larger.
It is generally considered impractical to make the storage capacity
that large and in most installations it is smaller by about an order
of magnitude, for example $50 - 100$ kg per m^2 collector. Consequently,
the storage volume is comparable to or smaller than the hourly collector
throughput, and hence the residence time of the fluid in the vessel
is one hour or less. As a result the tendency of the fluid to
stratify may be affected adversely because the relatively large flow
induces mixing. Apart from that, any stratification manifests itself
much more clearly at low flow rates (large residence times) than at
high flow rates, as is illustrated by the step responses of a ten-layer
vessel model shown in Figure 1. In short, at high circulation rates
the stratification is less developed and manifests itself more weakly,
and experiments are likely to show that stratification effects, although
clearly present, are not very pronounced.

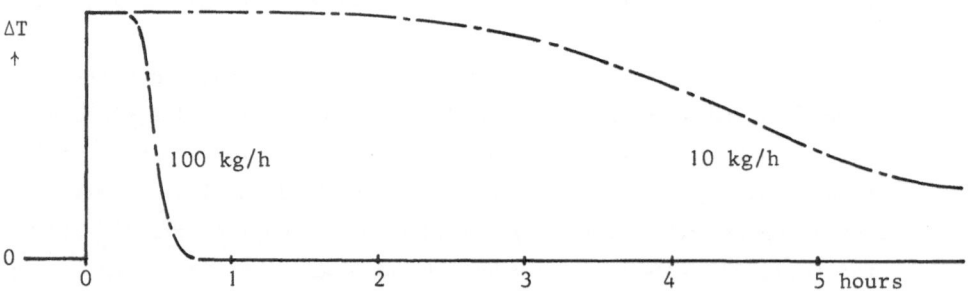

Figure 1 Greatest temperature difference (ΔT) present in ten-layer storage
as a function of time elapsed since the introduction of a stepwise inlet
temperature change for two flow rates (for details see table at the end)

From there it is but a small and apparently logical step to assume that
the effect of stratification is relatively insignificant and that the storage
vessel may be represented by a series of three or two well-stirred
volumes or even more simply: by a single well-mixed volume. An additional
motive may be that the basic equations for a layer in a storage vessel are
somewhat complicated (the generally favorite backwards-difference
approximation contains boolean expressions) and that the digital simulation
of a vessel containing many layers may become rather time-consuming, as
Veltkamp has already illustrated [1]*.

Thus, the basis is laid for a perfectly logical but misleading train
of thought. For if it is assumed that the vessel may be treated as a single
mixing unit, its energy balance will satisfy:

$$C_v \cdot dT_v/dt = Q_c - Q_d - Q_1, \tag{1}$$

where T_v represents the (uniform) vessel temperature, Q_c the heat flow
delivered by the collector, Q_d that to the demand subsystem and Q_1 the heat
loss to the surroundings. Assuming the collector inertia to be negligible
compared to that of the storage vessel, the well-known Hottel-Whillier
-Bliss equation may be employed [2]:

$$Q_c = c \cdot F_c \cdot (1 - e^{-U \cdot F'/(c \cdot F_c)}) \cdot (Q_p/U + T_a - T_v). \tag{2}$$

An elementary exercise in optimal control theory shows that the amount
of heat captured during any prescribed period may be maximised
by maximisation of the cooling medium flow rate when:

$$Q_p/U + T_a - T_v > 0, \tag{3}$$

whereas the flow rate should be zero if this condition is not satisfied.
This means that a simple on/off-control of the pump will accomplish
dynamic optimal control as well as optimisation of the momentary heat yield,
and that a high circulation rate is advantageous. As explained earlier,
the larger that flow rate is, the more the behaviour of a stratified
storage vessel will resemble that of a well-mixed volume, which would
seem to confirm the initial assumption of ideal mixing! Hence it
would appear to be quite logical to go one step further and to conclude
that the vessel may indeed be treated as being well-mixed.
As we shall see below, this apparently self-confirming train of
thought is fundamentally wrong.
-----
* References and symbols are listed at the end of this paper.

3. Exploitation of stratified storage dynamics

3.1 Conceptual tools

Let us consider the stratified storage of sensible heat in a water
vessel. Various arrangements – using internal and/or external heat
exchangers and movable inlets and outlets – are being used. While
numerical simulation of the dynamic behaviour of these systems is
not too difficult, it is often quite difficult to gain a clear
understanding of what is going on. We urgently need a good mental
tool, a simple way of reasoning about the behaviour of a stratified
storage system.

For this purpose we propose a simple concept that is the
very counterpart of ideal mixing (= perfectly uniform storage),
namely: _ideal nonmixing_, and as a standard of excellence we consider
perfect stratification , which implies that all heat is stored at
the temperature level at which it is supplied by the collector and
that the required heat is extracted at the required temperature
level (if available in the vessel). In this brief paper we shall
use only one of the attributes of such an ideal nonmixing storage
system, namely the concept of a _heat front_, which is an imaginary
horizontal boundary between the cold fluid coming from the water
supply and the fluid that has received collected heat. From now
on we shall reason as if such a heat front may indeed exist.

Suppose that at sunrise the vessel is filled completely with
cold water. When the collector starts to operate, and provided the
collector flow rate F_c is larger than the demand flow rate F_d, a heat
front comes into being in the top of the vessel and moves downwards.
If F_c becomes zero, or at least smaller than F_d, the front will reverse
its movement. Otherwise, it may eventually reach the bottom and,
so to speak, leave the vessel there; in that case, a new heat front
will come into being there as soon as F_c becomes smaller than F_d.
After sunset F_c will be zero and the heat front will rise when
heat is extracted. If enough hot water is consumed by the demand
subsystem, the heat front will leave the vessel through the top
and the whole vessel will be filled with cold water again.
Let us now consider the control of these movements by means of F_c.

3.2 An optimising control strategy

For ease of explanation, we shall consider a single-purpose solar
heating system having one loss-free storage vessel and further we
shall adopt an *a posteriori* approach: we look back upon one or more
days and investigate what would have been the best way of controlling
the solar system so as to maximise the amount of collected heat (which
might also be called: control based on perfect clairvoyance).

Suppose again that at sunrise the vessel is filled completely
with cold water and further that on/off - control of the collector
flow is used, the "on" flow rate ($F_{c,m}$) being so small that the heat
front stays in the upper half of the vessel. Let the integral amount
of heat collected be represented by IQ_c. All other circumstances being
equal, more heat could be collected by using a larger flow rate, for then
the collector would be cooled better and its heat loss would be smaller
(see Eqn. 2). However, if $F_{c,m}$ is increased beyond a certain value, the
heat front will eventually leave the vessel through the bottom and then
the collector inlet temperature will rise above the cold fluid temperature.
As follows from (2), this will adversely affect the amount of heat being
collected. The larger $F_{c,m}$ is made, the earlier the collector inlet temperature will start to rise and the higher it will rise; hence the
adverse effect will grow rapidly with increasing flow rate, whereas
the rise in collector efficiency levels off, as illustrated by Figure 2.
Obviously, the net heat yield will have a maximum for that flow rate
for which the two opposing marginal contributions are equal in absolute
value.

Figure 3 shows the behaviour of a simple system during one of a
series of clear days for three different values of $F_{c,m}$, case (b) representing the optimal choice, (a) a lower and (c) a higher value. The curves
represent the collector outlet temperature (T_c) and the temperatures
in the top (T_t) and the bottom (T_b) of the vessel. The desired temperature
T_d = 60 $^\circ$C. Note the behaviour of T_b for increasing collector throughput (earlier rise and higher peak, both having an adverse effect upon
the amount of heat collected, also because the collector is switched-off
earlier). It is interesting to note that case (b) is better than (a)
although the auxiliary heater has to work round the clock, while in case
(a) it can be switched-off during the early afternoon because T_c

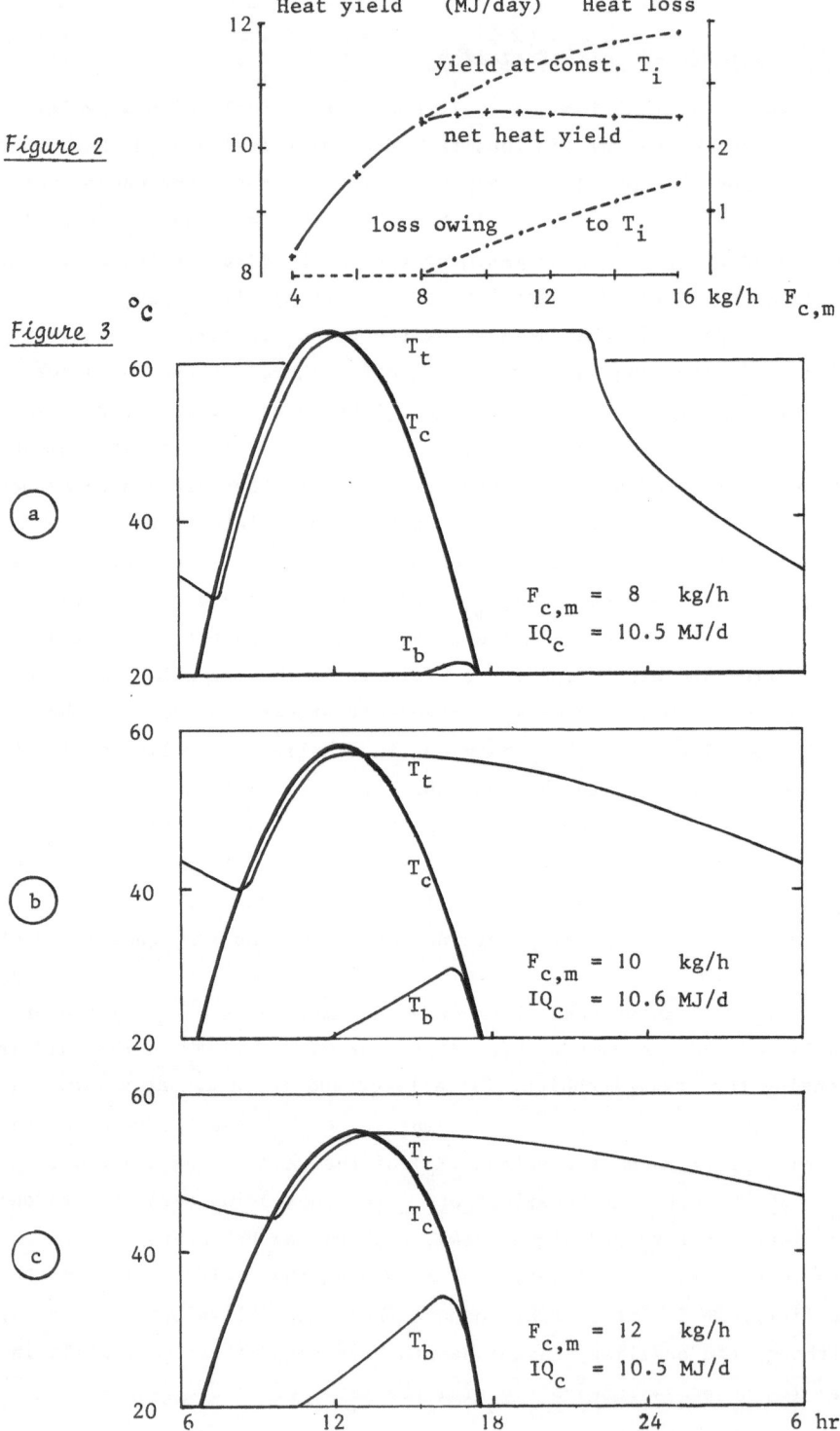

Figure 2

Figure 3

exceeds T_d. The explanation is that the collector flow rate is still below optimal. Note further that the optimal value of $F_{c,m}$ (10 kg/h/m^2) is small compared to customary values.

The optimal value of $F_{c,m}$ will be different for different days; during longer summer days it will even be lower than in the example considered above and during clouded days it will be larger. So the control strategy described here is one of (preferably adaptively) _modulated_ on/off-control.

We have given an on/off-control example here because it is simple and likely to appeal to those familiar with conventional solar-system operation. Further improvement is possible by replacing the on/off collector flow control by dynamically-optimal modulating control, but it seems rather unlikely that such a more advanced control strategy will be economically justifyable. Instead, it is probably more rewarding to look for an approach that is simpler and nearly as effective, but before discussing such a control strategy, let us consider the performance of the modulated on/off-control strategy somewhat more closely.

Figure 4 shows the total energy (IQ_c) collected during a clear, windless day as a function of $F_{c,m}$ in three cases: (a) well-mixed vessel, (b) stratified storage vessel, (c) constant collector inlet temperature (= infinitely large vessel). Curve (b) clearly shows that more heat can be collected at a low flow rate - and hence considerably less pumping energy - than if the system is operated in the customary high-flow fashion and that for all flow rates the energy gain is higher than with a well-mixed vessel (curve a). Curve (c) indicates the margin for further improvement by increasing the storage capacity.

Note that the shape of curve (b) and the location of the optimum depend on factors varying from day to day, like the duration of the insolation and the heat demand pattern. So, under _conventional_ on/off-control (i.e. _constant_ $F_{c,m}$) the yearly heat yield will be below optimal and one may even not be able to find a low-flow maximum at all.

3.3 A simplified, near-optimal control strategy

Numerous simulations have indicated that our original control strategy (called "_T.H.E. Strategy II_") may be considerably simplified with little loss of collected energy by choosing $F_{c,m}$ so that the heat front just does not leave the vessel through the bottom; this we call "_T.H.E. Strategy I_". It implies that the total amount of fluid pumped through

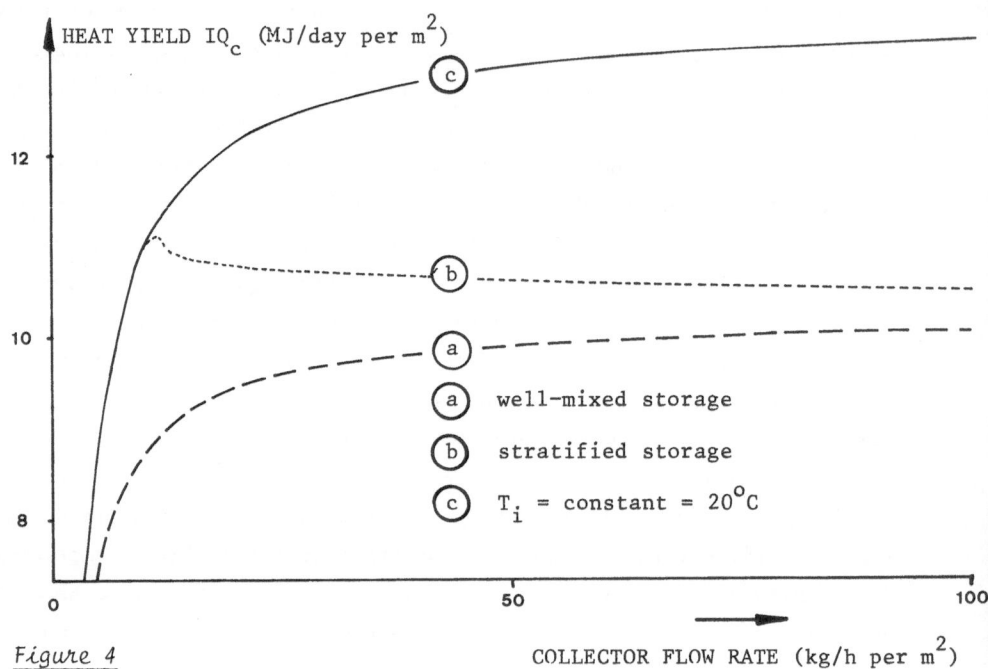

Figure 4   COLLECTOR FLOW RATE (kg/h per m^2)

the collector during a period of, say, one day has to be equal to
the amount of cold liquid present in the vessel at the beginning <u>plus</u>
the total amount of cold liquid supplied by the demand subsystem during
the operating period of the collector. In other words: all cold
liquid should pass once through the collector. Various simple schemes
may be envisaged for adjusting $F_{c,m}$ in such a way that the heat front
remains inside the vessel or leaves it only for a short time. The basic
rule is that the integral of the collector flow, IF_c, during the operating
period of the collector should be equal to the total amount of cold water.
The latter is determined by the integral of the demand subsystem flow,
IF_d. Hence the basic rule comes down to $IF_c = IF_d$. As already pointed
out by Veltkamp[1] and further explained in the preceding section,
a slightly higher value of IF_c is advantageous. The precise value of
IF_c is not very important because the optimum is not very sharp.

3.4 Influence of the heat demand pattern

Before concluding this brief survey, we should draw the attention to a
beneficial effect of the heat demand that may not yet be appreciated
sufficiently widely. Namely that the extraction of heat from the vessel
is accompanied by the supply in return of cold fluid that pushes
the heat front upwards or retards its downward movement, thus enabling
more heat to be collected by suitable adaptation of the collector
flow rate.

This effect is not insignificant. If the cooling effect of the return
stream is eliminated in the system considered in Figure 4, the daily
collector yield falls from 11.33 to 10.67 MJ and further to 10.37 MJ
if the demand flow is made zero (the corresponding increases in
the collector heat loss are +30% and +43%, respectively).

In the case of space heating, the heat demand is probably fairly
closely correlated to the heat collected by the building, and hence also
to the heat captured by the collector, whereas in the case of a
tap-water installation, the correlation is likely to be rather insignificant. This marks an important distinction between the two cases.
In the case of space heating, the correlation - whose characteristics
depend on the construction of the house and its heating installation -
has an impact on the control of the solar system as well as the
optimum size of the storage vessel. This should be investigated.

Additional control possibilities may be found in space heating
systems in which the values of F_d and T_d needed to satisfy a certain
heat demand may still be varied (subject to certain rules). As discussed above, F_d affects the system's performance; therefore, F_d can
be used to further optimise that performance. This should also be
investigated.

4. Closing remarks

It will be clear that our control strategies will be less effective
with systems using low-loss vacuum collectors than with those using
ordinary glass-covered collectors. It is to be expected that the
advantage is greatest with high-loss collectors, i.e. without any
cover. Therefore, systems using so called "energy roofs" may provide
particularly attractive applications.

Further it should be pointed out that the scope of the described control strategies is by no means limited to short-term storage systems. In fact, their applicability to large-scale seasonal storage systems may well be more profitable. Further, the scope is not limited to solar systems either. Instead of a collector, any other heat exchange process displaying the same general characteristics (heat gain rising with increasing throughput and falling with increasing inlet temperature used in combination with stratified storage, is a potential candidate and, perhaps, applications may even been found in entirely different processes in which an intensity variable other than temperature plays an analogous role.

References

1. Thermal stratification in heat storages.
   W.B. Veltkamp. This conference, same session.

2. Solar Energy Thermal Processes, Chapter 7.
   J.A. Duffie, W.A. Beckman. John Wiley 1974.

Symbols

c	specific heat of heat transport fluid	$J.K^{-1}.kg^{-1}$
C_v	total heat capacity of storage vessel	$J.K^{-1}.m^{-2}$
F'	collector efficiency factor [2]	
F_c	flow rate of heat transport fluid per m^2 collector	$kg.s^{-1}.m^{-2}$
$F_{c,m}$	maximum value of F_c under on/off-control	$kg.s^{-1}.m^{-2}$
F_d	demanded flow rate	$kg.s^{-1}.m^{-2}$
IF_c	integral of F_c over given time-interval	$kg.m^{-2}$
IF_d	integral of F_d over given time-interval	$kg.m^{-2}$
IQ_c	integral of heat gain per m^2 collector over given time-interval	$J.m^{-2}$
Q_c	heat flow captured per m^2 collector	$J.s^{-1}.m^{-2}$
Q_d	demanded heat flow, per m^2 collector	$J.s^{-1}.m^{-2}$
Q_l	heat loss flow from vessel, per m^2 collector	$J.s^{-1}.m^{-2}$
Q_p	solar power effectively absorbed by plate, per m^2	$J.s^{-1}.m^{-2}$
t	time	s
T_a	ambient temperature	K or C
T_b	temperature in vessel near bottom outlet	K or C
T_c	collector outlet temperature	K or C
T_d	demanded temperature	K or C
T_i	collector inlet temperature	K or C
T_t	temperature in vessel near top outlet	K or C
T_v	average temperature of C_v	K or C
U	plate heat loss coefficient	$W.K^{-1}.m^{-2}$

DATA USED IN THE CALCULATIONS OF THE FIGURES

Figure 1

Vessel holdup:	50 kg per m^2 collector
Inlet from collector:	in the top
Outlet to collector:	in the bottom
Collector flow rate:	10 or 100 kg/h per m^2
Demand flow rate F_d:	0
Vessel loss coefficient:	0

Vessel is initially filled with water of $20°C$. At t = 0 the collector outlet temperature changes from 20 to, say, $40°C$, which creates a temperature profile in the storage. The largest temperature difference in the vessel, ΔT, is plotted versus time. Eventually, ΔT tends to zero (uniform storage temperature of $40°C$) because the vessel heat-loss is neglected.

Figure:		2 & 3	4
$Q_{p,max}$	$W.m^{-2}$	640	640
T_a	K	0	0
U	$W.K^{-1}.m^{-2}$	5	5
T_d	K	60	40
Q_d	$W.m^{-2}$	156.2*	156.2*
T_{return}	K	20	20
Nr of segments		90	60
Vessel holdup	$kg.m^{-2}$	80	50
Solution type		periodic	1 day, cold start
Time step	min	0.2 - 2**	0.2 - 2**

* This constant heat demand leads to a daily consumption that is equal to the total amount of heat captured in one day if the heat removal factor equals F'.

** The usual time step is 2 min. At high values of F_c, smaller time steps may be necessary to ensure accuracy.

LATENT HEAT STORAGE

SUMMARY AND OVERVIEW

Prof. ir. C.J. Hoogendoorn
Delft University of Technology, The Netherlands

In this chapter eight papers are presented; two of them are general
papers on heat storage studies in the USA (Michaels) and in Sweden
(Ofverholm). They give an excellent review of work going on outside the
EEC and they will not be further reviewed in this summary. The other 6
papers are on research on latent heat storage and all were partly
sponsored by the commission of the European Communities. They show the
two main attractive points of latent heat storage:
1. high volumetric energy density
2. large heating effect at a nearly constant temperature.

Essentially two types of phase change materials (PCM's) have been
considered: salt hydrates (Furbo, Van Galen and Abhat) and organic
materials like paraffins (Mancini, Marshall, De Jong and Abhat). Each
have their distinctive problems in application like for salt hydrates:
1. segregation,
2. supercooling,
and for paraffins:
3. wide melting ranges,
4. low thermal conductivity,
whereas for all latent heat storage systems an important problem area
is:
5. temperature (exergy) losses in heat exchangers.

To overcome these problems the different papers indicate several
possibilities. Furbo uses the extra water principle and a stirred vessel
to solve the segregation problem for salt hydrates. Van Galen uses a
colloidal polymer matrix to reduce segregation when using sodium acetate
trihydrate. He also could overcome the supercooling problem by using
a permanent cold zone in his cylinders filled with PCM. Abhat stresses
the need to obtain experimentally thermophysical properties of PCM's
before testing them on larger scale. A differential scanning calorimeter

(DSC) can give the required calorimetric data. For paraffins Abhat and
De Jong show that they may have a wide melting range and secondary heat
effects instead of a constant melting temperature as quoted in literature.
De Jong indicated that a scond solid-solid phase transition in paraffinic
materials at temperatures 10 to 20 ^{o}C below the melting point may account
for 25 to 40% of the heat of fusion for these materials.

Heat exchanger development to obtain optimal results is required as
indicated by Abhat. This applies especially to the discharging (solidifying)
part of a cycle; the heat transfer rates are then limited due to solid
state thermal conduction and may reduce the discharge rate. He presented
a finned annulus heat exchanger to increase thermal performance. De Jong
also presented finned tube results, for paraffins. He shows that the use
of aluminium honeycombs or other thin strip metal matrix structures can
increase the thermal conductivities by a factor 4 for relative volume of
the metal in the paraffin of only 2%.

It is noteworthy to point out that in the paper by Fittipaldi in
the session on chemical storage essentially a latent heat storage
principle is discussed. Here a solid-solid phase transition in organo-
metallic chemical compounds has been used. Metal is incorporated in the
molecular structure and also gives rise to an increased thermal conductivity.
However, these materials are still expensive compared to paraffins or
fatty acids.

Other methods to improve heat transfer are encapsulation, formation
of small pellets that can directly be contacted with the heat transfer
fluid (Mancini). Finally in the USA-review paper Michaels indicated the
system where direct contact heat transfer has been used with an
immiscible heat transfer fluid.

Another important area touched upon by Marshall is the test procedure
for latent heat storage devices. He indicates that the ASHRAE Standard
Test Procedure shows anomalies and is not satisfactory in some respects.
He suggests a modified procedure. Also Van Galen found that the ASHRAE
test does not give sufficient information to predict actual storage
efficiency in a real solar system. He even suggests extended tests in a
computer simulated solar system coupled to a simulated load. It is clear
that for proper evaluation and optimisation of latent heat storage systems
an improved test method is required.

Commercially only a few latent heat storage devices are marketed. In the USA Michaels mentioned Dow Calcium Chloride rods (27 $^{\circ}$C) and a Calmac PCM (46 $^{\circ}$C). However, costs are still high compared to sensible heat storage. In general problems on corrosion, shrinkage and long term stability after many cycles are disadvantageous for PCM's.

Conclusions on latent heat storage devices from this overview are:
1. Costs higher or nearly equal to sensible heat storage
2. Reliability on a long term still low to average
3. Few commercial applications
4. Further work required on:
    a. Thermal physical properties
    b. Development of low cost new materials
    c. Encapsulation
    e. Test procedures

As a prediction of short term possibilities for PCM's it can be said that they will have a limited application versus the sensible heat devices. They are mainly attractive when a narrow temperature range is required by the solar and heat load system.

AN OVERVIEW OF THE U.S.A. PROGRAM FOR THE DEVELOPMENT OF THERMAL ENERGY
STORAGE FOR SOLAR ENERGY APPLICATIONS

ALLAN I. MICHAELS
ARGONNE NATIONAL LABORATORY
ARGONNE, ILLINOIS   60439, U.S.A.

1.   INTRODUCTION

The administrative responsibility for thermal energy storage (TES)
technology in the U.S.A. resides chiefly in the DOE Office of Solar Applications for Buildings (OSAB), and the Office of Advanced Conservation
Technology (OACT).   Management support and technical direction of DOE projects are provided by several U.S.A. National Laboratories.   The organizational structure related to TES programs is given in Figure 1.

FIGURE 1.   U.S.A. DOE ORGANIZATIONAL STRUCTURE DIRECTLY

RELATED TO THERMAL ENERGY STORAGE PROGRAMS

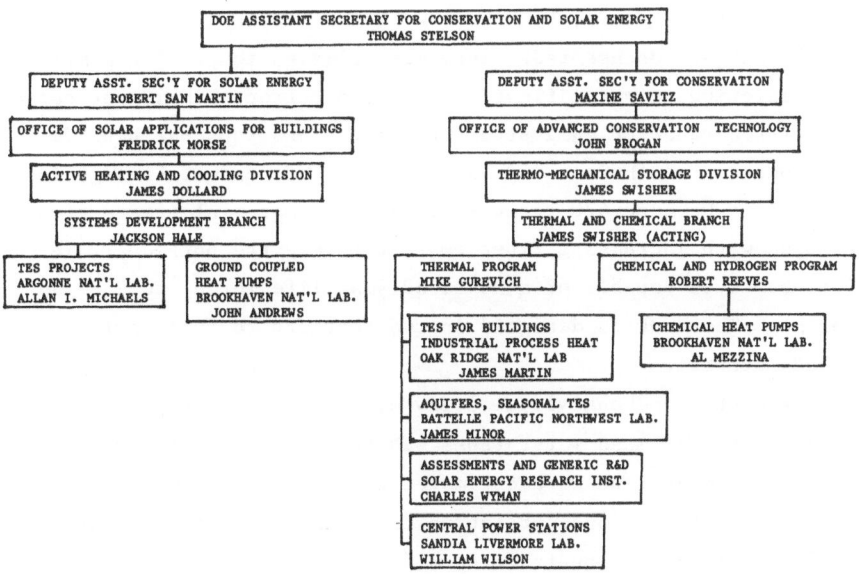

The OSAB program in TES is limited to solar heating and cooling.
The main thrusts of the program are to assess and disseminate information

on available TES methods and devices, to support innovative and engineering design improvements in conventional TES systems, and to develop new, higher density TES devices utilizing latent heat of fusion (LHS) and chemical reactions. The OACT has a broader TES responsibility, which includes: off-peak charging, waste heat recovery and solar energy for buildings; industrial process heat; and solar central power stations.

The following overview of U.S.A. TES technology will briefly describe projects of both Offices in sensible heat storage, latent heat storage and chemical heat pumps for solar heating and cooling; seasonal heat storage; and TES in industrial process heat and solar central power applications.

SENSIBLE HEAT STORAGE FOR DIURNAL HEATING AND COOLING

Water Tanks *

Argonne National Laboratory (ANL) has developed, Sha[1], and Lin[2], a 3-dimensional water tank model COMMIX-SA. Experimental data for validation of the code is being obtained from in-house experiments, Cole[1], and from measurements by Anderson[1] at the University of Nebraska.

Two improved one-dimensional water tank codes have been developed and experimentally validated, by Kuhn[1]. These models have been installed in TRNSYS and documented. They are believed to provide a more realistic prediction of stratification than the current TRNSYS subroutine.

An insulated sandwich wall structure fiberglass tank has been designed and constructed, Hudson and Jones[1], by Independent Living, Inc. The inner wall is Derkane 470 resin - glass fiber material (resin rich on the water side with increasing fiber content toward the outside). This material is believed capable of a 20 to 30 year life for up to 95°C hot water storage. The tank is designed in two sections to facilitate shipping and retrofit.

An experimental and analytical study of alternate support structures for flexible membrane lined water tanks was recently completed by Bourne[1]. On-site constructed designs were estimated to have installed costs of about 1/3 that of conventional prefabricated tanks. Two field installations are presently under construction.

Westinghouse Corp. is completing tests and analyses on multilayer, foamed concrete tanks, Buckman[1]. Three prototype tanks, two using molded semi-spherical sections, and the third rectangular panels have been

constructed and tested. Because of high shipping costs it was concluded
that on-site construction, using mobile plants for concrete foaming,
would be most economical.

Rock Beds

ANL is developing a 3-dimensional simulation code for rockbeds,
COMMIX-SA-RB, Sha[1]. Solaron Corp. has begun experimental studies to
provide required thermophysical parameters and data for validation of the
code, Jones [1]. Designs for uniform flow horizontal and vertical
rockbeds will be jointly developed.

An improved one-dimensional rockbed code, has been developed by Boeing
Computer Services, and installed in TRNSYS, Kuhn[1]. This code has been
experimentally validated and is considered to provide a more realistic prediction of performance than the current TRNSYS rockbed subroutine.

Miscellaneous

A Thermal Energy Storage Design and Installation Manual was published,
by Cole et. al[4]. Wachtell[1] has analyzed self-driven heat circulation
methods for solar energy systems. Colorado State University completed
tests on its prototype Direct Contact Liquid-Liquid Heat Exchange Water
Tank, Karaki[1]. Results indicate some degree of performance and cost
advantage over conventional heat exchange systems. Simulations are being
run to further evaluate performance.

2. LATENT HEAT STORAGE FOR HEATING AND COOLING

Commercial LHS Devices

One promising LHS entry to the market has run into difficulties. This
is the air heat transfer system manufactured by Valmont Industries of
Valley, NE. This system packages $Na_2SO_4.10H_2O$ (Glauber's Salt) dispersed
in a rigid suspension material in flat HDPE trays. Each tray has a rated
storage capacity of 1.60 MJ and costs \$12.50, equivalent to \$7.81/MJ.
Problems were encountered with leakage and a 40% decrease in storage capacity, after several hundred cycles, increasing storage cost to \$13.16/MJ.
Valmont has stopped production. A new formulation of Glauber's salt is
under test. Valmont expects to decide by January, 1981 if production will
be resumed.

Dow Chemical Co. is marketing a chemically modified form of $CaCl_2$.
$6H_2O$, called "Thermol 81" which is congruently melting and has minimal

super-cooling[5]. PSI Inc. of Fenton, MI, is selling "Energy Storage
Rods", which are carbon black coated HDPE cylinders filled with Thermol
81, with a capacity of 2.34 MJ, at $29.80/ rod, equivalent to $12.74/MJ.

Calmac Corp. of Englewood, NJ has introduced a line of "Heat Bank"
TES systems utilizing bulk containment of LHS media with internal spirals
of close-spaced plastic tubing for heat exchange, MacCracken[1]. Figure 2
shows a schematic of these units. The Heat Bank-115, which contains
$Na_2S_2O_3 \cdot 5H_2O$ (Hypo, $Tm = 46°C$), and an Ice Bank Unit, are being marketed.
Units with a 116°C salt ($MgCl_2 \cdot 6H_2O$) and a 7°C salt (a Glauber's salt
based eutectic) are available. The 46°C and 116°C units have been exten-
sively cycled with no loss in storage capacity. The 45°F unit, however,
has experienced capacity loss on cycling under some conditions. The Heat
Bank-115 unit has a storage capacity of 400.94 MJ for a ΔT of 19°C
(331.30 MJ LHS). A price of $3000 was quoted, equivalent to $7.48/MJ.

TES Inc. of San Diego, CA has marketed a similar LHS unit, but util-
izing aluminum double tube finned strips as the internal heat exchanger.
The medium is also hypo but with proprietary additives. The unit has a
LHS capacity of 263.78 MJ and costs $3500, equivalent to $13.27/MJ.

A bulk storage unit, with direct contact internal heat exchange, is
being marketed by OEM Products, Solarmatic Div. of Plant City, FL. This
"Heat Battery" unit uses Glauber's salt and pumps an immiscible mineral
oil out of arrays of nozzles located at various depths in the salt. Heat
exchange is via two copper coils located in the salt-oil mixture. The oil
stirs the salt and enhances heat exchange between the salt and the copper
coils. No significant loss in storage capacity with cycling is reported.
A schematic of this device is shown in Figure 3. The LHS capacity of the
unit is 258.73 MJ at 32°C. A unit was quoted at $2800, equivalent to
$7.81/MJ.

Engineering and Prototype Development And Testing Of LHS Devices

The Institute of Energy Conversion (IEC) at the University of Dela-
ware has developed a low cost multi-layer plastic film sausage or "chub"
shaped package for LHS materials. Tests by Frysinger[6] have established
adequate durability and impermeability. A prototype LHS unit has been
built with a 15°C, 93 kJ/kg Glauber's salt based eutectic plus a
thickening agent packaged in 0.051M diam., 0.51M long chubs stacked in
horizontal arrays[7] for a residential, storage assisted cooling system
under test by the New York State ERDA, Rizzuto[8]. IEC estimates a
production cost of $700 for a 285 MJ unit, equivalent to about $2.28/MJ.

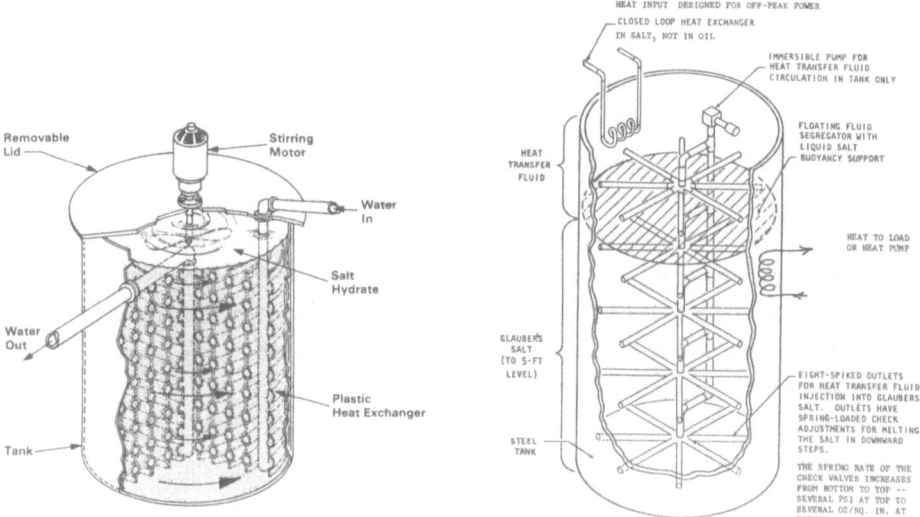

Figure 2.   CALMAC LHS Unit   Figure 3.   OEM LHS Unit

Form-stable, cross-linked HDPE pellets have been developed and
tested by Monsanto Corp[9]. This material melts at 130°C with 200 kJ/kg
and can withstand repeated cycling without loss in storage capacity or
pellet form-stability. The cost for a complete system was estimated to
be $8.90/MJ with ethylene glycol heat transfer fluid, or $7.48/MJ with air
heat transfer. A 6800 kg batch is being prepared by electron beam cross-
linking, Salyer[8], for use in LHS devices being developed at ANL.

A direct contact heat transfer LHS unit is under development by International Thermal Instrument Co. of Del Mar, CA[10]. A stirrer mixes
immiscible oil and salt to enhance heat transfer between the salt and a
heat exchanger coil located within the mixing region. A prototype 105 MJ
unit containing Sodium Acetate Trihydrate (Tm = 58°C) and "XTALOV #135C"
heat transfer oil has been built and is being tested.

A prototype system with a water tank enclosed in a paraffin layer
has been built and tested by Environmental Solar Systems, Inc. of Troy,
NY[11]. The paraffin serves to store additional energy and to provide
reserve heating in the discharge mode.

An ASHRAE Standard 94-77 for "Methods of Testing Thermal Storage Devices Based on Thermal Performance" was adopted in 1977, but has proven
to be difficult to apply, particularly for LHS devices[12]. ASHRAE is
convening a new committee to revise the standard. Improved test procedures for LHS devices are presently under development at ANL.

Annual Ice Formation And Storage Systems

The Annual Cycle Energy Storage (ACES) system has been under development at ORNL[13]. This system uses a water source heat pump for winter
heating, at the same time producing and storing ice for summer cooling. A
residential demonstration has been underway since October 1977, and technical feasibility has been fully established. The project is presently
focussing on developing lower cost components.

A passive system for winter formation and storage of ice in an insulated concrete underground tank is under development at ANL[14]. A
full-scale prototype has been built and a variety of heat pipes which
produce ice whenever ambient freezing conditions exist, and automatically
release ice from the freezing surface, are under test. A cost analysis
indicates that this system may be competitive with conventional A/C
systems.

R&D On LHS Materials And Packaging

Detailed mathematical modeling and computer simulation of melting/
freezing in LHS materials are being conducted by Solomon[1]. Thermal
convection is not included. Experimental verification has been obtained
by Deal[15]. Experimental melting rates, however, exceed predictions
by about 20% when convection is present.

Reduction in the loss of heat capacity with cycling for Glauber's
salt has been achieved by Marks[16]. Two methods for preventing
clustering of salts and loss in capacity were found. One is a crystal
habit modifier for both Glauber's salt and Na_2SO_4, and the other is a more
rigid or "spongy" thickener. Tests with each of these additives have resulted in retention of about 80% of the initial heat of fusion after 700
cycles, as compared to a 50% loss for the standard Glauber's salt formulation.

Pennwalt Corp. of King of Prussia, PA has begun a project for pelletization and roll coating encapsulation of LHS materials. Pelletization
will be attempted for a range of LHS materials with Tm's from 7°C to 115°C.

Candidate coating materials include acrylics, acrylonitriles, polyurethane, polyvinyl chloride, nylons, epoxies and fluorocarbons. Successfully pelletized and encapsulated samples, 0.003 m to 0.0127 m in diameter, will be thermally cycled in a small packed bed assembly.

Clemson Univ. has been experimenting with direct contact LHS systems for several years, and more recently completed a large scale test[17]. Best results were obtained with sodium thiosulfate pentahydrate, disodium hydrogen phosphate, sodium sulfate and calcium chloride, and with immiscible fluids Marcol 72 and Therminol 60. A 443 MJ unit was constructed using Marcol 72/disodium hydrogen phosphate. The cost of this system was $996, equivalent to $2.25/MJ. Measures taken to prevent salt carry-over were unsuccessful and eventually the system failed due to salt build up in a pump. It was concluded that substantial salt carry-over in the immiscible fluid is inevitable, and that a successful system design must allow for significant salt deposition.

The mechanisms of drop formation and heat transfer in direct contact, immiscible fluid-LHS heat exchange are under investigation at the Solar Energy Research Institute (SERI) in Golden, CO. Experiments on a single drop system, Wright[8], are exploring relationships among nozzle velocity, jet formation, drop size, and heat transfer rates. Tests are commencing on a multiple drop unit. Simulation studies have identified the drop size, the mechanism of heat transfer within the drop, and the dispersed phase hold-up as the significant parameters[18].

Several promising solid-solid transition LHS materials are currently under investigation by D. Benson at SERI. The following are the best candidates found to date:

1. <u>Pentaerythritol:</u> T_t = 188°C, ΔH_t = 323 kJ/kg, Cost = $1.28/kg, equivalent to $3.96/MJ.

2. <u>Pentaglycerine:</u> T_t = 81°C, ΔH_t = 216 kJ/kg, No cost given.

3. <u>Neopentyl Glycol:</u> T_t = 43°C, ΔH_t = 130 kJ/kg, cost = $0.80/kg, equivalent to $6.22/MJ.

3. CHEMICAL HEAT PUMPS FOR HEATING AND COOLING

Martin-Marietta Corp. has been developing a chemical heat pump (CHP) for solar heating and cooling, based on ammoniation/deammoniation of inorganic salts, Jaeger[1]. Operating characteristics and parameters

86

were determined for CHP's based on several paired salt reactions, plus salt reactions paired with condensed liquid ammonia. The two "best candidate" systems identified were liquid ammonia paired with liquid state ammoniates of ammonium nitrate, and of ammonium thiocyanate. A full scale prototype system using ammonium nitrate ammoniate/liquid ammonia was designed and constructed, but corrosion problems have prevented reliable experimental measurement of performance characteristics.

A Sulfuric Acid/Water CHP has been developed by Rocket Research Co., Clark et. al[8]. A 150 to 300 MJ, 25 MJ/hr engineering prototype has been designed and constructed, and is under test. Some 576 hours of operation, comprising 89 closed-loop charge/discharge cycles have been completed without any difficulties. Maximum acid temperature reached was 204°C. A 1000 MJ, 160 MJ/hr prototype unit has been designed for an industrial process heat application. This unit will largely be constructed of porcelain coated steel. It is expected to be used for upgrading 95°C industrial waste heat to 150 to 175°C process steam, with a return-on-investment potential of 1.5 to 2 years, Miller and Mezzina[8].

Tests have recently been completed on an engineering prototype of a solid state $CaCl_2$-CH_3OH (methanol) CHP, developed by EIC Laboratories, Inc. for solar heating and cooling applications, Offenhartz[8]. Experimental heat pumping rates appear to be adequate to meet requirements. The unit has been tested for 100 charge-discharge cycles with a 2% performance degradation, which appears to have stabilized. An economic analysis for the Washington, DC area, indicates that this CHP may be cost competitive with conventional heating and cooling systems.

For several years ANL has been developing a metal-hydride/hydrogen heat pump, Clinch[8]. An engineering prototype with a rapid cycling rate has been designed and tested. An analysis indicates a potential heating COP of 1.55. A DOE solicitation is currently in progress to select an industrial contractor for the design and engineering development of a full-scale prototype metal hydride heat pump.

4. SEASONAL HEAT STORAGE

The USA program for the development of large-scale, long-term TES subsystems has largely been concentrated on hot water storage in confined aquifers, and salt gradient ponds. However, there have been a number of analytical studies and small scale experiments on other seasonal heat

storage methods, including: undisturbed or moisture controlled earth;
rock or pebble-bed formations; fresh water ponds; and large, in-earth
water tanks. In addition, a substantial amount of analysis and experimentation is being conducted on ground coupled heat pump systems, which
may have relevance to solar seasonal storage applications.

The aquifer program is the most advanced. A large-scale experiment
has been successfully carried out by Auburn University in Alabama, through
two charge-discharge cycles[19]. Additional improved experiments are in
preparation. A three-dimensional computer code has been developed by
Lawrence Berkeley Laboratory and simulation predictions were in close
agreement with the Auburn experimental data[19]. An analysis of aquifer
storage in a planned district heating system established cost and energy
saving benefits[20]. Currently, the first phase of a demonstration program has begun, including projects for the utilization of aquifer TES in
an existing district heating system at the University of Minnesota in St.
Paul, in a low temperature district heating installation in Bethel, Alaska
and for winter coolness storage in Stony Brook, NY.

In the salt gradient pond program, a variety of small ponds have
been built and operated with a fairly high-degree of success, Nielson,
Wittenberg, Bryant[1]. Several one-dimensional computer codes have
been developed and utilized for analytical studies with a fair correlation
with experimental observations, Bryant[1], and Hull[21]. A more complex,
three-dimensional code is under development at ANL, Sha et. al[1]. An
analysis of a solar pond district heating system is underway at SERI[22].

There have been several analytical, Hooper[1], and cost[23] studies
of large scale water tanks in seasonal storage applications. One preliminary analysis of seasonal storage in earth was conducted by Yuan and
Bloom[24], and a small scale experiment is in progress, Walker[1]. An
analytical study conducted by Riaz[25] addressed seasonal heat storage
in various rock-earth configurations. Finally, several ground coupled
heat pump projects which are underway have relevance to seasonal heat
storage, Andrews, Bose[1].

5.   TES FOR SOLAR CENTRAL POWER STATIONS AND INDUSTRIAL PROCESS HEAT

SERI has carried out systems analyses, Copeland[8], and cost and
performance assessments, McKenzie[8], for a variety of high temperature
TES concepts. In addition, SERI is initiating R&D studies on high temperature thermochemical storage and transport, Wyman[8].

88

The bulk of the high temperature TES R&D work for Solar Power Systems and Industrial Process Heat is directed by Sandia Livermore Laboratories, Radovich[8], and Carling[8]. LHS in alkali and alkaline earth carbonate salts for temperatures of 704°C to 871°C is under investigation by Petri and Claar[8]. A 2 MWht energy storage boiler utilizing a $MgCl_2$, NaCl, KCl salt eutectic is under construction at the Naval Research Laboratory, Chubb and Veith[8]. TES in the latent heat of fusion of metal alloys with Tm from 325°C to 925°C is being experimentally studied by Birchenall[8]. EIC Laboratories is studying the interaction chemistry of various nitrates with water and atmospheric CO_2, White et. al.[8]. Corrosion problems are being addressed by De Van and Tortorelli[8], and by Osteryoung and Fernandez[8]. Experiments in high temperature sensible heat storage in thermocline storage tanks are being conducted by Gross and Harrigan[8], and Chaney et. al[8]. Finally, the utilization of LHS for thermal energy buffer storage for parabolic dish collector power systems is being studied by Manvi[8], Zimmerman et. al.[8], and Polzien[8].

REFERENCES

1. Proceedings, Active Solar Heating and Cooling Contractors Review Meeting, CONF-800340, March 1980.
2. E. I. H. Lin et. al., "COMMIX-SA: Validation, Application and Extension of a Solar Design Tool," Proceedings, 2nd Annual Systems Simulation and Economics Analysis Conference, San Diego, CA, January 1980.
3. W. T. Sha et. al., "COMMIX-SA-1: A Three-Dimensional Thermo-hydrodynamic Computer Program for Solar Applications," ANL-80-8, December 1979.
4. Cole, R. L. et. al., "Thermal Energy Storage Design and Installation Manual," ANL-79-15, January 1980.
5. Lane, G. A. et. al., "Macro-Encapsulation of PCM", Final Report, ORO/5217-8, November 1978.
6. Frysinger, G. R., "Life and Stability Testing of Packaged Low-Cost Energy Storage Materials", Proceedings, Fourth Annual Thermal Energy Storage Review Meeting, CONF-791232, 277-281, December, 1979.
7. Frysinger, G. R., "Storage Assisted Residential Heating/Cooling Using Solar Energy-Electric Heat Pump", Proceedings, Solar Energy Storage Options, Vol. 1, CONF-790328-P1, 221-229, March, 1979.
8. Proceedings, DOE Thermal and Chemical Storage Annual Contractor's Review Meeting, McLean, VA, October 14-16, 1980, 12-16.
7. Frysinger, G. R., "Storage Assisted Residential Heating/Cooling Using Solar Energy-Electric Heat Pump", Proceedings, Solar Energy Storage Options, Vol. 1, CONF-790328-P1, 221-229, March, 1979.
8. Proceedings, DOE Thermal and Chemical Storage Annual Contractor's Review Meeting, McLean, VA, October 14-16, 1980, 12-16.
9. Botham, R. A. et. al., "Form-Stable Crystalline Polymer Pellets for Thermal Energy Storage-High Density Polyethylene Intermediate Products", Final Report, ORNL/SUB-7398/4, January, 1978.

10. Greene, N. D. and Watson, W. K. R., "An Immiscible Fluid, Phase Change Heat Storage Battery", Proceedings, Solar Energy Storage Options, Vol. 1, CONF-790328-P2, 465-472, March 1979.

11. Scaringe, R. P., "A Variable Capacity Thermal Storage Device", Ibid, 609-618.

12. Jones, D. E. and Hill, J. E., "Testing of Pebble-Bed and Phase-Change Thermal Energy Storage Devices According to ASHRAE Standard 94-77", NBSIR 79-1737, May, 1979.

13. Holman, A. S. and Brantly, V. R., "ACES Demonstration Construction, Startup and Performance Report", ORNL/CON-26, October 1978.

14. Gorski, A. et. al., "Long-Term Ice Storage for Cooling Applications Using Passive Freezing Techniques", for presentation at the 1981 AS/ISES Annual Meeting, Philadelphia, PA, May 26-30, 1981.

15. Deal, R. P. and Solomon, A. D., "Heat Transfer Characteristics of Phase Change Materials", ORNL/MIT-286, February 1979.

16. Marks, S. B., "Thermal Energy Storage Using Glauber's Salt: Improved Storage Capacity with Thermal Cycling", Proceedings, 15th Intersociety Energy Conversion Engineering Conference (IECES), 1, Seattle,WA, August 18-22, 1980, 259-261.

17. Edie, D. D. et. al., "Immiscible Fluid-Heat of Fusion Heat Storage System", Proceedings, Fourth Annual Thermal Energy Storage Review Meeting, CONF-791232, 391-399, December 1979.

18. Cease, M., "A Model of Direct Contact Heat Transfer for Latent Heat Energy Storage", Proceedings, 15th IECES, 1, Seattle, WA, August 18-22, 1980, 624-629.

19. Tsang, C. F., Moltz, F. J. and Parr, A. D., "Experimental and Theoretical Studies of Thermal Energy Storage in Aquifers," Proceedings of the 15th IECES, 2, Seattle, WA, August 1980, 1245-1248.

20. Meyer, C. F., "Potential Benefits of Thermal Energy Storage in the Proposed Twin Cities District Heating - Cogeneration System" GE79TMP-32, GE - TEMPO, Santa Barbara, CA., July 1979.

21. Hull, J. R., "Computer Simulation of Solar Pond Thermal Behavior," Solar Energy, 25, 1980, 33-40.

22. Jayadev, J. S. and Edesess, M., "Solar Ponds", Solar Energy Research Institute, Golden, CO, SERI/TP-641-587, March 1980.

23. B. & A. Engineers, Ltd., "An Assessment of the Technoeconomic Feasibility of Seasonal Thermal Energy Storage," ANL, 1979.

24. Yuan, S. W. and Bloom, A. M., "Long Duration Earth Storage of Solar Energy," Final Report, ORNL, Oak Ridge, TN, January, 1978.

25. Riaz, M. et. al., "Long-Term Storage of Solar Energy in Native Rock," ISES, New Delhi, India, January 1978. Pergamon Press, Elmsford, NY, Vol I, 474.

THE SWEDISH SEASONAL THERMAL STORAGE PROGRAMME WITH APPLICATION TO THE
BUILDING SECTOR

E ÖFVERHOLM

1. INTRODUCTION

The Swedish climate is characterized by long and cold winters with little
or no sunshine. On the other hand solar insolation during the summer is
greater than in most parts of Europe. On a yearly basis the average insolation on a horisontal surface is a little less than 1 000 kWh/m^2. The average
insolation in December is 20 times less the average insolation in June.
Hence solar insolation for the months November, December and January is almost
negligible for solar energy plants. Heating demand for Stockholm is 3568
degree days and almost the double in the far north of Sweden. Average temperature during January in Stockholm is -2,9 ^{o}C.

Thus the need for seasonal storage in solar energy systems is essential
if these are going to be used for space heating. Both calculations and results from full scale experiments show that the cost for solar systems with
short time storage in Sweden is prohibitive, see Rosengren 1977.

Seasonal storage of solar energy is therefore one of the most important
parts of the energy research and development programme for the period 1981-
1984, recently presented to the government by the Swedish Council for Building Research. The programme contains three alternatives for seasonal storage:
storage in water, storage in the ground and thermochemical storage.

This paper will mainly deal with water storage systems, for which the
following goals have been set up for research, development, pilot studies
and full-scale experiments.

Storage in water:
- rock chambers are given priority in favour of cisterns above ground
- continued development of pits in ground
- two or three full-scale projects equivalent of approximately 500 dwellings
  Storage in the ground:
- increased research on geological, geohydrological and ecological aspects

- pilot studies and full-scale projects
- technical-economic evaluation of storage systems

Within the national energy R&D programme the Swedish Council for Building Research supports research, development, pilot studies and full-scale experiments within the sector Energy use for buildings. The Council has applied to the government for the following fundings.

Storage in water   96,7 million SEK 1981 - 1984

Storage in the ground,

including extraction

of heat from the ground   83,2 million SEK 1981 - 1984

2. STORAGE PROJECTS

Several types of storages and solar collectors are considered in Sweden. So far flat plate, CPC and line-focusing collectors have been used in full-scale experiments. Three seasonal water storage are now in operation, see fig. 1.

	Collector area m^2	Collector type	Storage volume m^3	Storage type	In operation
Studsvik	120	CPC	640	pit	Feb 1979
Växjö	1320	line, con- centrating	5000	cistern above ground	May 1979
Linköping	2875	flat plate selective	10000	pit in rock	April 1980

fig 1. Solar heating plants with seasonal water storages

Two ground storages have been built, one with plastic pipes in bore holes in rock and one with plastic pipes in clay. The clay storage is 30 m deep and operates between 12 and 20° C with a storage capacity of 70 000 kWh. Ongoing Swedish projects regarding chemical storage will be presented in another paper at this conference.

3. THE STUDSVIK PROJECT

This pilot project consists of a 640 m^3 pit dug into the ground. The pit is insulated with mineral wool and lined with rubber. Collectors are mounted on top of the floating and revolving pit lid, see fig 2.

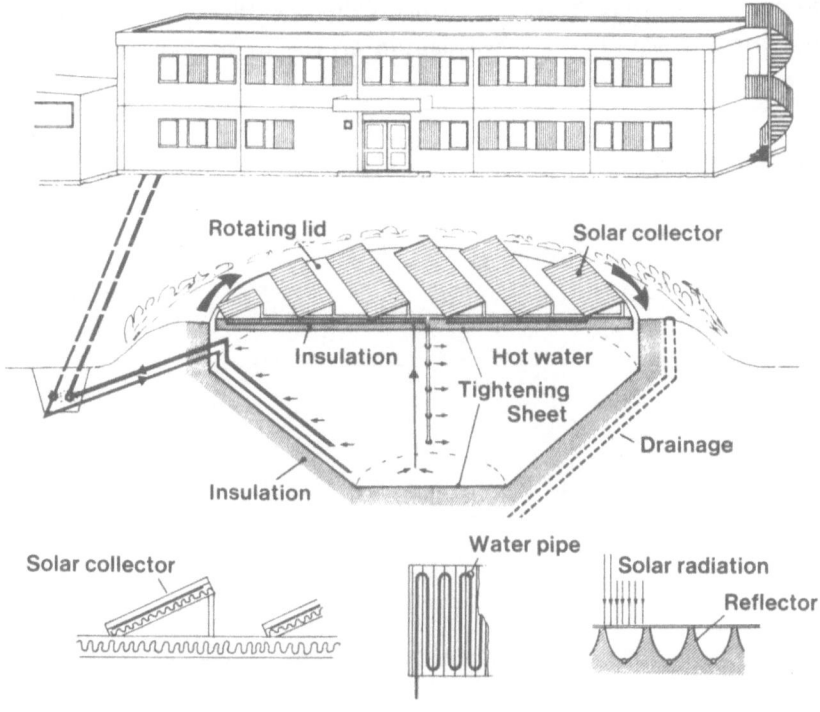

fig 2. Prototype installation; diagrammatic sketch

The system has been built for experimental purposes and therefore the storage volume is as small as 640 m^3. Heat losses are therefore 50 % of the stored energy. A future optimal plant would probably be a hundred times bigger. The technical specification of this plant is shown in fig 3.

Solar collector system		Heat magazine		Office building	
Type:	CPC intensification factor 4, reflector of aluminized plastic foil in a copper absorber painted black, 1 cover glass	Type:	Excavated magazine	Floor area:	500 m²
		Shape:	Truncated cone	Heat requirements:	22,5 MWh/yr
		Volume:	640 m³ (30°C)	Heating system:	Water/air heat exchanger
Total surface area:	120 m²	Temperature range:	30 — 70°C	Lowest distribution temperature	30°C (for the dimensioned case)
Inclination to the horizont:	25°	Insulation:	Bottom and conical surface, 40 cm mineral wool platters; cylindrical surface, lid, 40 cm expanded polyurethene	Project leader:	Rutger Roseen Studsvik Energiteknik AB
Positioning:	Rotaiton of magazine lid				
Annual production:	42 — 52 MWh	Heat losses:	19 MWh per annum		

fig 3. Technical specification of the Studsvik plant

The plant has been in operation since February 1979. A report on the design, building, measuring and evaluation of the plant has been made, Roseen 1980. The overall performance for the two first years of operation is in accordance with theoretical calculations if the actual solar insolation is considered, see fig 4.

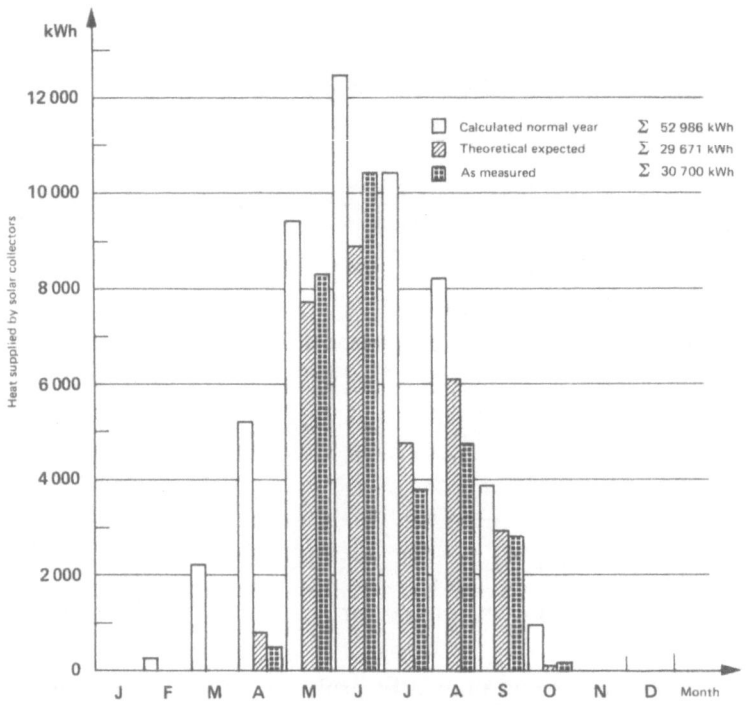

fig 4. Measured solar energy compared with calculated values for a normal year and for 1979 from Roseen 1980.

4. THE VÄXJÖ PROJECT

This plant has been in operation since May 1979. The basic layout and expected energy balance is shown in fig 5.

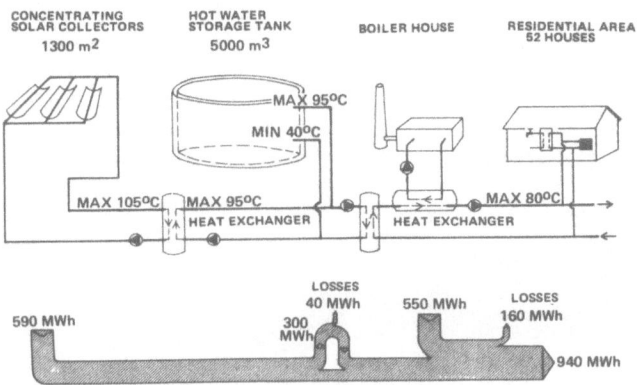

fig 5. Schematic diagram - energy flows, Växjö plant

The concrete cistern is situated on the ground and has a fixed water inlet at the top of the cistern. Mainly due to a much higher portion of diffuse
insolation even during clear days the performance has been much lower than
expected. For a report on the design and construcion phase, see Finn 1979.

5. THE LAMBOHOV PROJECT

This is the largest seasonal storage so far in Sweden, 10 000 m^3 of water.
It consists of a pit which has been blasted out in the rock and then insulated
with cement bound light-weight clay granules and lined with the same kind of
butyl rubber which was used in the Studsvik plant. The lid floats on the water
and the flat plate solar collectors are situated on the roofs of the 54 row-
houses which are furnished with the solar energy. Heat pumps are connected
to the storage in order to enhance the storage capacity. As the plant only
has been in operation since May 1980 and the monitoring system has mal-
functioned it is difficult to draw any conclusions of the operation of the
plant. However, the temperature increase in the accumulator has been close to
the predicted. See fig 6 for the calculated energy balance, fig 7 for technical data and Norbäck 1980 for full design report.

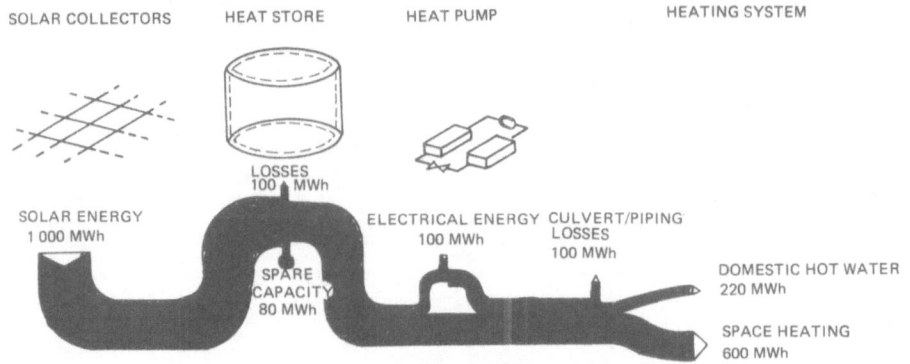

Fig 6. Schematic diagram of energy flows, Lambohov plant

Solar collectors   flatplate selective surface 2 568 m^2
    inclination   55°
    freeze protection   drainage

Storage   9 772 m^3
    diameter   32,2 m
    depth   12 m
    temperature range   20-70°C
    insulation   1,2 m
    U-value   0,1 W/m^2 ^{o}C (6 % moisture)

Heat pumps
    domestic hot water, output 82 kW 5/60° C
    spaceheating, output   432 kW 5/60° C

Distribution
    temperature: space heating 20-50° C (air system)
    temperature: domestic hot
    water   45° C

fig 7. Technical specification, Lambohov plant

6.   EXPERIENCE AND FUTURE PROJECTS

So far, lining and insulation have been one of the major problems with seasonal storage. Heat losses have been in the range of theoretical

values. Stratification has not always occured to the extent which was expected due to e.g. high water flows. However, during normal operating conditions temperature stratification has caused no problems. The measuring
equipment has had several malfunctions.

The main part of suggested new technical solutions to the problems with
seasonal storage of solar energy either have the ground as a container for
water, e.g. rock chambers, or use the ground as heat storage. These two solutions have in common lower costs than for cisterns above ground. However,
significant research and development work has to be accopmplished for these
technics.

Two projects with rock chamber storage are financed by the Council. One
is situated at Avesta and has a storage capacity of 15 000 m^3, the other
will be built in Uppsala with a storage capacity of 100 000 m^3.

To summarize, Sweden has several seasonal storage projects going on and
reasonable economy seems to be within reach, at least what large scale projects are concerned.

98

References

L Finn - A Swedish solar heating plant with seasonal storage. The Ingelstad project design and construction stage, Swedish Council for Building research, Document D 14:1979.

K Norbäck, J Hallenberg - A Swedish group solar heating plant with seasonal storage. Technical-economic description of the Lambohov project, Swedish Council for Building Research, Document D 36:1980.

R Roseen, B Perers - A Solar Heating Plant in Studsvik. Design and first year operational performance, Swedish Council for Building Research, Document D 21:1980.

B Rosengren - The Termoroc House - an experimental low-energy house in Sweden, Swedish Council for Building Research, Document D 8:1977.

INTERNATIONAL TNO SYMPOSIUM

"THERMAL STORAGE OF SOLAR ENERGY"

Novembre 1980

Amsterdam, The Netherlands

USE OF PARAFFINS FOR THERMAL STORAGE

N.A.Mancini

Istituto di Struttura della Materia, Università di Catania
Corso Italia,57 95129 Catania,Italy and FEP/SE-CNR, Roma,Italy

Actually many materials have been recognized that in
principle should be suitable to storage thermal energy at
low-medium temperature (30°÷70°C) by a phase-change.
Among these materials paraffins deserve particular attention
inasmuch they appear to exhibit the proper physical, chemical and thermophysical properties that are required so that
a substance may be utilized into a thermal energy stockage.
Let us recall the main characteristics of a potentially proper material:

a) high latent heat at the phase transition

b) good sensible heat in the thermal range of use

c) low thermal expansion coefficient at the phase transition

d) high thermal conductivity coefficient in the solid phase

e) high chemical stability

f) no appreciable supercooling

g) a congruous phase transition

h) low cost.

Some of these requirements appear to be satisfied to a large
extent by paraffins. Actually under this term are comprised
many compounds,saturated hydrocarbons, that although consti-
tued by the same chemical elements, Carbon and Hydrogen, dif

fer for the number of atoms and for their disposition resulting in largely different thermophysical properties. Looking
at Table I we can see how the melting temperature ranges from

Paraffin	Melting Point,°C	Latent Heat,Cal/gm	Density gm/cc
n-Dodecano ($C_{12}H_{26}$)	-12	-	0.750
n-Tridecano ($C_{13}H_{28}$)	-6	-	0.756
n-Tetradecano ($C_{14}H_{30}$)	5.5	54	0.771
n-Pentadecano ($C_{15}H_{32}$)	10	59	0.768
n-Esadecano ($C_{16}H_{34}$)	16.7	56.6	0.774
n-Eptadecano ($C_{17}H36$)	21.7	41	0.778
n-Octadecano ($C_{18}H_{38}$)	28.0	58	0.774
n-Nonadecano ($C_{19}H_{40}$)	32.0	-	
n-Eicosano ($C_{20}H_{42}$)	36.7	59	0.778
n-Eneicosano ($C_{21}H_{44}$)	40.2	48	0.758
n-Docosano ($C_{22}H_{46}$)	44.0	60	0.763
n-Tricosano ($C_{23}H_{48}$)	47.5	56	0.764
n-Pentocosano ($C_{25}H_{52}$)	49.4	-	0.769
n-Tetracosano ($C_{24}H_{50}$)	50.6	59.5	0.766
Cera Paraffina	54.4	35	0.88
n-Esacosano ($C_{26}H_{54}$)	56.3	61	0.770
n-Eptacosano ($C_{27}H_{56}$)	58.8	56.1	0.773
n-Octacosano ($C_{28}H_{58}$)	61.6	61	0.779
n-Nonacosano ($C_{29}H_{60}$)	63.4	57	-
n-Triacontano ($C_{30}H_{62}$)	65.4	60	-
n-Entriacontano ($C_{31}H_{64}$)	-	57.8	-
n-Dotricontano ($C_{32}H_{66}$)	70	40.7	0.782
n-Tritriacontano ($C_{33}H_{68}$)	71.0 .	-	-
n-Tetratriacontano ($C_{34}H_{70}$)	72.9	64	-
n-Esatricontano ($C_{36}H_{74}$)	76.1	56.2	-

Table I: Thermophysical properties of some paraffins

-12°C to +76°C and the melting latent heat from 35 cal/gram
to 64 cal/gram. On this respect we have to stress, just
from the outset, that a great care has to be exercised on ex
ploiting the possibility of utilization, looking at the actual paraffin one wants to use. Moreover we have to observe
that generally speaking paraffins have a density lesser than
unity (0.75÷0.80 grams/cc) so that their high mass thermal
capacity is contrasted by a lower volume thermal capacity.
Sometimes this results into an apparent poor economical usefulness for their use. The congruous phase transition, the
absence of supercooling and the large chemical inertia allow
a very large number of cycles for a single specimen, so that
these materials appear to be very advantageous for a recycled
use. Moreover these characteristics allow many materials
may be choiced as containers. However some disvantageous
properties of paraffin have to be recalled, mainly their ther
mal expansion coefficient when they become solid and their
low thermal conductivity in the solid phase. The first kind
of difficulty can be overcome in some cases utilizing elastic
containers or by foreseing safety volumes in the containers.
The latter difficulty, i.e. the low thermal conductivity in
the solid phase, is not so easily solvable and many research
efforts are now devoted to it. A method that was followed
was the inclusion of blocks of various size[1]). It is believed that the inclusion offers three advantages:
1) the raising of the chemical and physical stability, favou-
   ring the congruity and reproduction of the thermal charge and discharge process.
2) the easier production of storage units, also for materials different from paraffin.
3) the possibility to produce manufacts for structural and/

or architettonic uses.

With reference to paraffin a procedure of microenclusion was reported in[2]. For the various tests the employed specimens were of normal paraffin P116 by the Sun Oil Company envisaged as a potential phase change material. In Table II are reported their thermophysical properties

	- Density of Solid	0.63 gr/cc
	- Percent volume change	6÷7%
Tab.II	- Average Specific Heat	0.17 cal/gr°C
Paraffin 116	- Latent Heat of Fusion	35 cal/gr
Properties	- Melting Temperature Range	45÷49°C
	- Peak Latent Heat Temperature	46.7°C

A minor phase transition occurs at 34.4°C. This is low energy solid-to-solid transition which is present in hydrocarbons with an odd number of carbone atoms[3]. The measured latent heat of the P116 paraffins which combines the solid-to-solid transition and the solid-to-liquid transition was 35 cal/gr.

This values differs from the measured value reported in reference[4].

Fig.1 shows freezing if P116 Paraffin at room ambient conditions; to the measure the paraffin temperature a copper-constantan thermocouple junction was inserted in the center of the sample and was connected to a Hewlett-Packard 0÷10 Millvolt strip recorder.

Several Companies with encapsulation capability were interested, with the aim to obtain encapsulated material that don't realize paraffin and procedure to obtain convenient microcapsule size and coating thickness.

Table III summarize some encapsulated sample properties.

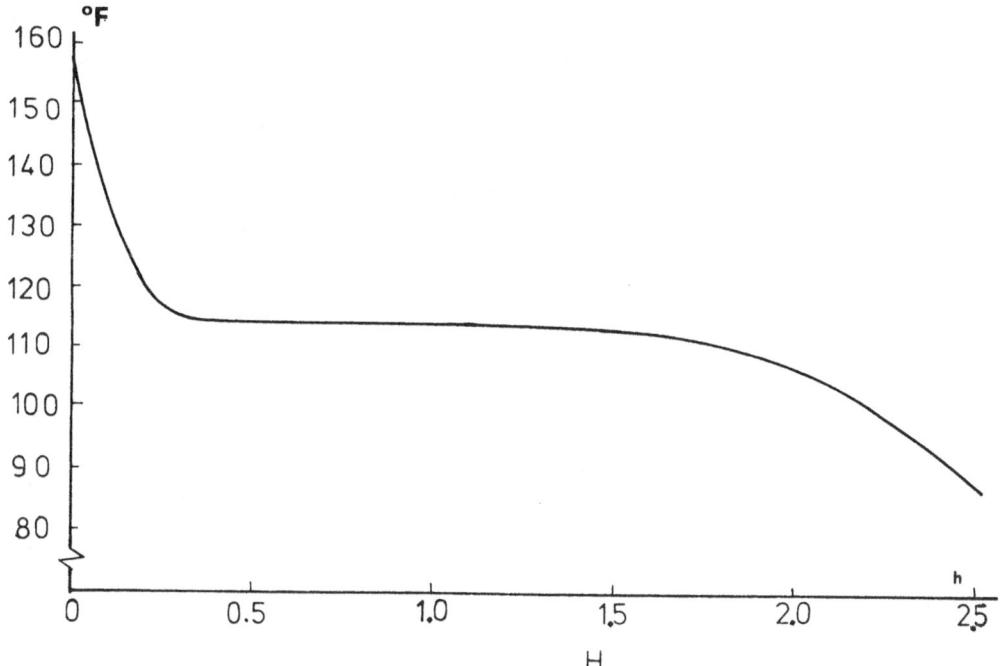

H

Fig.1: Freezing of P116 paraffin at room conditions.
Each sample has been subjected to as many as 300 temperature
cycles from 15.6°C to 54.5°C, at a rate of 6 cycles per hour.
The results show that all the samples using gelatin wall
failed within 60 melting cycles but only P116-2 samples had
small traces of paraffin leakage after 150 cycles, with a
percentage of paraffin released ranged from 1 to 25 percent.
Some specimens exhibited also a small amount of wall break-
down. Capsules with a urea formaldehyde wall material were
subjected to similar thermal cycling tests through the melting temperature of the paraffin with little paraffin leakage.
A water slurry of each encapsulated paraffin samples was also subjected to temperatures approaching the water boiling
point.
All the specimens show wall degradation or paraffin leakage
where water boiling temperature occurred.

SAMPLE DESIGNATION	SUPPLIER	BASIC WALL MATERIAL	PARTICLE DIAMETERS	% WALL WEIGHT	REMARKS
FJ 716	NCR	Gelatin	100-500 μ	15%	
FJ 717	NCR	Gelatin	100-500 μ	15%	Same process as 716 but post treated
FJ 718	NCR	Gelatin	100-500 μ	15%	
FJ 719	NCR	Gelatin	100-500 μ	15%	Same process as 718 but post treated
FJ 680	NCR	Gelatin	50 μ	25%	Additional post treatment
FJ 681	NCR	Gelatin	50 μ	25%	Specific Gravity 1.0
P116-1	Pennwalt	Modified Nylon	50 μ	10%	25% wall crosslinking
P116-2	Pennwalt	Modified Nylon	1000-2000 μ	10%	0 wall crosslinking
P116-3	Pennwalt	Modified Nylon	50 μ	10%	0 wall crosslinking
P116-4	Pennwalt	Modified Nylon	50 μ	20%	10% wall crosslinking

Table 3:Summary of encapsulated sample properties

Stirring agitation, thermal capacity and pumping tests were also performed. The measured heat storage capacity was within 10 percent of the theoretical one, and it can be increased by a factor of 2 and as high as a 3 times the equivalent water system.

The criterion on which the microencapsulation implicitly rests is traying to obtain the highest possible exchange coefficient, i.e. an emission rate so large that the storage energy may be released in a very short time.

If the storage system has not to be operating instantaneously, i.e. it has not to contrast brief and sudden black out of the primary source, but on the contrary it has to be a source me-

diated in time, so to average the phase shifts between source and utilization, such criterion appears to be of poor use_
fulness.

Actually in the storage systems that operate at low-medium
temperature the latter of the quoted criteria is the most interesting.

The use of macroscopic containers appears so to be more advan
tageous also if we take into account the more complete elimi-
nation of paraffin losses. Indeed a paraffin loss results
into an immission of the paraffin itself into the head distri
bution circuit and this could be responsible of the occlusion
of the circuit due to the possibile solidifiication of the
paraffin in the coldest point. The complex as a whole could
be rest.

However to use a macroscopic container the obstacle of the
low thermal conductivity of the paraffin has to be circumven-
ted. Indeed during the heat withdraw from the melted paraf
fin, the liquid which is nearer to the container walls become to be solid so that this region acts as a ther a insulator due to its low termal conductivity. This difficulty can
be avoided by the introduction of proper thermal shorts inside the paraffin. Several attempts have been performed utilizing different geometrical shapes. Among these let us recall
the first attempts to introduce metallic powders into the paraffin mass and the utilization of coaxial tubular containers
finned in the mass of the paraffin[5]).

The system containing metallic powders appeared uncongruous
due to the deposition of the powders themselves on the bottom
of the macrocontainer after few thermal cycles.

The container realized by coaxial tubes, utilized under diffe
rent versions for a building air conditioning was capable to

106

level the phase shift between utilization and furnishing,at least for a period of 24 hours operating in the thermal range 17 - 28 °C,(Fig.2).

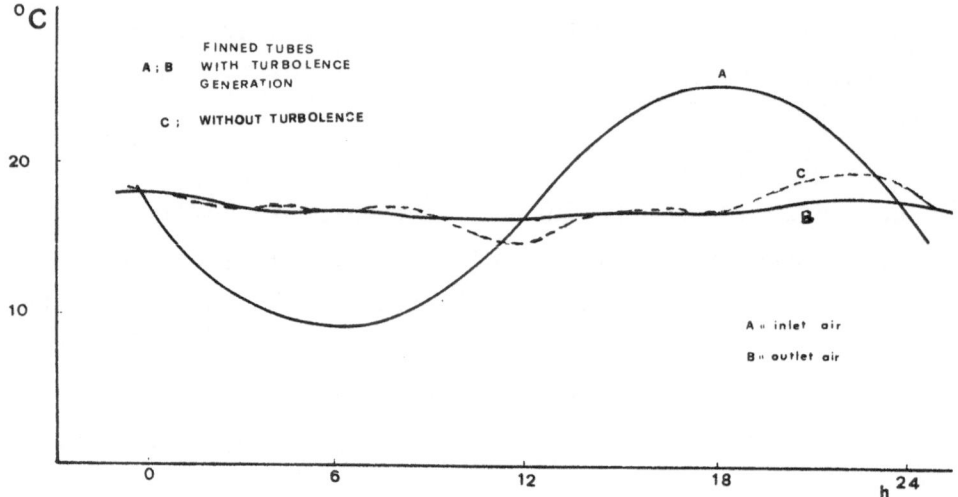

Fig.2: Flattening of outlet air.

Fig.3:Effect of ther mal conductivity on fluid outlet tempera ture.

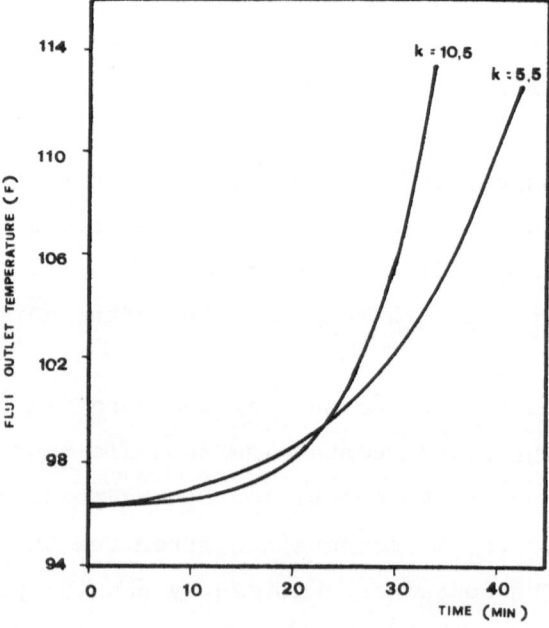

A theoretical analysis on thermal accumulators with planar symmetry with alluminum hexagonal honeycomb structure as ther mal short has been performed in the range between 35°C and 45°C[6]. The theoretical treatment was pursued both under the simplifying assumption that a negligible temperature gradient exists across the fluid (Asymptotic Expansion technique[7]) and assuming a finite value for the convective heat transfer coefficient at the capacitor fluid channel wall (Goodman technique)[8]. The calculations show that the effective conductivity and that the release time of the storaged energy depend on the honeycomb cell sizes and that a factor as large as two can be obtained (Fig.3).

It has been shown too[9] that a finned container largely redu ces the withdraw times of the thermal energy from the paraffin. In the perspective of cost reduction the finning of the tube was replaced by a honeycomb structure ($\Delta V/V = 2 \div 15\%$) or by alluminum thin-strip matrices ($\Delta V/V = 0.4 \div 2\%$).

Melting and solidification times shortened from two up to fifteen times and from two up to seven times have been observed as function of the adopted metallic volume percentage respectively.

A Finned Heat Pipe Heat Exchanger (FHPHE) has been developed too[10]. The dispositive consists of two zones, the first runned along by the warm fluid coming from the source, the latter by the fluid coming from the user. Both water and air can be adopted as fluids. The two zones are connected by an alluminum finned heat pipe inclosed by the change pha se material (PCM). Several FHPHE can be connected in parallel and enclosed into a modular container or can be disposed according to the required geometries, for example in box or in stack arrangements. Let us stress the relevance of the temperature gradient needed to get heat from the material

undergoing the phase-change and of the exchanger geometry,
i.e. fin height (DFIN), fin thickness (SR), fin spacing (DR)
and the void fraction (SF). Typical results are shown in
Table III and Fig.4.

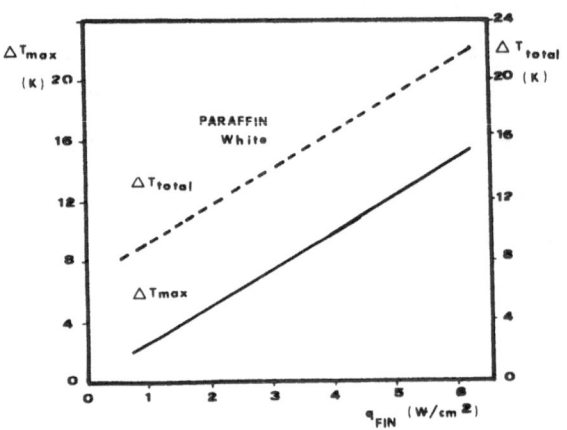

Fig.4: Influence of Melting Range on the Maximum Temperature
Gradient.

The method of a container realized by two coaxial tubes,
linked through longitudinal fins made from a good heat conducting material (e.g. alluminum) has been recently restored
and some possibilities of use are reported[11].
It has also been realized the possibility to use a passive
heat exchanger operating on the conditions of inversion[12].
A passive heat exchanger heat storage (PHEHS) was realized
introducing PVC tubes containing white paraffin and thermal
shorts into a thermically insulating tank containing water
as exchange fluid[13]. The radii of the PVC tubes are larger
than the Porter critical radius corresponding to the iron
thermal conductivity so that the thermal shorts were realized by this material. Experimental tests on dispositives

with realistic dimensions are in progress. A reduction of
the exchange time as large as a factor 4 was predicted and
verified as a function of an uniform withdraw into a period
of 24 hours. As about its cost, the dispositive appears to
be convenient for a commercial use inasmuch it is possible
to utilize cheap materials, the life time of the system being
longer than ten-twenty years.

Few words about the problem of the costs.

The economical use of a thermal accumulator is often calculated simply on its cost, life time and requirements. These
last need of a proper definition so to avoid physical misun-
derstandings[14]).Indeed if the storage system has to be utilized for solar energy we have to take into account the smaller
thermal excursion required for a system utilizing phase-change materials with respect to sensible heat systems.

The working of solar collectors in a narrower and better defined thermal range results into a larger efficiency inasmuch high working temperatures become unnecessary so that
smaller collection surfaces are needed. To such occurrence
corresponds a lower cost, i.e. a benefit for the system as
a whole. So even if the cost of the heat storage part
was higher,the complessive system can be cheaper. These
considerations are of the outmost relevance for a global
economical analysis.

Thanks are due to Mrs.G.Giuffrida for typing the manuscript.

REFERENCES:

1) G.A.Lane et al., Dow chemical company Contract NSF-C 906, 1976

2) E.M.Mehalich, A.T. Tweedie, General Electric Co NSF Grant AER 74-01986, NSF/RANN/SE/ AER 74-09186,1975

3) Broadhurst, Martin,G. Journ. of Research, of the D.B.S., A Phys. and Chem.66A,3,1962

4) NSF/RANN/SE/GI 27976/PR73/5 University of Pennsylvania, 1973

5) J.Chevalier, Activite du CSTB,1978

6) J.A.Bailey et al., Proceeding if the workshop on Solar Energy, Charlottesville, Virginia,1975

7) A.Erdelyi, Asymptotic Expansions, Dover, N.Y.1956

8) T.R.Goodman, ASME Trans., 80,335-342,1958

9) A.G.De Jong, C.G.Hoogendoorn, International TNO Symposium, Thermal Storage of Solar Energy, Amsterdam 1980

10) A.Abwat, Proceeding of First Seminar on Solar Energy Storage, Trieste, Italy,1978

12) A.W.Porter, Phil.Mag.20,511,1910

11) A.Abbat et al. International TNO Symposium, Thermal Storarage of Solar Energy, Amsterdam,1980

13) N.A.Mancini et al. to be published

14) R.H.Marshall, International TNO Symposium, Thermal Storage of Solar Energy, Amsterdam 1980.

A THEORETICAL STUDY OF ASHRAE STANDARD 94-77 FOR TESTING THERMAL
STORAGE DEVICES

R. H. MARSHALL

1. INTRODUCTION

In the solar field the ASHRAE Standard 94-77, ref.1, has formed the basis
for the testing of thermal storage devices on the basis of thermal performance.
However, the application of this Standard on a prototype phase change storage
device has been found to lead to meaningless results. Therefore it is the
intention of this article to review the procedure as applied to an ideal fully
mixed sensible heat store powered by an imbedded heat exchanger, without internal capacity and thereby demonstrate that the Standard

   (a) fails to correctly identify the loss coefficient,

   (b) is confused and confusing as to the form of the loss term
       during a charge test,

NOMENCLATURE

C_p	specific heat $(kJ\ kg^{-1}\ K^{-1})$	x	distance; $x' = x/L$	
$\dot{e}$	error rate (W)	COP	coefficient of performance	
E	energy stored/discharged (J)	NTU;	$NTU = 1-\exp(-UAHTX/\dot{m}C_p)$	
L	length of heat exchanger (m)	T_{in}	inlet stepped value at $t=0$	(^oC)
$\dot{m}$	mass flow rate $(kg\ s^{-1})$	TSC	theoretical storage capacity over $T_z - T_{in}$	(J)
M	mass (kg)			
N	number of fill periods	T_z	reference temperature; $T_z = T_{in}(t=0^-)$	(^oC)
R	ratio of heat capacities	UA	loss coefficient	$(W\ K^{-1})$
t	time; $t' = t/t_f$	UAHTX;	overall heat transfer product	$(W\ K^{-1})$
t_d	delay time $t_d = R/(1+R)$	ξ	heat exchanger effectiveness; $\xi = 1 - \exp(-NTU)$	-
t_f	fill time (reference) $t_f = (1+R)MC_{ps}/\dot{m}C_{pf}$	β	loss coefficient ratio; $\beta = \xi/\Sigma$	-
T	temperature; $T' = (T-T_z)/(T_{in}-T_z)$	Σ	sum; $\xi + UA/\dot{m}C_p$	-
		λ	store time constant; $\lambda = (1+R)\Sigma$	-

<u>Subscripts:</u> c charge, d discharge, f fluid, in inlet, o outlet, s store,
            AR ASHRAE, NBS National Bureau of Standards, ACT actual

(c) fails to identify the heat exchanger effectiveness necessary for modelling and system performance,

(d) is incorrect in its assumption of zero loss during a discharge test.

After discussing the sources of error for the zero capacity heat exchanger, a "bandage" procedure is discussed which can alleviate some of the problems with the present procedure. This study concludes that a full scale study is necessary with emphasis on realistic stores in which the identified parameters serve both the needs of the theoretical modeller and the design engineer.

2. THE MODEL

2.1 Two equation model

The mathematical description of the thermal storage behaviour of the idealized storage device seen in Fig.1 is given by the solution to the governing equations

$$M_s \, C_{p_s} \, \dot{T}_s \; = \; \dot{m} \, C_p \, (\overline{T}_f - T_s) + UA(T_a - T_s) \qquad \ldots \ldots (1)$$

$$M_f \, C_{p_f} \, T_{f_t} \; = \; UAHTX \, (T_s - T_f) - \dot{m} \, C_{pL} \, T_{f_x} \qquad \ldots \ldots (2)$$

$$\overline{T}_f \; = \; \frac{1}{L} \int_0^L T_f(x) \; dx \qquad \ldots \ldots (3)$$

The two parameters, UA and UAHTX, are the parameters whose identification ought to be the objective of the ASHRAE test procedure. However, in the following sections it will be shown that even for this idealized case the procedure leads to only an approximate form of the first parameter, UA, and to a complicated function of the second, UAHTX.

Using the notation listed in the Nomenclature the non-dimensional forms of equations 1-3 become, after dropping primes,

$$\dot{T}_s \; = \; (1+R) \, NTU(\overline{T}_f - T_s) + (1+R) \, \frac{UA}{\dot{m} C_p} \, (T_a - T_s) \qquad \ldots \ldots (4)$$

$$T_{f_t} \; = \; (\frac{1+R}{R}) \, NTU(T_s - T_f) - (\frac{1+R}{R}) \, T_{f_x} \qquad \ldots \ldots (5)$$

$$T_f \; = \; \int_0^1 T_f(x) \; dx \qquad \ldots \ldots (6)$$

Equation (4) is an ordinary differential equation whose solution depends upon the solution to the linear partial differential equation (5) and the evaluation of the integral, equation (6). The particular solution to equation (5) for a unit step change in inlet temperature at t=0 is given by

$$T_f(x,t) \; = \; e^{-NTU \, x} {}_{<T_{in}(t-t_d x)>} + e^{-NTU \, t/t_d} \int_{t-t_d x}^{t} \exp(NTU \, \tau/t_d) \, \frac{NTU}{t_d} T_s \, (\tau) \; d\tau \qquad \ldots \ldots (7)$$

where the unit step function is given by

$$<T_{in}(t-t_d x)> \equiv T_{IN} \quad ; \quad t > t_d x \qquad \dots \text{(8)}$$

and $\quad <T_{in}(t-t_d x)> \equiv T_{IN}(t<0); \quad t \leq t_d x. \qquad \dots \text{(9)}$

The outlet temperature is therefore given by

$$T_o(x=1,t) = e^{-NTU}<T_{in}(t-t_d)> + \int_0^1 e^{-NTU\,z} NTU\,T_s(t-t_d z)\,dz \qquad \dots \text{(10)}$$

The mean fluid temperature, then, is given by

$$\overline{T}_f = e^{-NTU\,x}T_{in}(t-t_d x)\,dx + \int_0^1 e^{-NTU\,t/td}\int_{t-t_d x}^t e^{NTU\,\tau/td}\frac{NTU}{t_d}T_s(\tau)\,d\tau \text{(11)}$$

Note that the consequence of a finite heat exchanger, R, is to introduce a "memory" function, $<T_{in}(t-t_d)>$, such that the storage material responds to an event depending on what has previously occurred. Because the solution to equations (10) and (11) cannot be presented easily in closed form for a finite capacity heat exchanger this author would prefer to see the zero capacity, zero loss fully mixed store replace the stratified store as the reference performance case because the analytic closed form solution then exists.

In the next section, therefore, we review the analytic solution in closed form for the zero capacity heat exchanger case with the attention focussed on the identification of the governing parameters by the ASHRAE/NBS procedures.

2.2 Case I: Ideal zero capacity heat exchanger (R=0)

In the zero capacity case, R=0, equations (4-6) can be readily integrated to yield the familiar solution

$$T_s(t) = T_z + (1-e^{-\lambda t})(T_{in}-T_z - \frac{UA}{\Sigma}(T_{in}-T_a)) \qquad \dots \text{(12)}$$

where $\quad \Sigma = \dot{m}C_p\,\xi + UA \quad$ and $\quad \lambda = \xi + \frac{UA}{\dot{m}C_p}$

with $T_o(t) = \xi T_{in} + (1-\xi)\,T_s(t) \qquad \dots \text{(13)}$

The steady state equation, $t \to \infty$, then reads

$$T_s = T_{in} - \frac{UA}{\Sigma}(T_{in} - T_a) \qquad \dots \text{(14)}$$

As an example, a plot of the transient traces is shown in Fig.2 for two cases: $\xi = 1$ (a fully mixed store with a perfect heat exchanger) and $\xi = .5$ (a fully mixed store with a poorer heat exchanger). The plots of both $T_s(t)$ and $T_o(t)$ are shown for an initial step from ambient temperature to a value which represents $T_a + 25^o$ as required for a heat loss test, Region I. Then the inlet is raised to a second level, Region II, which may represent the charge test. Finally the inlet is decreased, representing the discharge test, Region III.

2.3 Heat loss coefficient identification

The procedure states that a single heat loss test shall be performed in which the inlet temperature is raised 25° above ambient and the steady state loss coefficient, UA_{AR}, see Fig.2, Region I, shall be determined from

$$UA_{AR}(T_{in}-T_a) = \dot{m}_{H.L.}C_p (T_{in}-T_o) \qquad \ldots \ldots (15)$$

Because we are treating the ideal case with no experimental error, only the steady state rate equation, equation (15) needs to be considered.

The flow rate required, $\dot{m}_{H.L.}$, depends on which standard is being used, however. In Table 1 are listed the choices in dimensional units for the heat loss test as applied to the particular store and the flow rates associated with each charge/discharge test.

	Heat Loss	Charge/Discharge t_f = 2 hours	Charge/Discharge t_f = 4 hours
Ref(2)	$\dot{m}_{H.L.} = \dfrac{MC_{ps}}{C_{pf}} \dfrac{25^{o}C}{25^{o}C \, 1\,hr}$	$\dot{m} = 1/2 \, \dot{m}_{H.L.}$	$\dot{m} = 1/4 \, \dot{m}_{H.L.}$
Ref(1,3,4)	$\dot{m}_{H.L.} = \dfrac{MC_{ps}}{C_{pf}} \dfrac{25^{o}C}{25^{o}C \, 4\,hr}$	$\dot{m} = 2 \, \dot{m}_{H.L.}$	$\dot{m} = \dot{m}_{H.L.}$
Ref(5)	$\dot{m}_{H.L.} = 1 \, 2 \, \dfrac{1/hr}{MJ}$	$\dot{m} = \dot{m}_{H.L.}$	six trajectories

Table 1.   Flow rate for heat loss test

It is seen from Table 1 that the heat loss flow rate is the minimum for the majority of the references. Therefore, in the following all references to $\dot{m}$ refer to the use of $\dot{m}_{H.L.}$ as the flow rate because the errors diminish for higher flow rates.

Turning back to equation (15), after inserting the steady state equation: (13 and 14), the parameter actually being identified by the heat loss test, UA_{AR}, is related to the real heat loss coefficient, UA, by the equation

$$UA_{AR} (T_{in}-T_a) = \dot{m}C_p \, \xi^* \, \frac{UA}{\dot{m}C_p \, \xi + UA} (T_{in}-T_a) \qquad \ldots \ldots (16)$$

i.e. $UA_{AR} = \beta \, UA; \; \beta = \dfrac{\dot{m}C_p \, \xi}{\dot{m}C \, \xi + UA} < 1$ $\ldots \ldots (17)$

Thus the parameter, UA_{AR}, identified by the procedure, is less than the theoretical loss coefficient and it is a function of the second unknown parameter, On the other hand for a low loss coefficient, UA, and/or a high enough flow rate, $\dot{m}C_p$, the apparent heat loss coefficient will closely approximate the true one. The practical experience of this author has indicated that the value β lies in the range $.9 < \beta < 1$.

Note that the expression for β is a strong function of flow rate so that an increased flow rate should be used in the heat loss test in order that the error will be minimal.

Interestingly, equations (15-17) form the basis for an alternative test procedure discussed later.

2.4   The charge test

Having established a heat loss coefficient, albeit in error, the test procedure requires a charge test, Region II Fig.2, best seen in a close-up over a single fill period, Fig.3. The test is performed to determine the "effective capacity for storage", EC, i.e. the amount of energy stored in one time period t_f; and coefficient of performance COP_c, which is the ratio of the energy stored to theoretical maximum energy which could be stored over the temperature trajectory $T_{in}(t<0) \to T_{in}$.   Under the present non-dimensionalisation the parameters required are given by

$$COP_c = \frac{EC}{TSC} = \left\{ \int_0^1 (T_{in}-T_0) \; dt - \frac{UA}{\dot{m}C_p} (T_z + T? - T_a) \right\} TSC^{-1} \qquad \cdots \quad (18)$$

Reference	$T_z + T? - T_a$	Remarks
Ref.1	$T_z + \dfrac{T_{in}-T_0}{2} - T_a$	Text unclear if $T_{in} - T_{out} = \dfrac{1}{t_f} \int_0^{t_f} (T_{in}(t)-T_0(t)) \; dt$
Ref.2,5	$T_z + \dfrac{T_{in}-T_z}{2} - T_a$	Same remark
Ref.3 p.6	$T_z + \dfrac{T_{in}+T_0}{2} - T_a$	Text unclear if $\dfrac{}{T_{in} - T_{out}} = \dfrac{1}{t_f} \int_0^{t_f} (T_{in}(t)-T_0(t)) \; dt$
Ref.3 p.29	$T_z + \dfrac{T_{in}-T_0}{2} - T_a$	Same remark
Ref.4	$T_z + \dfrac{\overline{T_{in}-T_0}}{2} - T_a$	Same remark

Table 2.   Definitions of heat loss term, $T_z + T? - T_a$

116

The term T? is intentional as it represents both a source of confusion in interpreting the test procedure and a source of error in computing the values of EC and COP_C. Table 2 summarises the confusion. It would appear that the expression $(T_z + T? - T_a)$ is intended to be an approximation to $T_s(t)$. However, the plot of these values (seen as the dashed lines in Fig.3) illustrates the fallacy of this presumption. The reference (2,5) form clearly is too large over the test period while the reference (1,3,4) form not only is too large initially but also decreases with time. These errors are best enumerated by determining the instantaneous error rate during charging and then, by integration, the accumulated error.

2.5 The ASHRAE/NBS error rates

Subtracting the expression for the charging rate using the ASHRAE heat loss expression in eqn (18) from the actual charging rate using the actual store temperature, eqn (12), results in the relative storage error rate for the ASHRAE (Ref.1) assumptions, $\dot{e}_{AR}$,

$$\dot{e}_{AR} = \frac{\dot{Q}_{AR} - \dot{Q}_{ACT}}{TSC} = UA\{T_s - T_a\} - UA_{AR}\{T_z + \frac{T_{in} - T_o}{2} - T_a\} \quad \ldots \quad (19)$$

By inserting eqns (12,16) into eqn (19) for the T_o and UA_{AR} terms, simplifying with $\beta \approx 1$, then integrating (neglecting $UA/\dot{m}Cp$ in exponential expressions), the total accumulated relative error during charging over N fill periods can be shown to yield

$$E_{AR} = \int_0^N \frac{\dot{e}_{AR}}{TSC} \approx \frac{UA}{\dot{m}Cp} \{(1 + \frac{\xi}{2})\left[N + (\frac{e^{-N\xi} - 1}{\xi})(1 - \frac{UA}{\Sigma})\right] - N\frac{\xi}{2}\}$$

$$- \frac{UA}{\dot{m}Cp}(1 + \frac{\xi}{2})(\frac{UA}{\Sigma})(N + \frac{e^{-N\xi} - 1}{\xi} - \frac{UA}{\Sigma}\frac{(e^{-N\xi} - 1)}{\xi})(1 + \frac{T_z - T_a}{T_{in} - T_z}) \quad \ldots \quad (20)$$

so that over a single fill period, $N = 1$, the accumulated error is

$$E_{AR} = \frac{UA}{\dot{m}Cp}\left[g(\xi) + (1 + \frac{\xi}{2})e(\xi)\frac{UA}{\Sigma}\right] - \frac{UA}{\dot{m}Cp}(1 + \frac{\xi}{2})(f(\xi) + e(\xi))(\frac{UA}{\Sigma})(1 + \frac{T_z - T_a}{T_{in} - T_z})$$

$$\ldots \quad (21)$$

where $d(\xi) = 1 - e^{-\xi}$, $\quad e(\xi) = \frac{d(\xi)}{\xi}$

$$f(\xi) = 1 - e(\xi), \quad g(\xi) = (1 + \frac{\xi}{2})f(\xi) - \frac{\xi}{2} \quad \ldots \quad (22)$$

Following a similar derivation using the NBS definitions seen in Table 2 the error rate and hence the accumulated error for the NBS assumptions, is given by

$$E_{NBS} = \int_0^N \frac{\dot{e}_{NBS}}{TSC} \approx \frac{UA}{\dot{m}C_p} \left[(N + \frac{e^{-N\xi}-1}{\xi})(1 - \frac{UA}{\Sigma}) - \frac{N}{2} \right]$$

$$- \frac{UA}{\dot{m}C_p} (\frac{UA}{\Sigma}) (N + \frac{e^{-N\xi}-1}{\xi} - \frac{UA}{\Sigma} (\frac{e^{-N\xi}-1}{\xi})) (1 + \frac{T_z - T_a}{T_{in} - T_z}) \quad \cdots \cdots (23)$$

so that over a single fill period the accumulated error is

$$E_{NBS} = \frac{UA}{\dot{m}C_p} \left[h(\xi) + \frac{UA}{\Sigma} e(\xi) \right] - \frac{UA}{\dot{m}C_p} (f(\xi) + \frac{UA}{\Sigma} e(\xi)) \frac{UA}{\Sigma} (1 + \frac{T_z - T_a}{T_{in} - T_z}). \quad \cdots (24)$$

where $h(\xi) = f(\xi) - \frac{1}{2}$. $\quad \cdots \cdots (25)$

The functions defined by eqns (22) and (25) have been plotted as Figure 4.
Surprisingly the dominant error coefficient, $g(\xi)$, used in the ASHRAE Ref.1
procedure over a single fill period is minimal despite its physically meaningless form (see Fig.3). The negative value $h(\xi)$ arising from the NBS (ref.2→4)
form could be anticipated (see Fig.3) because the loss term is clearly too large
over the first fill period. However, the error expressions, eqns (20 and 23)
grow linearly when the analysis is extended to longer periods.

2.6 The inexorable build-up of error

Extending the experiment to N multiples of the fill time such that
$\exp(-N\xi) \to 0$, taking only leading terms and ignoring all UA/Σ leads to the
approximate accumulated errors:

$$E_{AR} \to \frac{UA}{\dot{m}C_p} \left[N - 1/2 - 1/\xi \right] \quad \cdots \cdots (26)$$

$$E_{NBS} \to \frac{UA}{\dot{m}C_p} \left[\frac{N}{2} - \frac{1}{\xi} \right] \quad \cdots \cdots (27)$$

Thus an inexorable build-up in error begins to accumulate after just a few fill
periods, well within the normal data acquisition period.

In physical terms the ASHRAE and NBS approximate values of the store
temperature T_s, lead to too low a heat loss which, when integrated over some
4 to 5 fill times leads to an inexorable build-up of error in the amount of
energy stored during charging. Clearly then if the analysis of test data is
carried out over N fill periods where N is greater than a critical value, N_C,
when eqns (26 and 27) become positive, the amount of computed stored energy will
be greater than the theoretical maximum. It was, in fact, this discovery which
precipitated this study. However, even if the test is restricted to a single
fill period the procedure is still unclear, because of the lack of a model, as
to what parameter is being identified from the test.

2.7   <u>Heat exchanger effectiveness - identification</u>

Having found the error rates which lead to an error in accumulated energy stored we examine now the parameter "identified" by the procedures. The integration of eqn (18) using the ASHRAE, NBS, or actual heat loss expression leads to the Coefficient of Performance as a function of N multiples of the fill time and given by

$$
\begin{aligned}
COP_{AR}(N) &\approx \left[(1-e^{-N\xi})(1 - \frac{UA}{\Sigma}) + \frac{UA}{\Sigma} (1 + \frac{T_z - T_a}{T_{in}-T_a})\{N-(1 - \frac{UA}{\Sigma})(1-e^{-N\xi})\} \right] (1-\beta/2 \ \frac{UA}{\Sigma}) \\
&\quad - \beta \ \frac{UA}{\dot{m}C_p} \left(N \frac{T_z - T_a}{T_{in}-T_Z} \right)
\end{aligned}
$$
$$\dots \dots (28)$$

$$
\begin{aligned}
COP_{NBS}(N) &\approx \left[(1-e^{-N\xi})(1 - \frac{UA}{\Sigma}) + \frac{UA}{\Sigma} (1 + \frac{T_z - T_a}{T_{in}-T_z})\{N - (1 - \frac{UA}{\Sigma})(1-e^{-N\xi})\} \right] (1-\beta/2 \ N\frac{UA}{\Sigma}) \\
&\quad - \beta \ \frac{UA}{\dot{m}C_p} \left(N \frac{T_z - T_a}{T_{in}-T_z} \right)
\end{aligned}
$$
$$\dots \dots (29)$$

$$
\begin{aligned}
COP_{ACT}(N) &\approx \left[(1-e^{-N\xi})(1 - \frac{UA}{\Sigma}) + \frac{UA}{\Sigma} (1 + \frac{T_z - T_a}{T_{in}-T_a})\{ N\xi-(1 - \frac{UA}{\Sigma})(1-e^{-N\xi})\} \right] \left(\frac{\Sigma}{\dot{m}C_p\xi} \right) \\
&\quad - \frac{UA}{\dot{m}C_p} \left(N + N\frac{T_z - T_a}{T_{in}-T_z} \right)
\end{aligned}
$$
$$\dots \dots (30)$$

where an approximate exponential term which neglects UA/Σ has been used.

In all cases the leading term is the same so that the performance coefficient is given by

$$COP_C \cong (1 - e^{-N\xi}) \quad\quad\quad \dots \dots (31)$$

which for N = 1 appears as $d(\epsilon)$ in Fig.5.

As is apparent from the rather complicated expressions (28) and (29), and even the "actual" expression (eqn.30), the parameter being identified, COP_C, does not lead directly to the modelling parameter, ξ, the heat exchanger effectiveness. Further, the result is clouded by the appearance of the ratio $(T_z-T_a)/(T_{in}-T_z)$, the appearance of the loss ratio UA/Σ, and the heat loss coefficient error ratio, β.

These results show that even when no experimental error is present the parameter identification of UA and COP_C is either in error or does not yield the parameters necessary for system modelling or system design. In short, the ASHRAE/NBS identification procedure is unsatisfactory. The analysis in the discharge case is similar and the result is worse because the ASHRAE/NBS procedures ignore a heat loss expression altogether. As a result the discharge coefficient will be larger than it truly is over any number of test fill periods. In fact, it too may exceed unity, thereby indicating (fallaciously) that more energy is extracted than was stored.

We outline next, therefore, a "bandage" procedure which may alleviate the problems noted above, at least for the case of the fully mixed store.

3.   THE "BANDAGE" PROCEDURE

For the very special case of a perfectly mixed store powered by a zero capacity heat exchanger the identification of the ASHRAE/NBS loss coefficient was seen as eqn(17).

Combining eqns (15 and 17) to eliminate UA_{AR}, then inverting, yields

$$\frac{T_{in}-T_a}{T_{in}-T_o} = \dot{m}C_p \left(\frac{1}{UA}\right) + \frac{1}{\xi} \qquad \qquad \ldots \ldots (32)$$

With precise measurements of the left hand side for various flow rates, $\dot{m}C_p$, at various inlet temperatures, T_{in}, in the steady state one can "identify" the two unknown parameters; the loss coefficient, UA; and the heat exchanger efficiency ξ. Provided that the store indeed behaves as a fully mixed store during the transient, one can identify the governing parameters without recourse to a transient technique.   Further, eqn(32) is also valid for storage devices powered by heat exchangers of arbitrarily large finite capacity. Although eqn (32) in this form is only valid for a limited range of flow rates near the design point, an expression based on the effectiveness, ξ, as a function of flow rate is easily derived.   What remains, however, is to extend the above analysis to include other storage devices, particularly those governed by non-linear time or temperature dependent equations.   This requires a much deeper study of both modelling and system identification procedures.

4.   CONCLUSION

We have shown that the present ASHRAE/NBS procedure is unsatisfactory. The procedure leads to an error in the identification of the heat loss para-meter UA for an idealized fully mixed store powered by a zero capacity heat exchanger. We then have shown that the definition of the heat loss expression for the charge test is not only confused and confusing but incorrect if it intends to represent the store temperature.   We have shown, further, that this incorrect loss expression leads to an inexorable build-up in error after just a few fill times in the computed value of the energy stored and hence the performance coefficient so that more energy is (falsely) stored than is possible We also indicated that the zero loss assumption for the discharge case leads, again incorrectly, to a greater energy extraction and a greater performance coefficient than is possible.   We then showed that the computed performance coefficient is not equal to but is a complicated function of the heat exchanger effectiveness. We were, however, able to derive an expression, eqn (32),based on steady state measurements, which leads directly to the two governing parameters, thus obviating the need for a transient procedure altogether.

4. REFERENCES

1. Hill, J.E. et al. ASHRAE Standard 94-77, Methods of Testing Thermal Storage Devices based on Thermal Performance, ANSI B 199.1-1977, American Soc. of Heat, Refrig. and Air Con.Eng. Inc., 345 E 47th St., New York, N.Y. 10017, USS.

2. Hill, J.E. and Kelley, I.E. 1974. Method of Testing for Rating Thermal Storage Devices based on Thermal Performance, NBSIR 74-644, National Bureau of Standards, Washington, DC.

3. Hunt, B.J., Richtmeyer, T.E., Hill, J.E. 1978. An Evaluation of ASHRAE Standard 94-77 for Testing Water Tanks for Thermal Storage, NBSIR 78-1548, National Bureau of Standards, Washington, DC.

4. Jones, D.E. and Hill, J.E. 1979. Testing of Pebble Bed and Phase Change Storage Devices according to ASHRAE Standard 94-77, NBSIR 79-1737, National Bureau of Standards, Washington, DC.

5. Van Galen.E.1979. Recommendations for Testing Thermal Storage Devices, CEC Coordinator on Storage, TNOTPD, Delft, Holland. Private communication

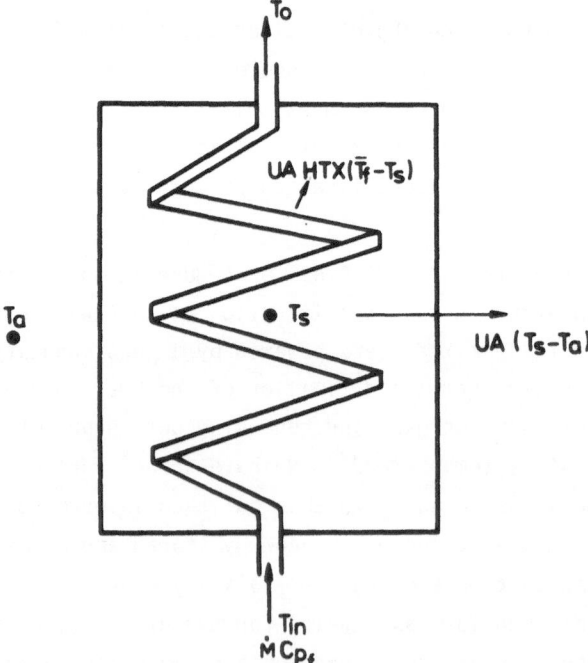

Figure 1. The Ideal Fully Mixed Store

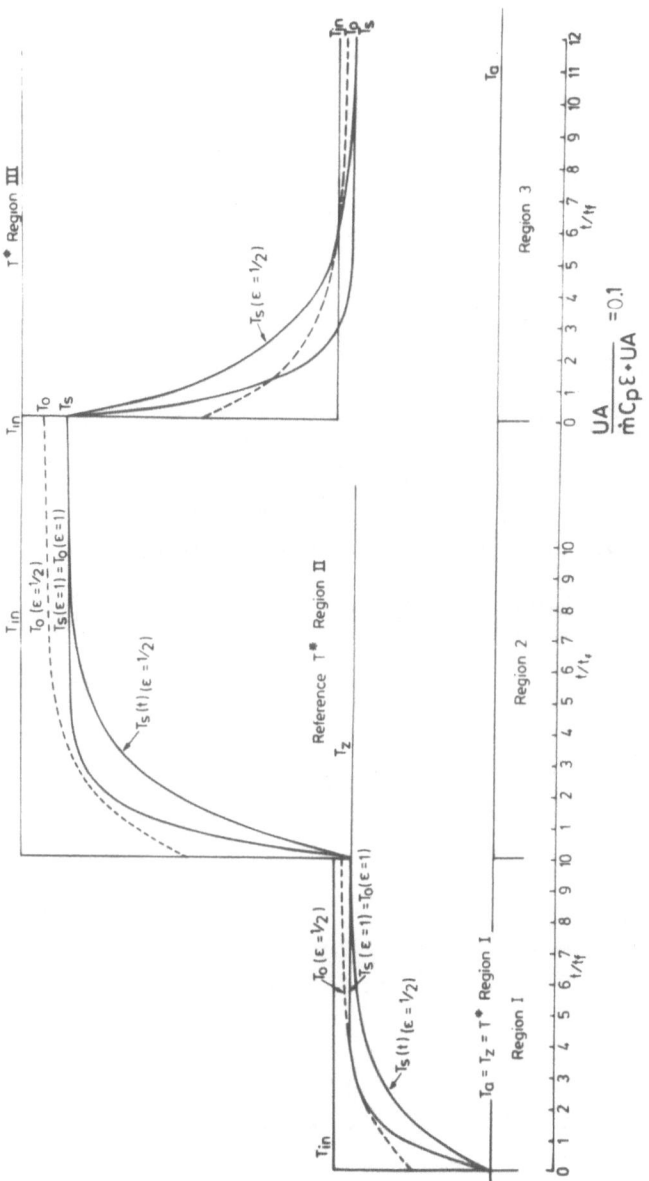

Figure 2. The Transient Behaviour

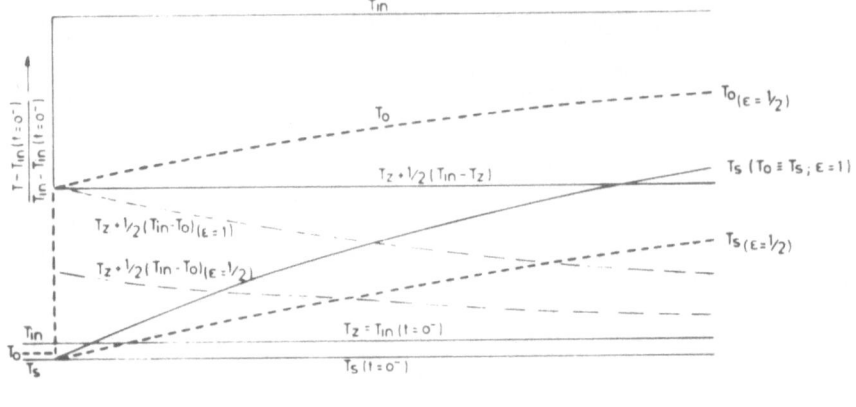

Figure 3.   Detail of Transient during Charging

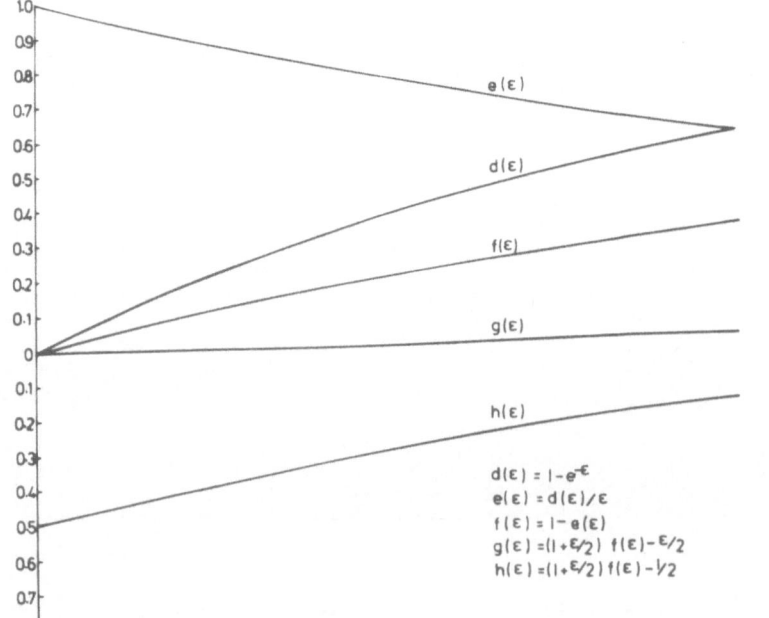

$$d(\varepsilon) = 1 - e^{-\varepsilon}$$
$$e(\varepsilon) = d(\varepsilon)/\varepsilon$$
$$f(\varepsilon) = 1 - e(\varepsilon)$$
$$g(\varepsilon) = (1 + \varepsilon/2) f(\varepsilon) - \varepsilon/2$$
$$h(\varepsilon) = (1 + \varepsilon/2) f(\varepsilon) - 1/2$$

Figure 4.   Error Rate Functions

IMPROVEMENT OF HEAT TRANSPORT IN PARAFFINES FOR LATENT HEAT
STORAGE SYSTEMS.

A.G. DE JONG [x)] AND C.J. HOOGENDOORN

Applied Physics Department, Delft University of Technology,
The Netherlands.

1. INTRODUCTION

For an efficient use of solar energy a certain heat storage
capacity is always required. Often sensible heat has been used
requiring large vessels filled with water or a rock bed.
For reduce the storage volume latent heat systems are well
known. Materials showing a solid/liquid phase change at a
suitable temperature are used. Two types of materials are
under consideration for this phase change behaviour. One the
hydrates of different anorganic salts, like the well known
Glauber's salt having a melting point near $32,4^{\circ}C$. They have
relatively high heats of fusion, however after going through
a number of cycles they show segregation of salt and liquid.
Also a considerable subcooling during solidification may occur.
These factors often reduce the storage capacity considerable.
Next to these materials organic homogeneous materials having
the right melting temperature can be considered. For the
temperature range 30 to $70^{\circ}C$ used in solar heating of buildings paraffine waxes can be used as reviewed by Stahl [1]
and Bailey [2]. In general they have a lower heat of fusion
(180 - 210 kJ/kg versus about 250 kJ/kg for hydrates).
However they don't show any segregation or subcooling effects
even after repeated cycling. Moreover for higher temperature
applications (70 - $130^{\circ}C$) there are several other organic
materials like polyethylene with the right melting point.
Especially for these applications latent heat storage at a
fixed required temperature level is advantageous.

[x)] Now at Grontmij, De Bilt.

It prevents too high temperatures when storing and ensures reasonable collector efficiencies. Moreover when used for absorption cooling to extract sensible heat at lower temperatures has no use.

A drawback of organic phase change materials are their lower thermal conductivity, which gives a reduced rate of heat extraction when solidifying the material. During the heat uptake and melting of the material convection in the melt is often more important and the low thermal conductivity of the liquid is a smaller problem.

In our study we aimed at improving the heat transport through solidified paraffine layers as formed during a heat extraction cycle.

A possible method is to use finned tubes as reported on by Abhat (3). Next to this way of improving heat transfer rates we used metal matrix structures in the paraffinic material to enlarge the thermal conductivity.

In our paper we will first discuss the physical properties of the two paraffines used and further report on the effect of metal structures on thermal conductivity and on solidification time.

2. PHYSICAL PROPERTIES

2.1. Heat capacities

Two paraffinic materials have been used in our study. One a pure substance: n - Eicosane ($C_{20} H_{42}$) with a melting temperature of $37^{\circ}C$ and the other a commercial wax with a melting range of $42/44^{\circ}C$. Heat capacities for the solid and liquid phase as well as heat of fusion have been measured. Both a caloric method and a modified dynamic DTA (Differential Thermal Analyser) method have been used. The latter as developed by TPD/TNO (4,5). By using heat-flux meters this modified DTA method also yields caloric data.

The results showed that the in literature reported heats of fusion at the melting temperature (range) are higher than the measured ones. In fact we found for n-eicosane 125 kJ/kg versus

211 kJ/kg from literature and for the commercial wax 146 kJ/kg.
This effect has also been reported by Abhat (3).

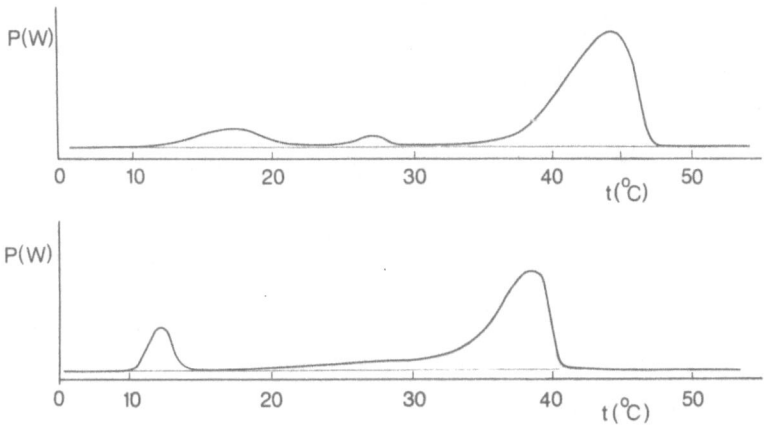

FIGURE 1. DTA-results for commercial wax (upper) n-eicosane
(lower).

The reason is that 10 to $25^{\circ}K$ below the melting point there
are secondary latent heat effects due to recrystallization
of the paraffinic material. This can be clearly seen in the
DTA-curves as given in fig. 1. Often the secondary heat effects
below the melting point are included in literature data on
specific heats of the solid material. However we assumed the
solid phase specific heat to be the value as found below $10^{\circ}C$.
By integration from $10^{\circ}C$ to the melting range the total latent
heat value could be calculated including the secondary effects.
This values comes nearer to the reported literature values.
Table 1 gives the results for both materials from the two
measuring methods.
Also in the caloric method the solid phase c_p value showed
higher c_p values due to secondary heat effects above $10^{\circ}C$.
A conclusion from these data is that to make full use of the
heat of fusion of paraffinic materials the secondary recrys-
talization temperature should be brought near to the melting
point by adding suitable crystallization aids.

TABLE 1. Heat capacities of paraffines.

		n-eicosane		comm. wax	
		Caloric	DTA	Caloric	DTA
LIQUID c_p	kJ kg^{-1}K^{-1}	2.3	2.3	2.4	2.3
SOLID c_p (<10°C)	kJ kg^{-1}K^{-1}	-	2.4	-	2.2
LATENT HEAT AT MELT RANGE	kJ kg^{-1}	126	124	128	146
SECONDARY HEAT EFFECTS	kJ kg^{-1}	-	51	-	50

2.2. Thermal conductivity.

Thermal conductivity for both the liquid and solid phase
have been measured by the transient hot-wire method as
described by Tye (6). For the liquid phase short measuring
times could be used to prevent convection. In the solid phase
special care has to be taken that the solidified material has
good contact with the wire. A local air gap may occur leading
to too low λ-values. On the other hand in the solid phase the
secondary latent heat effects will lead to too high λ-values
in this method. So the accuracy of the method is less good
for the solid than for the liquid phase.
As an alternative method for the solid phase the transient
cooling or heating of a cylindrical sample at temperatures
well below the melting point have been used. From these data
the thermal diffusivity a could be found and using the known
c_p values also the λ. This method has been used because in the
case of metal matrix structures applied in the paraffinic
material the hot-wire can not be used. With this method
somewhat higher λ-values are found, see table 2. As an average
a value of 0,12 Wm^{-1}K^{-1} for the solid phase and 0,20 Wm^{-1}K^{-1}
has been found. These values are a factor 5 below those for
hydrates. It means that a 12 mm solid paraffine layer gives
an effective heat transfer coefficient of only 10W/m^{2}K.

TABLE 2. Thermal conductivities ($Wm^{-1}K^{-1}$)

	n-eicosane	comm. wax
Hotwire, liq. ph.	0,20	0,19
Hotwire, solid ph.	0,08	0,12
Transient cyc. solid	0,11	0,16

3. SOLIDIFICATION TIMES

3.1. Measurement method.

   To measure the solidification time cooling down tests have
been done in cylindrical vessels of 47 mm diameter and 200 mm
length. Cooling down was obtained by a water stream outside
the thin metal wall the bottom and top were well insulated.
The paraffinic material was cooled down from a temperature
of about 5 to 10°C above the melting point by a water stream
of about 10 to 20°C below the melting point.
With thermocouples in the cylindrical vessel the solidification
front could be determined as a function of time (see fig. 2).

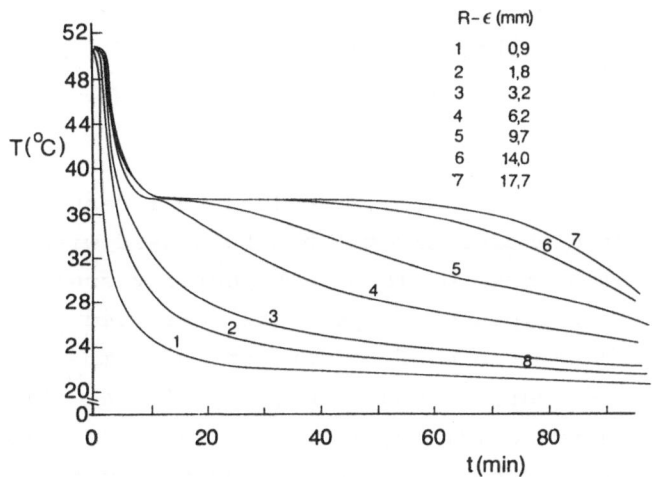

FIGURE 2. Typical solidification curves for n-eicosane,
R-ε distance relative to wall.

The solidification plateau as seen in the curves are not in
all cases very clear. To use a standard method we measured

128

the time to reach a temperature 0,5°C below the melting points
(44 and 37°C).Fig. 3 gives the measured solidification times
as function of the solid layer thickness. Eicosane has a
somewhat longer solidification time due to its lower λ, but
also due to the cooling ΔT being lower (14°C instead of 20°C).
On the other hand its primary heat of fusion is lower.

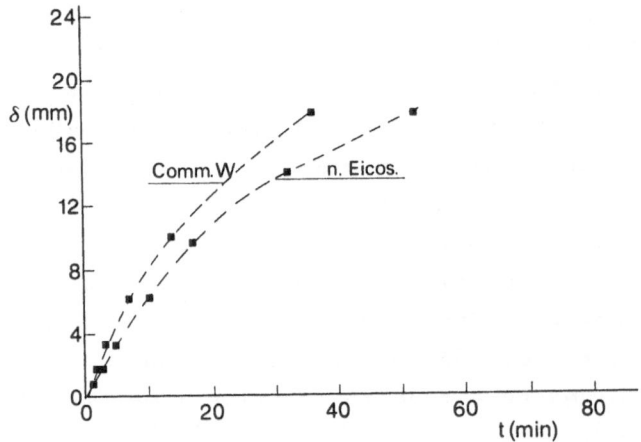

FIGURE 3. Solidified layer thickness δ as function of time t.

3.2. Finned tubes.

   To increase the heat transport a few finned tube tests
have been done. In these cases a similar cylindrical vessel
with a somewhat larger diameter (50 mm) has been used with
An internal copper cooling tube of 10 mm. The outer vessel wall
now being isolated. Two types of copper fins were used
extending to the outer wall. One of 0.5 and one of 1 mm
thickness, fin distances were 10 and 20 mm, see fig. 4.
These finned tubes led to solidification times for commercial
wax being up to a factor 10 better than for the cylinder
discussed under 3.1. The solidification times were slightly
dependent on the heat transfer coefficient in the cooling
pipe. Fig. 5 gives the solidification time for about 90% of
the material as a function of the Reynolds number of the
cooling water flow for two different fin geometries.

FIGURE 4. Finned tube, with fin distance of 10 mm.

These times are now between 3 and 8 minutes, whereas 40 min.
were found for the empty cylinder cooled from the outside,
and 80 min. if cooled from the internal cooling tube without
fins.

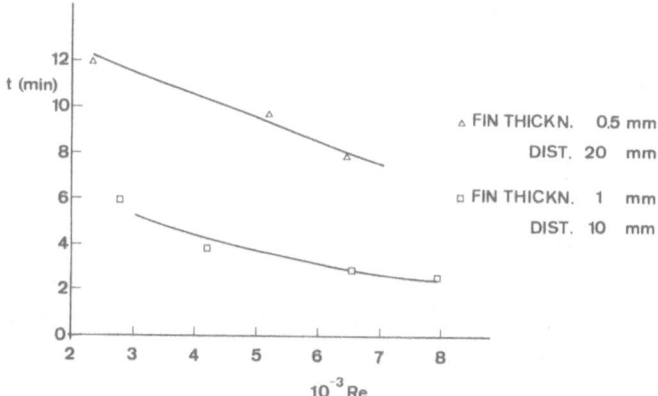

FIGURE 5. Solidification times for two different finned tube
geometries for n-eicosane.

3.3. Metal matrix structures.

Two types of metal matrix structures have been used in
our tests.

1. Aluminium honey-combs as used in the aircraft industry (fig.6a)

FIGURE 6a. Honeycomb.

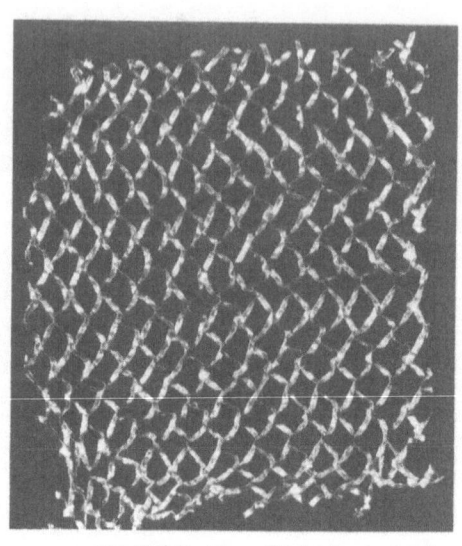

FIGURE 6b. Metal matrix

2. Aluminium thin-strip matrices as known under the trade name
Explofoil(see fig. 6b). This cheap material is used as explo-
sion prevention material for gasoline tanks.

The first material is available in a wide range of packing
densities: 2 - 20% volume, we used 10% (volume). The second
material can be compacted as required, we used in our tests
0,4, 0,8 and 1,6% vol. The same cylinder and method as described in 3.1 has been used. The obtained solidification
times were much shorter than those without metal structures.
The solidification times at a position in the vessel were
compared for the cases with and without the metal matrix.
Fig. 7 gives the factor of the reduction in solidification
time t averaged for the different positions in the vessel as
a function of volume fraction (%) of the metal stuctures.
For eicosane and the commercial wax about the same results
were obtained. As can be seen a reduction by a factor 4 can
be obtained by using only 1,6% of explofoil. The honeycomb
structure with 10% volume resulted in a factor 7 (not included
in fig. 7).

In fig. 7 also the data for the thermal conductivity as

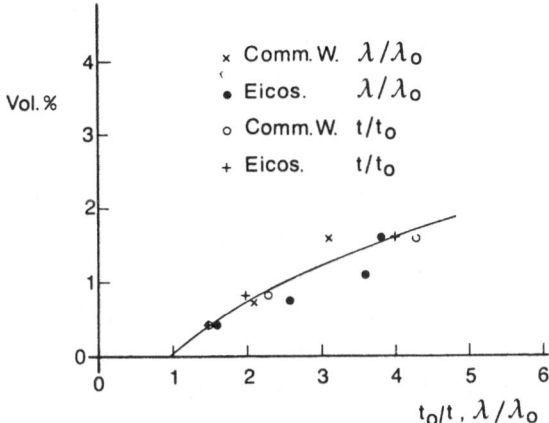

FIGURE 7. Relative decrease of solidification time t and
increase of thermal conductivity λ as function of volume
fraction metal matrix to and λ_0 without metal.

as measured by the transient cylinder method (2.2) have been
given. The data are relative to the λ-value for the solid
material without metal structure. The decrease in solidification time is approximately equal to the increase of λ.
Except for the honeycomb structure where the λ increase is
much larger, a factor 15 versus 7 for the reduction in t.
The finned tube data given in fig. 5 could also be expressed
as a decreased solidification time versus the volume fraction
taken by the fins. For the 0.5 mm. fin this gives a factor
5 at 2.5% for the 1 mm fin a factor 12 at 10%. These values
are in rough agreement with the data for the metal matrices.
The reduction of a factor 4 of the solidification time for
paraffines using 1.6% of metal matrix structure gives them
equal or better heat extraction efficiencies than obtained
with hydrates.

4. COMPARISON WITH COMPUTATIONS

For the mathematical simulation of the heat transfer during
solidification in our experiments the so-called Enthalpy Method
(7) has been used. In this method the interface between solid
and liquid state is eliminated from consideration in the

formulation of the model. We described our model elsewhere (8). For paraffines we have modified it by including a melting range instead of a melting point. The DTA-heat values were simplified by assuming a uniform distribution of the latent heat over the melting ranges (2-4°C), whereas the secondary effects were included in an increased specific heat. In general it has been possible to get an agreement within $\pm$ 10% in the solidification times calculated and measured. For the liquid phase using only conduction resulted in a too slow cooling down of the melt. This due to natural convection in the melt. By using an equivalent higher λ for the melt this could be corrected for.

5. CONCLUSIONS

Measured latent heat values at the melting points of paraffines are 30 - 45% lower than expected due to secondary heat effects at lower temperatures. A mathematical model taking this into account gives a good prediction of the solidification process. Thermal conductivities of paraffines are low and restrict their use. Improvement can be obtained with finned tubes and by metal structures. In the latter use of a 1,6% (vol) aluminium thin strip matrix gives a four fold reduction in solidification time, bringing them in this respect in the range of the more suitable salt-hydrates.

ACKNOWLEDGEMENT

The authors are indebted to the students mr. Burger, Dalhuijsen, de Geus and Koffeman for their accurate measurements.

REFERENCES

1. Stahl J. 1980. Kompakter Latentwärmespeicher Statusbericht Sonnen-energie I, p.395 - 402, VDI-Verlag.
2. Bailey J.A, Mulligan J.C, Liao C.K. 1978. Research on Solar Energy Storage subsystems utilizing the latent heat of phase change of certain organic materials. Dep. Mech. & Aerosp. En. N.Carol. State Un. Raleigh.
3. Abhat A, Aboul-Enein S, Malatidis N and Never G. 1980. Latentwärmespeicher für Solare Heizingssysteme. Statusbericht Sonneneenergie I p.375 - 394, VDI-Verlag.
4. Van der Graaf F. 1978. Description of a calorimeter for quantitative differential thermal analysis.

Report Inst. Appl. Physics, TNO-TPD, Delft.
5. De Waal H. 1965. Quantitative Differential Thermal Analysis
   with an isothermal microcalorimeter.Instr. Pract. 19,
   p.1022 - 1028.
6. Tye R.P. 1969. Thermal conductivity. Vol. 2, p.127, Ac.Press.
7. Shamsundar N and Sparrow E.M. 1975. Analysis of multidimensional conductions phase change via the Enthalpy
   Model. Journ. of Heat Transfer, Trans ASME, Vol. 97 C,
   p.333 - 340.
8. De Jong A.G and Bergmeijer P.W. 1979. Short term latent
   heat storage. Proceedings 18th Int. Conference USA.
   Solar Energy, New Prospects,Milan.

HEAT STORAGE WITH AN INCONGRUENTLY MELTING SALT HYDRATE AS STORAGE MEDIUM BASED ON THE EXTRA WATER PRINCIPLE

SIMON FURBO

Storage material

Heat of fusion storage material candidates are inorganic congruently melting salt hydrates, incongruently melting salt hydrates where phase separation is avoided, organic compounds and organic and inorganic eutectic mixtures. As a rule the organic compounds have small densities compared with the inorganic salt hydrates resulting in a small storage density per unit volume. This makes the salt hydrates favourable as heat storage materials. Harmless inexpensive salt hydrates with melting points in the temperature interval 25-70°C and high heat of fusions are attractive as heat of fusion storage materials in connection with solar heating systems. Examples of such salt hydrates are listed in table 1. The melting point, the heat of fusion, the price for large quantities as well as the melting behaviour are indicated. Since almost all salt hydrates which are attractive as heat storage materials melt incongruently, it is important to solve the phase separation problem which is connected to the use of incongruently melting salt hydrates.

Formula	Melting point, °C	Heat of fusion kJ/kg salt hydrate	Price Dkr/kg salt hydrate 20 t delivery 1 $ = 5.8 Dkr	Congruent melting
$Na_3PO_4 \cdot 12H_2O$	65	190	1.80	no
$NaCH_3COO \cdot 3H_2O$	58	265	2.10	no
$Na_2S_2O_3 \cdot 5H_2O$	48	209	1.45	no
$Ca(NO_3)_2 \cdot 4H_2O$	43	153	0,75	no
$Na_2HPO_4 \cdot 12H_2O$	35	266	1,90	no
$Na_2CO_3 \cdot 10H_2O$	33	247	0.45	no
$Na_2SO_4 \cdot 10H_2O$	32	251	0.40	no
$CaCl_2 \cdot 6H_2O$	29	174	1.10	no

Table 1. Data for salt hydrates which are attractive as heat storage materials in connection with solar heating systems.

Phase separation problem

An incongruently melting salt hydrate consists of an anhydrous salt with corresponding crystal water. The solubility of the anhydrous salt in water at the melting point is not great enough to dissolve all the anhydrous salt in the corresponding crystal water. The molten salt hydrate therefore consists of a saturated solution and some anhydrous salt undissolved in the water. When nothing is done to prevent it this settles down as sediment in the storage tank due to its higher density. By cooling salt hydrate crystals are formed in the dividing line between sediment and solution, by which a solid crust is formed. This solid crust prevents the anhydrous salt at the bottom and the upper layer of saturated solution in getting in contact with each other. Only a part of the anhydrous salt in the solution is active in the phase change, and the salt hydrate consists of three parts at temperatures lower than the melting point: at the bottom the anhydrous salt, then a layer of solid salt hydrate crystals, and at the top a layer of saturated salt solution. If the salt hydrate is not stirred, the salt hydrate crystals will melt by heating and form a supersaturated solution. The amount of sediment increases, and the heat storage capacity decreases by each melting/crystallization cycle, and therefore the phase separation has to be avoided before it is possible to make use of an incongruently melting salt hydrate as a reliable heat storage medium.

Extra water principle

The extra water principle is a method which prevents phase separation resulting in a stable heat of fusion storage making use of an incongruently melting salt hydrate as storage material.

The method consists in adding extra water to the salt hydrate so that all the anhydrous salt can be dissolved in the water at the melting point, that is, the storage medium is a saturated salt solution at the melting point. The storage medium is stirred softly while it is cooled or heated. Since crystallization only takes place from a saturated solution, and the solubility of the salt in water for temperatures below the melting point decreases for decreasing temperatures, solidification takes place by

decreasing temperatures. For temperatures below the melting
point the storage medium consists of a salt hydrate solid phase
and a saturated solution. By heating, a part of the solid salt
hydrate melts, and due to the soft stirring the solution will
still be saturated also at the higher temperature, so that the
mixture will remain stable. The soft stirring is necessary due
to differences in density between the salt solution and the
melted salt hydrates. The stirring could be as well natural
convection caused by temperature differences inside the heat
storage as forced convection produced by a stirrer. Melting as
well as solidification take place in the temperature interval
from the melting point and downwards. The heat of fusion of the
storage medium is situated in the same temperature interval.
The heat storage capacity of the storage medium is less than the
heat storage capacity of the ideal working incongruently melting
salt hydrate, since only a part of the anhydrous salt in the
salt water mixture is active in the phase change. On the other
hand the heat storage remains stable.

Figure 1 shows in the temperature interval 0-100°C the theoretical heat storage capacity of an incongruently melting salt hydrate, Glauber's salt, if it could work stably. The heat storage capacity of a salt water mixture consisting of 33% Na_2SO_4 and 67% of water based on weight, that is a storage medium making use of Glauber's salt and the extra water principle, and the heat capacity of water is shown too. The ratio between the heat storage capacity of a salt water mixture and of water depends on the temperature

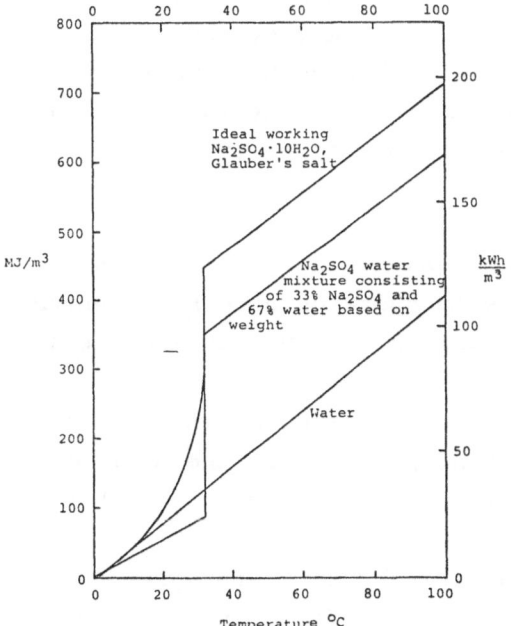

Figure 1. Heat storage capacity in the temperature interval
0-100°C of an ideal working incongruently melting
salt hydrate, a salt water mixture based on the
extra water principle, and water. The anhydrous
salt is Na_2SO_4.

138

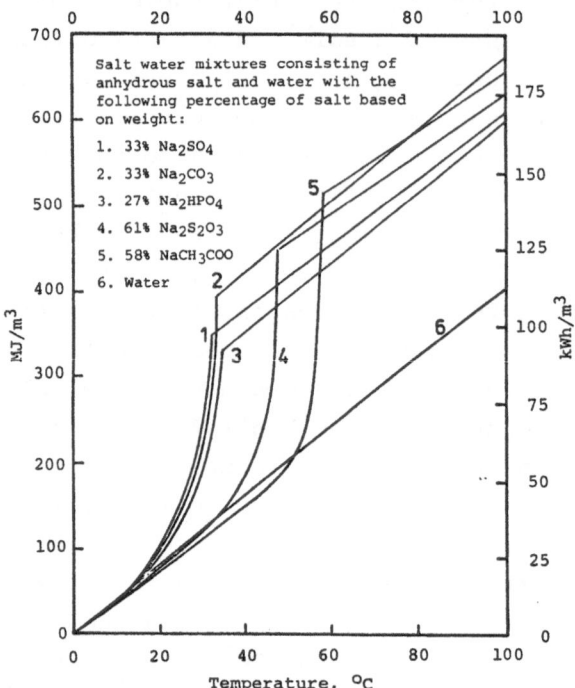

Figure 2. Heat storage capacity of different
heat storage materials.

interval which is chosen
for the comparison. For
small temperature intervals
around the melting point
the ratio is great, and
for increasing temperature
interval the ratio decrease

Figure 2 shows the heat
storage capacity in the
temperature interval 0-100°C
for five inexpensive incongruently melting salt
hydrates from table 1 with
different melting points
making use of the extra
water principle. The salt
hydrates are attractive
as storage medium in different systems: salt hydrates
with high melting points
in systems demanding high temperatures, salt hydrates with low
melting points in systems demanding low temperatures.

Experiments

Different full scale heat storages making use of the extra
water principle with different salt hydrates have been tested by
means of as well long as short term experiments during the last
5 years.

The main result of the long term experiments with heat storages
making use of the extra water principle is that the unstability
caused by phase separation is avoided resulting in reliable and
stable heat of fusion storages.

Important problems concerning salt hydrate storages based on
the extra water principle are: heat transfer problems caused by
the poor thermal conductivity of the solid salt hydrate crystals,
super cooling problems and salt water mixture stirring problems.

These probelms must be solved by an inexpensive and proper heat storage design.

Figure 3 shows schematical illustrations of 3 storage types which have been tested. In the first heat storage the heat

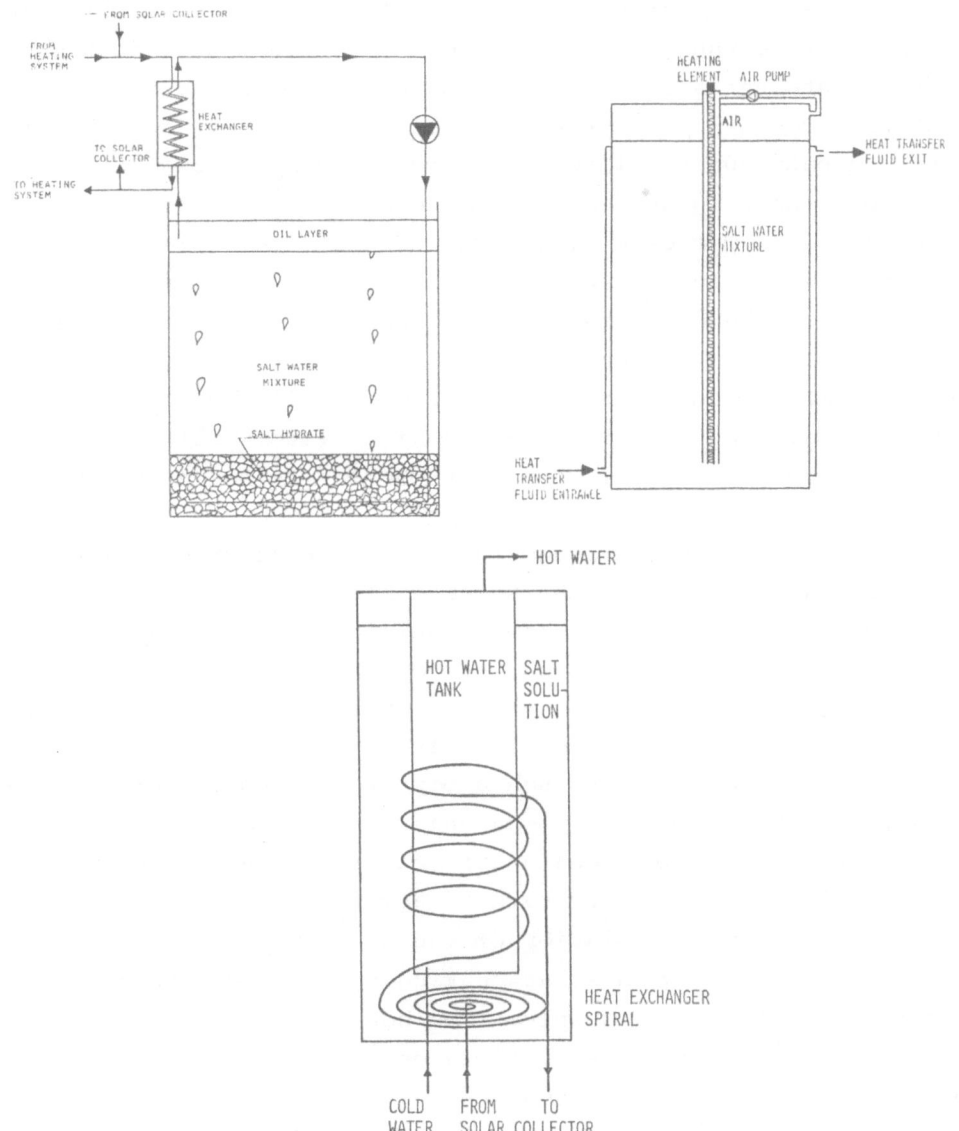

Figure 3. Schematical illustration of 3 heat storages.

transfer problems are solved due to the direct contact between
the heat transfer and the heat storage medium. Oil is the heat
transfer medium, which is circulated in an inner cirquit through
a heat exchanger where heat is transferred to or from the storage.
The oil must be completely immiscible with the salt water mixture.
Due to density differences the oil will form a layer at the top
of the storage tank. The oil is conducted in a pipe to the bottom
of the container and through a nozzle system to the salt water
mixture. The oil drops take care of the stirring, and the stirring prevents supercooling. The storage can be used in connection
with solar heating systems for space heating and domestic hot
water supply. The experiments showed that it is difficult to
keep the separation between the oil layer and the salt water
mixture when the oil flow rate is great. Since additionally
crystal growth in the nozzles appears for low temperatures the
heat transfer from the storage to the oil will be poor for low
temperatures. Further the heat storage system will be rather
expensive due to the oil, the heat exchanger, the nozzle system
and the oil pump.

In the second heat storage the fluid, which transfers heat to
and from the storage, is the solar collector fluid. The fluid
is circulated in a mantle around the surface of the storage tank.
The heat is conducted through the container wall to and from the
storage medium. The top of the tank serves as expansion possibility, and the tank is closed. Air situated at the top of the
container is conducted through a pipe to the bottom of the container by means of an air pump, and the air bubbling through the
salt water mixture produces a soft stirring. A heating element
situated in the pipe takes care of any salt hydrate crystals
which prevent the circulation of the air. The storage can be
used in connection with solar heating systems for space heating
and domestic hot water supply. The heat transfer from the
storage to the heat transfer fluid and the price of the heat
storage are of reasonable amounts. Therefore this storage
system is favourable compared with the storage with oil as the
heat transfer fluid.

In the third heat storage a hot water tank is situated
inside a tank in which the salt water mixture is situated.
Heat from the solar collector is transferred to the storage
by means of a heat exchanger spiral in which the solar collector
fluid is conducted. The heat exchanger spiral is situated in
the bottom of the storage tank and around the bottom of the hot
water tank. While cold water enters the bottom of the hot
water tank a part of the solution is cooled, and a soft stirring
of the salt solution caused by temperature differences occurs.
The heat transfer area between the salt solution and the hot
water tank is large resulting in good heat transfer from the
salt solution to the water. The storage can be used in connection with solar heating systems for domestic hot water supply.
The heat storage worked as planned, when the heat exchanger
spiral is situated in the bottom of the storage tank, so that
all the salt hydrate crystals during heating melt before the
temperature of the upper part of the salt solution has risen.
During the next year this storage type will be studied carefully with different salt water mixtures as heat storage materials.

An example of the advantages by using heat of fusion storages
in solar heating systems for domestic hot water supply

The yearly thermal performance of small solar heating systems
for domestic hot water supply is found by means of computer
calculations. The solar heating systems which are used in the
calculations are shown schematically in figure 4. Two different
storage types are used: A traditional hot water storage and a
heat of fusion storage making use of a $Na_2S_2O_3$ water mixture as
storage material based on the extra water principle.

The computer program simulates the solar heating system.
On the basis of the data of the Danish Test Reference Year
(1980 version) the solar radiation and the heat balance for the
heat storage is calculated every half hour. The three parts in
the heat balance are: the useful solar energy from the solar
collector cirquit, the heat for the hot water consumption and the
thermal loss from the heat storage to the environment. For
every half hour these amounts of heat are calculated, and they

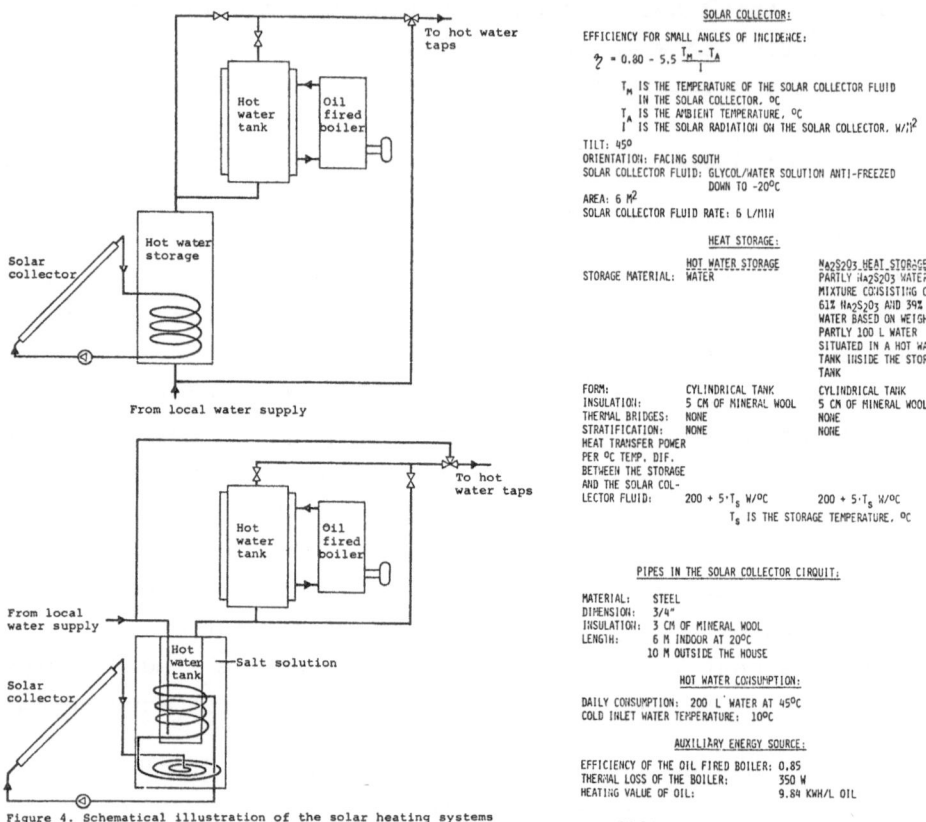

Figure 4. Schematical illustration of the solar heating systems with a hot water storage and a heat of fusion storage.

TABLE 2. ASSUMPTIONS USED IN THE CALCULATIONS

are summed up for the year, that is the yearly thermal performance of the solar heating system is found.

Some of the results of the calculations, with the assumptions
listed in table 2, are given in figure 5. The total utilized
yearly solar energy from the solar heating system consists of
the net utilized solar energy and the saved thermal loss from
the oil fired boiler in the summertime, where the oil burner
can be turned off.

The difference between the yearly net utilized solar energy
from solar heating systems with $Na_2S_2O_3$ water mixture storages
and with hot water storages is small. For solar heating systems
with the same yearly net utilized solar energy the volume of a
hot water storage will be about twice as big as for a $Na_2S_2O_3$
water mixture storage. Almost all the increase in the yearly

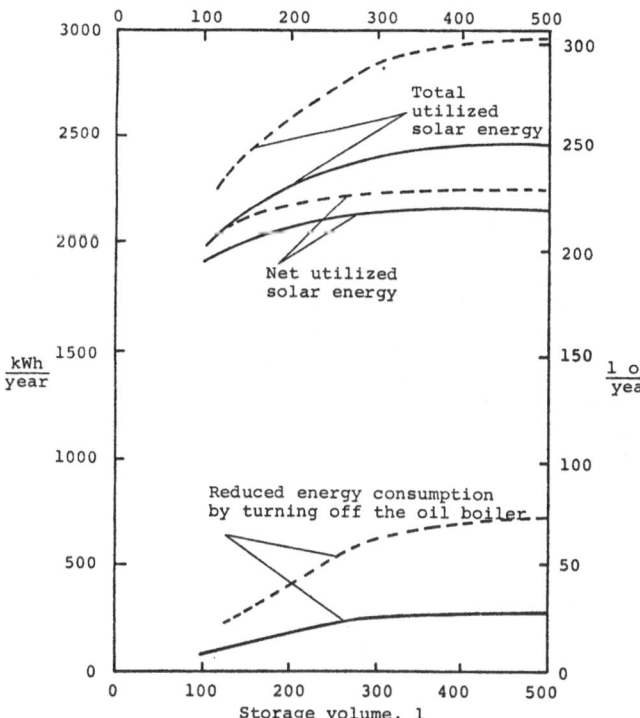

Figure 5. Utilized yearly solar energy from 6 m^2
solar heating systems for domestic hot
water supply as a function of storage
type and storage volume.

Hot water storage: ——————

Na$_2$S$_2$O$_3$ water storage: - - - - - - - -

net utilized solar energy by using a Na$_2$S$_2$O$_3$ water storage instead of a water storage is found in the summer months, where well insulated houses have no space heating demands. When the auxiliary energy source is an oil burner this fact is important since it is possible to turn off the oil burner in periods where the solar heating system supplies all the wanted amounts of water at 45°C. In this example the number of days with the oil burner turned off is chosen as the number of days in unbroken periods longer than one week, where the solar heating system produces all the wanted amounts of water at 45°C. The number of days with the oil burner turned off and with that the reduced energy consumption from the oil fired boiler increases for increasing storage capacity. Therefore the saved energy increases for increasing storage volume, and therefore the Na$_2$S$_2$O$_3$ water storage is favourable compared with the hot water storage as it appears from figure 5.

In this example the total yearly solar energy from the solar heating systems increases for increasing storage volume, and the total solar energy from the Na$_2$S$_2$O$_3$ water storage solar heating system is greater than from the hot water storage solar heating system. The advantage by using the salt water storage depends

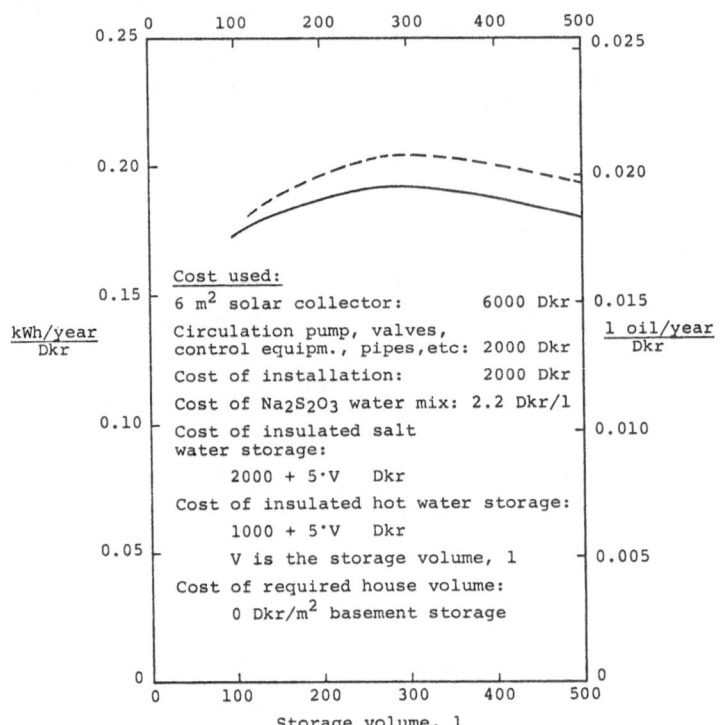

Figure 6. Yearly utilized solar energy per cost of investment for a 6 m^2 solar heating system as a function of storage type and storage volume.
Na$_2$S$_2$O$_3$ water storage: ----------
Hot water storage: ―――――

additionally of the costs for the different solar heating
systems. Figure 6 shows an example of the total yearly utilized
solar energy per cost of investment for the solar heating
systems as functions of the storage type and the storage volume.
The maximum of each curve gives with the used assumptions the
best volume of the storage. In this example the best storage
volume for both storages is 300 l, and the total yearly utilized
solar energy per cost of investment is about 6% greater for the
Na$_2$S$_2$O$_3$ water storage than for the hot water storage.

Conclusion

With the extra water principle a reliable, stable and relatively inexpensive heat of fusion storage making use of an
incongruently melting salt hydrate as storage material can be

constructed.   The advantage by using a salt water storage
instead of a hot water storage in solar heating systems is the
great heat storage capacity which involves small demands for
storage volume.   For solar heating systems with the same yearly
net utilized solar energy, the salt water storage volume is
about twice as small as the water storage volume.   For solar
heating systems for domestic hot water supply, salt water
storages with a suitable melting point are able to supply all
the wanted hot water in the summertime during longer periods
than ordinary hot water storages.   This fact makes the salt
water storages particularly favourable in domestic hot water
supply systems with an auxiliary energy source which during
the summer have a large energy consumption compared with the
energy demands for the hot water supply.

EXPERIMENTAL RESULTS OF A LATENT HEAT STORAGE SYSTEM BASED ON SODIUM
ACETATE TRIHYDRATE IN A STABILIZING COLLOIDAL POLYMER MATRIX, TESTED AS
A COMPONENT OF A SOLAR HEATING SYSTEM

E. VAN GALEN

Institute of Applied Physics TNO-TH, Delft, The Netherlands

INTRODUCTION

The design and development of thermal heat storage systems, providing
an optimum tuning between heat demand and heat supply in a solar heating
installation has become one of the greatest efforts in solar research.
In 1976 our Institute started a research work on this item for the
Commission of the European Communities and in co-operation with Meyhall
Chemical AG.

The purpose of this work was twofold. Firstly the development of a
small and economically feasible latent heat storage system. Small systems
are of importance for a simple integration in dwellings!

Secondly the development of a testmethod, which gives information on
the performance of the storage system as a component of a complete solar
installation.

The work started with the selection and optimization of heat storage
materials by both theoretical and practical means. At the same time
effort was given towards the design of the storage system and the test-
circuit. Theoretical support was given by computer models, modelling
the store in detail or as a component of a solar heating installation.
Now, at the end of 1980, the first experimental results of the heat
storage system tested as a component of a solar heating system are presented
in this paper.

THE HEAT STORAGE MATERIAL

Salt hydrates were considered to be the basic materials for our
investigations, offering the best aspects with respect to a number of
selection criteria i.e.:
- a high energy density (heat of fusion, specific heat, specific weight)
  over the operating temperature range (20 - 80 $^\circ$C)

- a melting temperature at a certain level, low enough to allow a good efficiency of the solar collector, but still high enough for the heating system
- sufficient heat transfer properties (thermal conductivity)
- chemical stability and insensitivity to ageing processes
- toxity, flammability, corrosiveness
- easy availability at reasonable prices.

Specific problems connected with salt hydrates are the segregation of the basic salt and its water of hydration in the molten phase (incongruently melting) and their tendency for supercooling. For every possible candidate a compatible additive must be found to prevent this segregation and a suitable nucleating compound must be found to prevent supercooling.

After a first selection and optimization the thermal properties of 9 storage materials have been measured. Table 1 gives the results for the two most promising materials. Because of its favourable melting point sodium acetate trihydrate has been chosen as a filling mass for the storage system.

	SODIUM PHOSPHATE 12 HYDRATE	SODIUM ACETATE 3 HYDRATE
latent heat of fusion (kJ/kg)	200	270
transition temp. range (°C)	30 - 50	56 - 58
'crystallization temp.' (measured in prototype) (°C)	30 - 33	
specific heat unmolten (J/kg K)	1850	1900
specific heat molten (J/kg K)	3100	2500
specific weight at 20 °C (kg/m³)	1425	1150
expansion between 20 -80 °C	4 - 5%	4 - 5%
thermal conductivity molten (W/mK)	0.7	0.5
thermal conductivity unmolten (W/mK)	0.6	0.5

TABLE 1: Thermodynamic properties of two promising storage materials.

For both materials based on respectively a sodium phosphate 12 hydrate
and on sodium acetate trihydrate a stabilizing and a nucleating compound
were found. However, both additives for sodium acetate trihydrate do not
function ideally. Although it was seen in experiments that the recrystal-
lisation behaviour has been improved, a factor of uncertainty remained.
This fact has been considered in the design of the storage system. With
respect to segregation, the ageing experiment still showed segregation
after 400 cycles in a 1 metre long vertically placed cylinder resulting
in a 20% energy content decrease.
A shorter test of 50 cycles in a cylinder of 30 cm showed a positive
result. New ageing tests in various geometries are underway.

THE HEAT STORAGE SYSTEM

The designer of the storage system must be aware of function and
place of the storage system within the solar heating installation. From
figure 1, a diagram of a for Dutch circumstances normal solar system, it
is clear that the storage system is part of both the collector circuit (I)
and the heat distribution circuit (II).

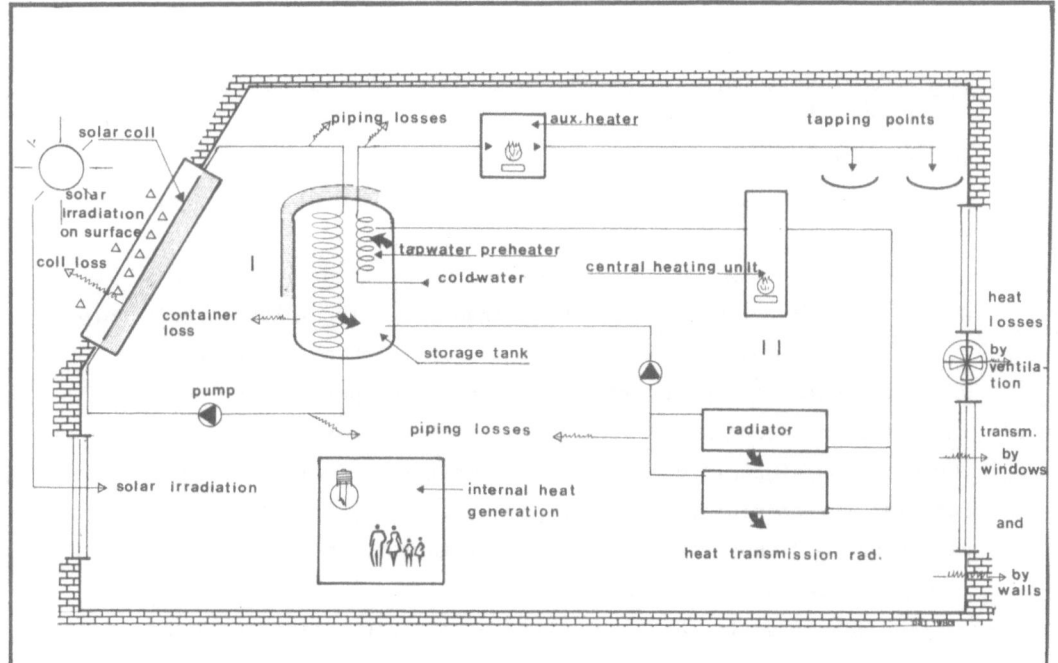

Fig. 1.: Diagram of a solar system, designed for Dutch circumstances

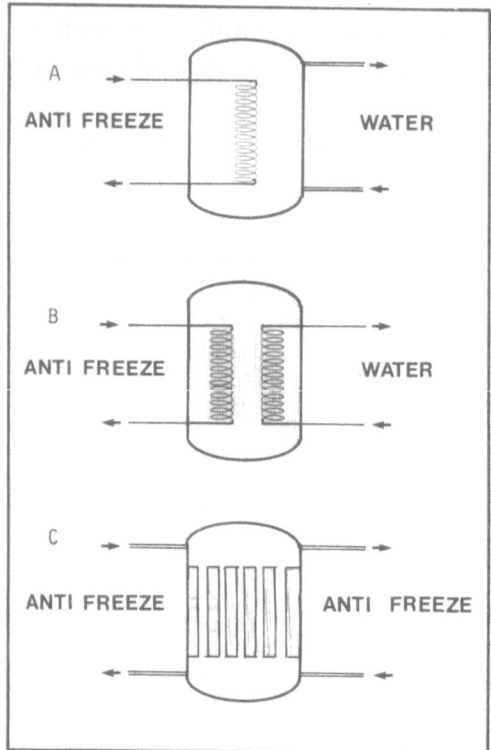

The heat exchanger in the collector circuit, being the partition between the liquids of both circuits (anti-freeze and water) is integrated in the storage tank (see fig. 2A).

When instead of water a salt hydrate is used as storage material, a second heat exchanger seems to be necessary in this system (see fig. 2B)

However, if anti-freeze is used in the heat distribution system as well and if the heat exchanger is used for charging and discharging the solution of figure 2C is possible.

Fig. 2.: Various solutions for the integration of the heat exchanger and the storage tank.

This last solution has several advantages:

- Dependent on the construction (geometry) of the heat exchanger, the mean value of the heat transfer coefficient (k)
  under operating conditions will be in the order of 10-100 W/m^{2}K. For a well designed water store, a heat transfer rate of 3000 W/K is very common. This means a surface F of 6 m^2 for the heat exchanger (k $\simeq$ 500 W/m^{2}K). Apparently relatively large heat exchangers are needed to install in a latent heat store the same heat transfer rate. Therefore, the solution of 2C is preferred. As a matter of fact the storage material is incorporated in a large number of cylinders, piled up against each other and thus forming the heat exchanger. The anti-freeze fluid flows in between (see figure 3).
- Another advantage is the possibility of a direct use of collected heat in the heat distribution system.
  This short-circuit takes care of:

a. higher storage outlet temperatures to the demand system.

b. lower storage outlet temperatures to the collector system, resulting in a higher system efficiency.

· The use of a more expensive heat transfer fluid in this store is not so important, because of the rather small fluid contents. The diameter of the containers is determined by the heat transfer rate required in house heating systems. (The average heat demand for well-insulated dwellings is 5 kW in the Netherlands).

Computations for a 1 m^3 store show that a diameter in the order of 5 cm will meet this demand (see figure 4).

Fig. 3.: The containers of the storage material piled up against each other in the storage tank.

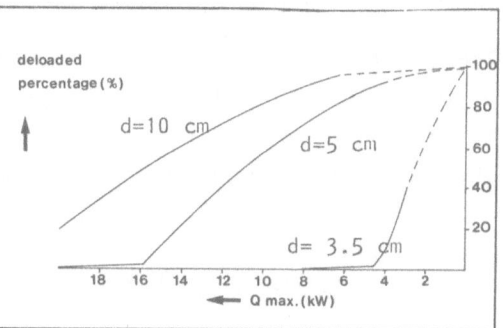

Fig. 4.: This figure shows till which percentage a storage system of 1 m^3 (0.7 m^3 storage material) can be discharged for various container diameters and as a function of the heat exchange rate. The calculations have been made for a heat transfer coefficient k of 20 W/m^2K.

These computations have been carried out for a mean heat transfer coefficient of 20 W/m^2K. This value has been deduced from measurements. The problem of supercooling has been solved by creating a little zone with a permanent cooling at the cylinder feet in which a very small part of the material remains unmolten. Such an auto-nucleating zone guarantees a prompt rehydration of the whole storage mass if the temperature falls below the fusion point again. A metal needle stabs an insulating foot and rests on a cold surface, which keeps the temperature of the needle always below the crystallisation temperature of the storage material. The form of the feet is hexagonal and they fit into each other insulating the heat transfer fluid from the cold surface (see figure 5)

<u>Fig. 5.</u>: Cooling foot of storage material container.

The thickness of the feet is 60 mm, reducing the heat losses through the bottom to an acceptable level. The bottomplate can be cooled in a useful way either by cold tap water or by mechanical ventilation air.
The data of the storage system realized are:
- a diameter of 44 mm and a length of 1500 mm for the 200 salt hydrate containers
- a total volume of 0.64 m^3 (70.7% storage mass, 19.6% water, 9.7% heat exchanger)
- a tank diameter of 707 mm
- a heat capacity between 20 ^{o}C and 80 ^{o}C of 250 MJ
- a heat loss coefficient of 3.5 W/ m^2K

THE TESTPROCEDURE

The only well known test for thermal storage devices is the ASHRAE 94-77 test procedure. However, the usefulness of these tests is limited. The response of the storage system is measured after a sudden temperature change of the heat transfer fluid (ΔT positive charging, ΔT negative discharging). Especially for these types of storage systems the results are very much determined by the start conditions and the size of the temperature step.
The real performance of the store finding expression in the solar contribution to the heating load can only be found by doing a test in which the boundary conditions are continuously changing in a way that would be met in practice.

Such a simulation test circuit requires two loops: one loop which simulates
the collector circuit and a second loop which simulates the heat distribution
system of a solar heating installation (see fig. 6).

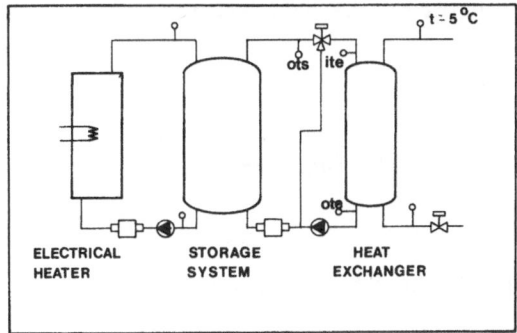

OTS : outlet temperature storage

ITE : required inlet temperature
emitters

OTE : outlet temperature emitters

Fig. 6.: Test circuit for storage systems.

The collector is simulated by an electrical heater, the heat emitters
are simulated by a 'cooling unit', This 'cooling unit' consists of a heat
exchanger which transfers the heat to a cooling fluid ($T = 5\ ^{\circ}C$). The
required cooling load can be met by a control of the flow rate of the
cooling fluid by means of a motorized two-way valve.
A three-way valve controlling the temperature level in the heat distribution loop is also part of the test circuit.

- The heating load for the store is calculated by a simulation model
  describing the heat balances of the collector and the heat losses of the
  piping of the collector circuit. The result of these calculations is
  determined by the outlet temperature of the store in the collector loop,
  the insolation on the tilt of the collector, the open air temperature
  and the characteristics of the collector.

- This heating load is delivered by the electrical heater.

- The cooling load for the store is calculated with a model describing
  the heat balance of a dwelling inclusive of the behaviour of occupants,
  internal gains etc., resulting in a heat demand of the dwelling. For
  a fixed value of the flow rate in the heat distribution system and a
  certain heat demand, the requested outlet temperature of the heat emitters
  can be calculated. The 'cooling unit' is controlled in such a way that
  this temperature is effected.

- The state of the three-way valve is determined by three temperatures:
  1. the temperature at the outlet of the storage system OTS
  2. the temperature at the outlet of the heat emitters OTE
  3. the required temperature at the inlet of the heat emitters
     (= the 'cooling unit') ITE

The following three situations can arise:
  a. OTS > ITE
     The position of the valve is such that the required temperature ITE
     is effected.
  b. OTE < OTS < ITE
     The complete flow rate goes through the storage system
  c. OTS < OTE
     The complete flow rate goes through the bypass

Every 5 minutes the mircro-computer receives signals from the datalogging
system, giving the average values over this five minutes' time periode
temperature differences, absolute temperatures and flows in the test
circuit. Next to every hour the microprocessor reads some weather data
like open air temperature, insolation etc. Then the microprocessor
calculates the new setpoints for the three-way valve and the heating and
cooling unit from the measured data and the given weather data. Finally
every five minutes signals are sent by the microprocessor to adjust the
different devices, while the most interesting data are stored on a
cassette. These data are used for a comparison of the actual performance
of this storage system with the performance of an unstratified water storage
system as a component of the same total system. The performance of an
unstratified water storage system can be simulated with a computer model
with enough accuracy for a realistic comparison because of its simple
physical behaviour.

RESULTS OF THE FIRST SIMULATION TESTS

Because of the limited powers of the heating and cooling unit and
the relatively small volume of the store (0.64 m^3 with 70% storage mass),
a relatively small solar collector area (15 m^2) and solar house (4.5 kW)
have been simulated. All other system parameters have adjusted, realistic
values.

The simulation period of the first tests was one week. The accuracy of the simulation was 5%. The test results were reproduceable.

A first test has been carried out with a start temperature of 60 °C for the storage system which means that the heat of fusion has been stored (melting point = 57 °C). The useful output of the store was measured. This result is compared in figure 7 with the calculated useful output of unstratified water stores of different volumes (T_s = 60 °C).

For equal volumes of both stores a 30% higher useful output is found for the latent heat store.

For equal outputs, it is found that the volume of the water store must be 1.8 times the volume of the latent heat store.

A second test has been done with the same weather data, but starting with a storage temperature of 40 °C, which means that the storage mass was unmolten. During the test the storage mass stayed unmolten, so actually a sensible heat storage system was used with a lower heat capacity per degree centigrade than a water store.

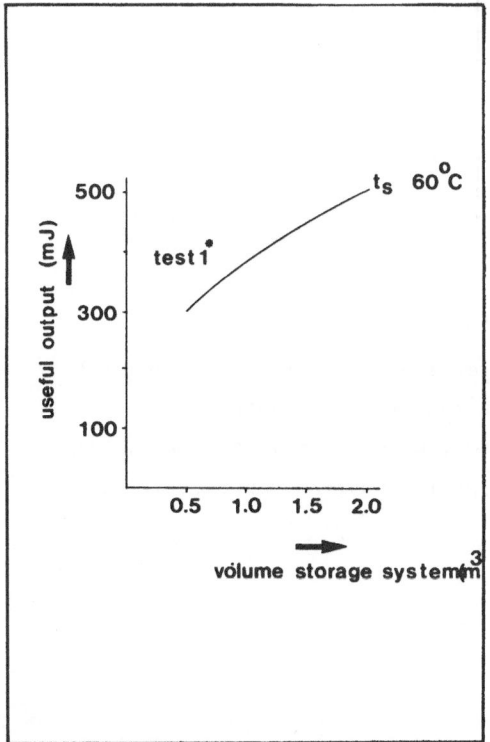

Fig. 7.: Comparison of the result of a first simulation test with the theoretical curve of a water store.

Indeed it is found that the same useful output is realized with a smaller water store (figure 8). It is obvious that the final result of a test over a week's period is completely determined by the start conditions.

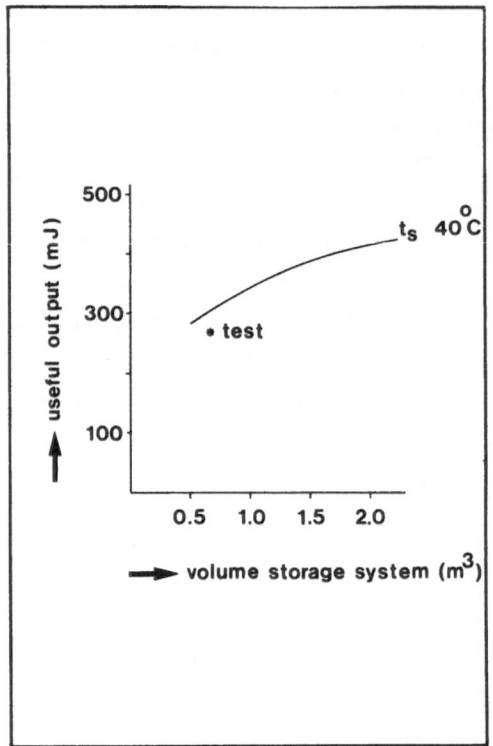

<u>Fig. 8.</u>: Comparison of the result of a second simulation test with the theoretical curve of a water store.

But the results of these tests answered the expectations, which will finally end in a volume decrease of a factor 2 or higher.
Nevertheless attention must be paid to the required length of the test period and its nature. The test period must be representative for the whole year.

HEAT-OF-FUSION STORAGE SYSTEMS FOR SOLAR HEATING APPLICATIONS

A. ABHAT, S. ABOUL-ENEIN, N.A. MALATIDIS

1.   INTRODUCTION

The present paper addresses itself to the development of heat-of-fusion storage systems for low temperature solar heating applications, such as space heating and domestic hot water production. Results from investigations on the two major aspects of heat-of-fusion storage technology - heat storage materials and heat exchangers - are presented.

Differential scanning calorimetry proved itself to be a useful technique for the investigation of heat-of-fusion storage materials. Calorimetric measurements of thermophysical properties of 14 selected organic and inorganic heat storage materials melting within the temperature range 20 - 80 °C, were undertaken with a precision differential scanning calorimeter (DSC). The frequently needed, but often missing, data pertaining to the phase transition temperature and enthalpy, as well as the specific heat of the heat storage materials were measured. The DSC furthermore provided significant information on the influence of thermal cycling on the melting and freezing behaviour of the storage substances. Changes in the thermophysical properties of the thermally cycled materials served as an indication of the change in their melting and freezing characteristics. Results from the DSC measurements thus delivered conditions under which the heat-of-fusion storage substances should be used in latent heat stores or what should be expected of them when used in the heat stores.

A new concept of a passive heat exchanger design for a heat-of-fusion store is presented. The heat exchanger comprises of a number of "finned-annulus" heat exchanger elements bundled together in a conventional shell-and-tube pattern. Results obtained from experimentation with one finned-annulus heat exchanger element are discussed in the paper. For the sake of comparing performance of the heat exchanger with different heat-of-fusion storage substances, a paraffin, a fatty acid and a salt hydrate are employed as the storage materials. Extrapolation of the results to larger systems for solar heating applications is provided.

2. HEAT STORAGE MATERIALS

Table 1 provides a list of the heat storage materials selected for the investigation of their thermophysical properties, along with the manufacturers' data or literature value of their melting point and cost. The commercial paraffins are characterized by the chain length of their carbon atoms and their oil content. For the sake of comparison, a research grade paraffin - octadecane - was included in the tests. The salt hydrates listed in Table 1 are rather well known in the literature and two of these, $CaCl_2.6H_2O$ and $Na_2S_2O_3.5H_2O$, have attracted special interest due to their low cost.

The Perkin-Elmer Differential Scanning Calorimeter, Model DSC-2, was used for the investigations. The evaluation of data was carried out either manually or with the aid of a computer. Further details of the evaluation techniques and precision may be found in Ref. 1.

Table 1. Heat Storage Materials Selected for the Investigations.

Heat Storage Material Group		Storage Material		Details		Freezing Point/ Range Manufacturers' Data or Literature Values	approx Cost
Group No.	Name	No.	Name/ Formula	Chain Structure	Oil Content /%/	/°C/	DM/kg
I	Paraffins	1	Type 6106	n	5	42 - 44	0,60
		2	Type 5838	n	<0.5	48 - 50	1,00
		3	Type 6035	iso	4	58 - 60	0,60
		4	Type 6403	iso	<0.5	62 - 64	1,00
		5	Type 6499	iso	3	66 - 68	0,70
II	Fatty Acid	7	Lauric Acid			42 - 44	2,55
		8	Palmitic Acid			63	2,30
III	Organic	9	25,1 % Propionamide + 74,9 % Palmitic Acid			50	1)
IV	Anorganic Salt Hydrates	10	$CaCl_2.6H_2O$			29,7	0,36
		11	$Zn(NO_3)_2.6H_2O$			36,4	2,39
		12	$Na_2S_2O_3.5H_2O$			48,0	0,30
		13	$Ni(NO_3)_2.6H_2O$			56,7	3,00
		14	$Ba(OH)_2.8H_2O$			78,0	1,75

1) Data not available

(a) Paraffins

Due to their long-term stability and ability to melt and freeze reproducibly without supercooling, paraffins form an important group of latent heat storage materials. Cost considerations, however, necessitate the use of cheap paraffins in heat-of-fusion stores, which are essentially paraffin mixtures and not totally refined of oil.

Typical thermograms for n- and iso-paraffins are presented through the example of Paraffin 5838 and Paraffin 6403 in Figs. 1 and 2, respectively. The n-paraffin exhibits a sharp peak around 46 °C and secondary peaks at temperatures between 12 °C and 46 °C, whereas the iso-paraffin exhibits one peak which is, however, spread over a large temperature range of about 43 K. Evaluation of the peaks shows that both paraffins undergo solid-liquid and solid-solid phase transitions. The DSC technique additionally provides a quantitative measurement of the energy content associated with each of the phase transitions.

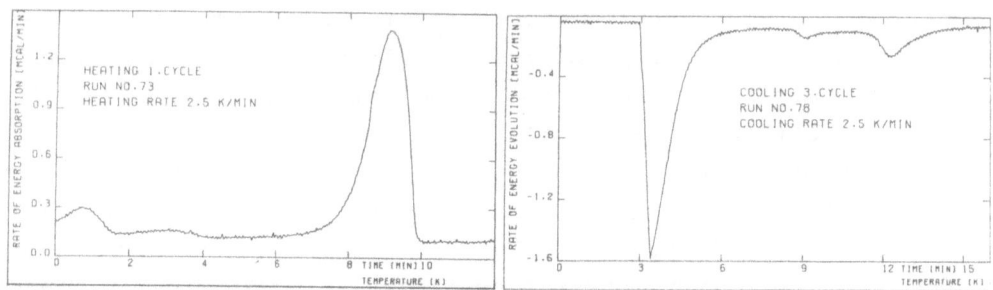

FIGURE 1. DSC-Thermograms for n-Paraffin 5838 (a) Heating, 1st Cycle (b) Cooling, 3rd Cycle

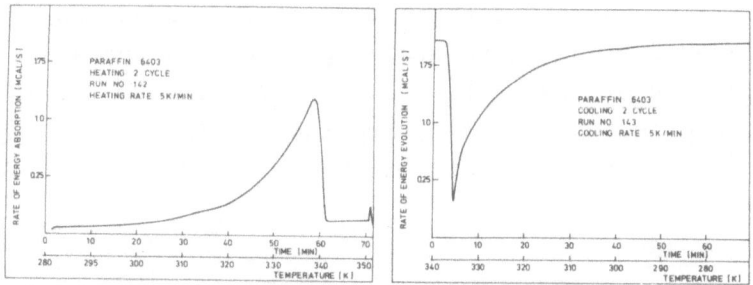

FIGURE 2. DSC-Thermograms for iso-Paraffin 6403 (a) Heating, 2nd Cycle (b) Cooling, 2nd Cycle

160

Table 2. Measured Values of Phase Transition Temperatures and Associated
Phase Transition Enthalpies for Paraffins

Paraffin			Manufacturer's Data		Measured Values							
Type	Chain structure	Oil content	Freezing Range	Heat of Fusion	Phase Transition Temperature Range	solid-liquid phase transition		solid-solid phase transition		Phase Transition Enthalpy		
					PT	PT_1	ΔH_1 [1]	PT_2	ΔH_2 [1]	Total $\Delta H = \Delta H_1 + \Delta H$	Proportions $\frac{\Delta H_1}{\Delta H}$	$\frac{\Delta H_2}{\Delta H}$
		/%/	/°C/	/kJ/kg/	/°C/	/°C/	/kJ/kg/	/°C/	/kJ/kg/	/kJ/kg/	/%/	/%/
6106	n	5	42-44	189	19-44	40.7-44	129.8	19-40.7	49.2	179	72.5	27.5
5838	n	<0.5	48-50	189	12.7-48.3	46.2-48.3	134.4	12.7-46.2	63.0	197.4	68.0	32.0
6035	iso-	4	58-60	189	-8-64.4	39-64.5	85.7	-8-39	83.2	168.9	50.7	49.3
6403	iso-	<0.5	62-64	189	23-65.6	51.7-65.6	129.8	23-51.7	59.2	189	68.7	31.3
6499	iso	3	66-68	189	-6-71.6	2)	2)	2)	2)	145	2)	2)
Octadecane [3]			28-29	246	27-28.5	27-28.5	230	-	-	230	100	-

[1] ΔH_1 and ΔH_2 represent the enthalpies associated with the solid-liquid and solid-solid phase transitions respectively

[2] Paraffin 6499 exhibited a peak spread over a large temperature range of about 78 K, so that the two phase transitions could not be distinguished from each other

[3] n-Octadecane (99 % pure) was employed as a reference paraffin during the investigations

Some results from the measurements of the phase transition
temperatures and enthalpies of paraffins are summarized in Table
2. For all commercial paraffins tested, the deviation between the
manufacturers' data and the measured values is noteworthy. In
contrast to the manufacturer's data, all paraffins, with the
exception of research grade octadecane, exhibited two phase transitions with significant energy contents in the solid-solid phase
transition (ca. 30 to 50 %). Moreover, while n-paraffins experienced the primary phase transition (solid-liquid) in a narrow
temperature range of about 2 K, the corresponding temperature
range for iso-paraffins was large (≥ 14 K). These results are of
particular importance for the choice of the paraffins in heat-of-
fusion stores for low temperature applications where the temperature excursions of the store are generally limited to 10 K about
the melting point. N-paraffins are thus preferred in comparison
to their iso-counterparts and may be selected irrespective of
their oil content.

(b) Fatty Acids

Fatty acids possess good melting and freezing characteristics, as may be seen in Fig. 3 from the sharp peak tracing on the

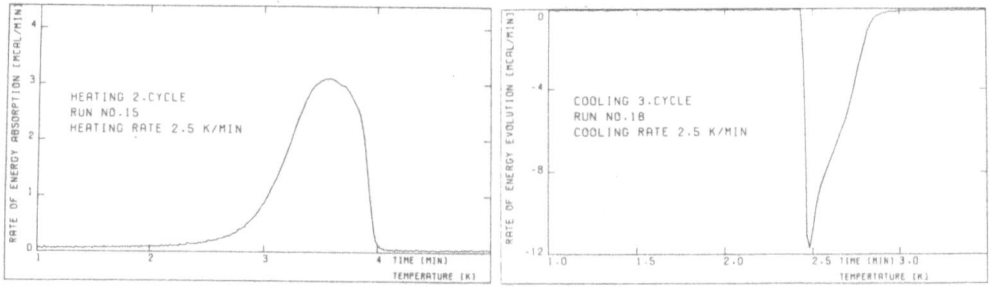

FIGURE 3. DSC-Thermograms for Lauric Acid (a) Heating, 2nd Cycle
(b) Cooling, 2nd Cycle

DSC-thermograms during the heating and cooling of lauric acid.
Fatty acids are excellent heat storage substances but are somewhat
too expensive for large scale solar applications. They are, however, much cheaper than research grade paraffins and are recommended for the function tests of heat-of-fusion storage systems.

(c)   Salt Hydrates

The calorimetric measurements with salt hydrates gave very
interesting results which are illustrated here through the example
of the incongruently melting calcium chloride 6-hydrate. Measurements with this substance were undertaken in two operation modes:
(a) hermetically sealed pans and (b) non-hermetically sealed or
'open' pans, wherein a hole was punched in the hermetically sealed
pans. Fig. 4 shows four typical thermograms obtained during the
heating and cooling of a $CaCl_2 \cdot 6H_2O$ sample in open pans. Operation
in open pans results in decomposition of the substance, as indicated by the occurrence of two or more peaks following the very
first melting process. During the heating process in the 5th cycle
- Fig. 4(d) - the partial decomposition of the hexahydrate into
the tetrahydrate is observed clearly by the second peak at ca.
45 °C. Samples tested in hermetically sealed pans, however, did not
experience any decomposition upto the 11th cycle.

Similar results were obtained with some of the other salt
hydrates tested, so that the use of salt hydrates only in sealed
or encapsulated systems is recommended. The DSC-measurements of
samples in hermetically sealed pans yielded phase transition tem-

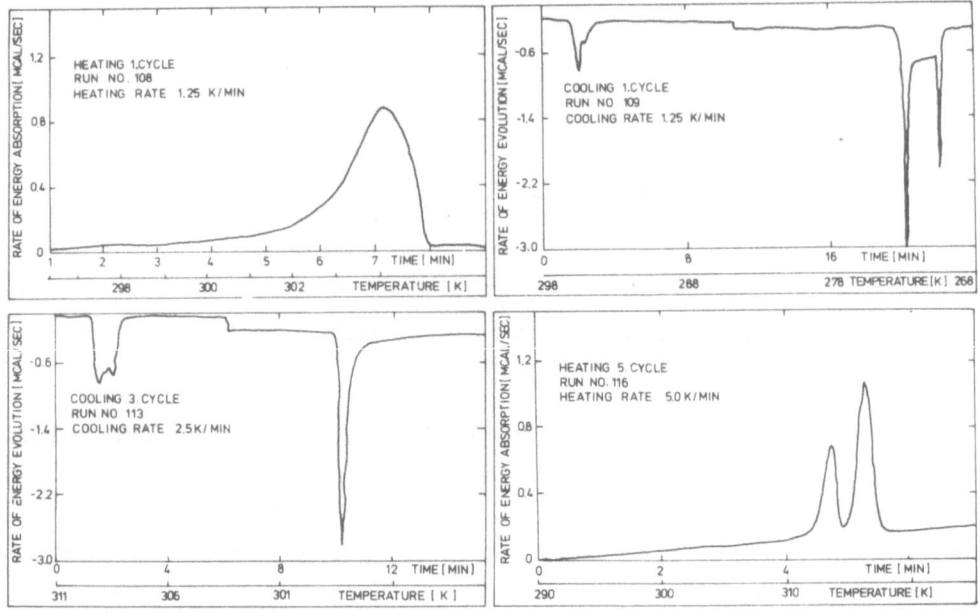

FIGURE 4. DSC-Thermograms for CaCl$_2$.6H$_2$O Samples in Non-Hermetically Sealed or 'Open' Pans (a) Heating, 1st Cycle (b) Cooling, 1st Cycle (c) Cooling, 3rd Cycle (d) Heating, 5th Cycle

perature and enthalpy values in good agreement with data referenced in the literature /2/. The degree of supercooling during freezing was however very high (ca. 50 K), as a consequence of which the phase transition temperatures during cooling of the samples were rather low. The unrealistic value of the supercooling is due to the extremely unfavourable nucleating condition in the small test pans.

3. HEAT EXCHANGER DEVELOPMENT

The commercial realization of heat-of-fusion storage systems is dependent on the development of suitable heat exchanger geometries for charging and discharging of the heat storage medium. The development of a new concept of a passive heat exchanger design, called the 'Finned Annulus Heat Exchanger' is presented in this paper. For most low temperature applications, particularly where high performance systems are needed, the freezing case is the more

critical due to the absence of thermal convection effects during
heat flow out of the system. The finned-annulus concept is particularly suited for these conditions.

3.1   The Finned-Annulus Heat Exchanger

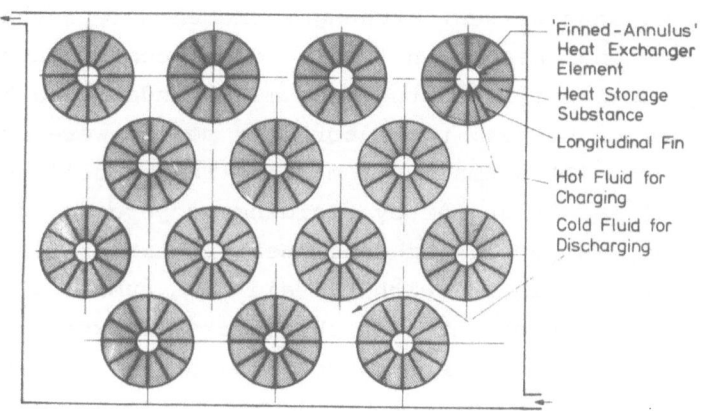

'Finned-Annulus'
Heat Exchanger
Element

Heat Storage
Substance

Longitudinal Fin

Hot Fluid for
Charging

Cold Fluid for
Discharging

FIGURE 5. Schematic of the Finned-Annulus Heat Exchanger Concept for a
Heat-of-Fusion Store

Fig. 5 shows a schematic of the finned-annulus heat exchanger
concept. The heat exchanger consists of a number of 'finned-annulus' heat exchanger elements bundled together in a conventional
shell-and-tube pattern. Each element comprises of an inner tube and
an outer tube maintained in thermal contact with each other
through longitudinal fins made from a good heat conducting material, e.g. aluminum. The region in the annulus and between the
fins is filled with a heat-of-fusion storage material. Hot liquid
from a heat source, e.g. solar collector, flows within the inner
tube of an element, whereas the cooler liquid from a heat sink,
e.g. the load of a heating system, flows around the outer tube.

The finned-annulus geometry offers a number of advantages for
use in a heat-of-fusion store. The fins provide an extended heat
transfer surface, they ensure mechanical stability, and they
restrict the storage medium to small volumes. Depending on the
thermal conductivity of the fins and of the inner/outer tubes,
the tube surface may itself serve as a fin to the longitudinal
fin. The heat exchanger thus permits not only charging and dis-

charging, but also simultaneous charging and discharging.Further-
more, the finned-annulus heat exchanger concept permits the operation of the heat storage system as a hybrid latent heat-sensible
heat store, whereby the sensible heat is stored in the liquid
surrounding the finned-annulus elements in the containment.

The finned-annulus heat exchanger elements represent a case
of three dimensional heat transfer in r-∅-z geometry accompanied
with melting and freezing of the medium. To understand the heat
transfer phenomena better, two laboratory models were constructed
and investigated: (A) a small 0.1 m long 'Test Model' to simulate
two-dimensional heat transfer by avoiding temperature gradient in
the z-direction, (B) a ca. 1 m long 'Heat Exchanger Element'
having a geometry optimized with the aid of a numerical analysis
program. Some results of interest obtained from experiments with
these laboratory-scale systems and extrapolations to larger
systems are presented here.

A. The Test Model

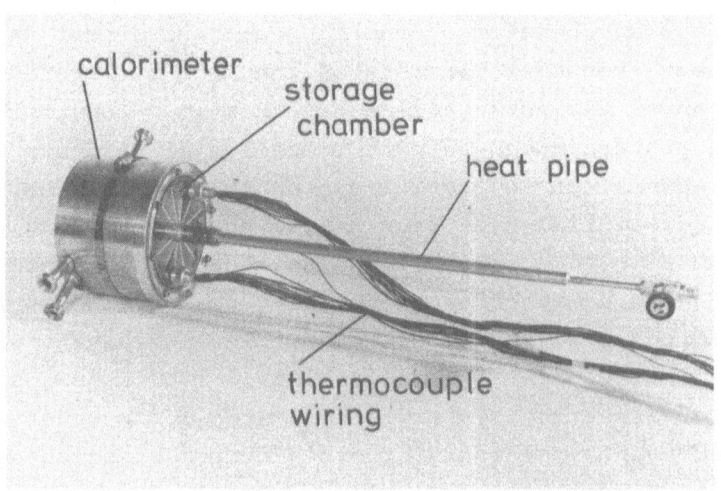

FIGURE 6. Photograph of the Test Model

Fig. 6 presents a photograph of the Test Model. It is instru
mented with 30 NiCr-Ni thermocouples for measurements of the temperature distribution. Both ends of the cylindrical chamber are
closed with plexiglas flanges to allow for visual observations an

for photographing the motion of the solid-liquid interface. Eicosane, lauric acid and $CaCl_2 \cdot 6H_2O$ were employed as the heat storage substances during the tests.

FIGURE 7. Sequence of Photographs Depicting Charging of the Test Model with Eicosane

Figs. 7 and 8 respectively present a sequence of pictures taken during charging and discharging of the test model. The heat storage medium is eicosane, a paraffin melting at 37 °C. Charging of the heat store is performed by flowing hot liquid at 50 °C at the evaporator end of the heat pipe, while discharging takes place by flowing cold liquid at 15 °C around the outer tube of the element. The pictures indicate clearly that thermal convection effects eventually set in within the segments of the top hemisphere during charging, inspite of the fact that the fins assist in suppressing convection. During discharging, on the other hand, the heat transfer mechanism is that of pure heat conduction and the frozen front moves symmetrically in all segments. A quantitative assessment of the rate of motion of the phase change interface could be obtained from an evaluation of the photographs /1/.

FIGURE 8. Sequnence of Photographs Depicting Discharging of the Test
Model with Eicosane

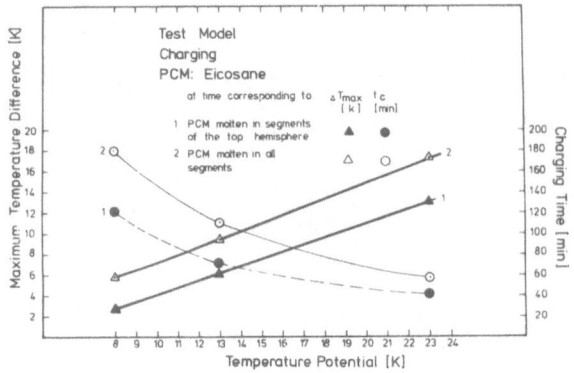

FIGURE 9. Charging Time, t_c, and the Maximum Temperature Difference Required
for Charging, ΔT_{max}, corresponding to the Melting of the Heat Storage
Substance (Eicosane) in: 1) Segments of the Top Hemisphere
2) All Segments in the Test Model

Fig. 9 shows the charging time, t_C, and maximum temperature
differences, ΔT_{max}, required for charging of the test model
filled with eicosane as a function of the temperature potential
available for charging, $|\Delta T_p|$. The curve '1' corresponds to com-
plete charging, whereas the curve '2' corresponds to the case whe
the storage medium only in the segments of the top hemisphere has
changed phase (see Fig. 7). The difference is substantial, e.g.
for $|\Delta T_p|$ = 10 K the storage medium in the segments of the top
hemisphere melts in 100 min while that in the bottom hemisphere in
150 min. The corresponding ΔT_{max}-values are 4 K and 7 K respec-

tively. The large differences are due to the fact that no convection effects can occur in the lower hemisphere,thereby resulting in poorer heat transfer. At t = 100 min, more than 70 percent of the medium has however melted, and in some application cases it may be worthwhile to stop the charging process at this time and commence with the discharging.

Fig. 10 shows the temporal variation of the heat transfer rate during charging for three different heat storage substances. The driving temperature potential for these measurements is 13 - 15 K. Initially the heat transfer rate is high but falls rapidly during the sensible heating of the solid phase. The phase transition range is marked by a plateau, which is seen more clearly during tests with $CaCl_2.6H_2O$ as this substance melts at a point rather than in a range. Interestingly enough, the heat transfer rate experiences an increase for a short while when the organic substances are undergoing a phase transition. Visual observations through the plexiglas flange at this time showed that due to gravity effects the heavier solid phase floating in the liquid phase fell on the hot walls of the fins or tubes (Picture 4, Fig. 7), resulting in improvement in heat transfer rate from the heat transfer surface to the storage medium.

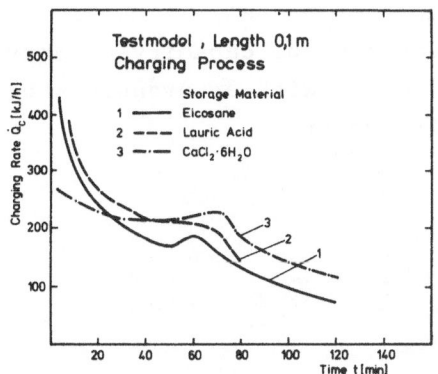

FIGURE 10. Charging Rates Measured with the Test Model for Three Different Heat Storage Substances

The maximum temperature differences. ΔT_{max}, during discharging are compared for the three storage substances in Fig. 11.

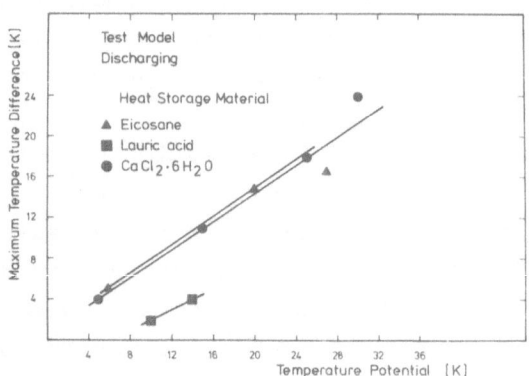

FIGURE 11. Maximum Temperature Differences For the Complete Discharging of the Test Model with Eicosane, Lauric Acid and $CaCl_2.6H_2O$

$CaCl_2.6H_2O$ with a thermal conductivity value almost three times as large as the organics obviously exhibits significantly small values of ΔT_{max} for the same temperature potential.

B. Heat Exchanger Element

Heat exchanger economics demand that each finned-annulus element have the largest possible diameter and heat storage capacity while meeting the heat transfer and technological constraints imposed upon it. The results of investigations with the Test Model and optimization studies carried out with a numerical analysis /1, led to the development of a 'Heat Exchanger Element' that has an estimated latent heat storage capacity of approximately 0.5 kWh (1.8 MJ) per meter length with an organic substance and 1 kWh (3.6 MJ) with a salt hydrate.

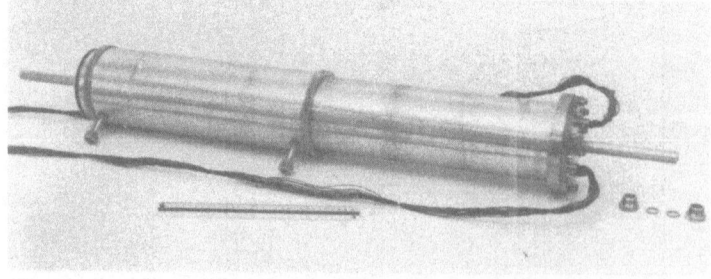

FIGURE 12. Photograph showing the Heat Exchanger Element

Fig. 12 shows a photograph of the Heat Exchanger Element. Measurements with the Element with myristic acid (Melting range 52 - 54 °C) showed that for an allowable temperature potential of 10 K, a heat transfer rate of 350 W per meter length of the Element could be drawn during charging and discharging.

3.2 Extrapolation of Results to Larger Systems

The heat transfer rate of 350 W per meter length of the element is more than necessary, if the elements are to be used for a latent heat store for solar space heating or combined space heating and domestic hot water production applications. The heat transfer rate is, however, insufficient for pure hot water production applications. In the latter case, however, the heat storage system can be easily designed in a manner that sufficient heat is stored as sensible heat in the water surrounding the elements (see Fig. 5) to meet the peak load demands. For a pure hot water application in a 4-person household, it is, in fact, advisable to design a two-day heat store, wherein the finned heat exhanger elements provide a one-day latent heat storage capacity and the surrounding water an additional one-day sensible heat storage capacity.

4. ECONOMICS OF HEAT-OF-FUSION STORAGE SYSTEMS

The economics of a heat-of-fusion store and its cost in comparison with conventional heat storage systems strongly influence its applications in solar heating systems. As an example cost estimates of the heat-of-fusion store employing finned-annulus heat exchanger elements are presented for the domestic hot water (DHW) applications. The cost calculations are made on the assumption that paraffins or salt hydrates may serve as potential heat storage substances.

Fig. 13 shows the costs of a heat store for this application. Specific geometrical and thermal data related to the heat store were assumed for the purposes of cost estimations. These are:

o Number of finned-annulus heat exchanger elements   7
o Diameter of outer containment (shell), m   0.43
o Length of outer containment, m   2

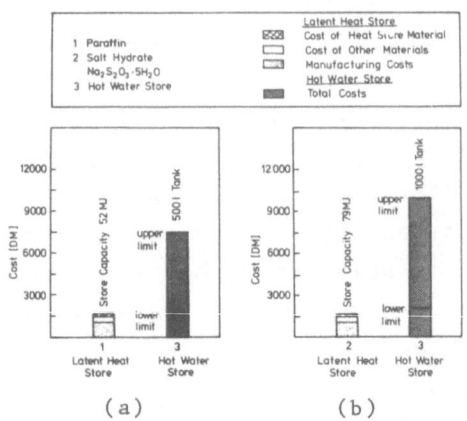

FIGURE 13. Cost of the Heat-of-Fusion Store, based on the Finned-Annulus Heat Exchanger Concept, for the Domestic Hot Water Application. Costs are shown for: (a) Paraffin as heat storage substance (b) The salt hydrate $Na_2S_2O_3 \cdot 5H_2O$ as heat storage substance. Costs for an equivalent hot water store obtained from Ref. 3 are also shown for the sake of comparison.

o Latent heat storage capacity of the storage
   substance, MJ
   - with salt hydrate filling   49
   - with paraffin filling   22
o Additional sensible heat storage capacity of the
   water filling the free volume between the heat
   exchanger elements (with $\Delta T = 10$ K), MJ   30

Results indicate that a hybrid latent heat-sensible heat store of the type described above can be cost competitive with conventional hot water stores for the DHW application, even when employing an organic heat-of-fusion storage substance. Applications requiring larger heat storage capacities, e.g. as in residential space heating systems, would, however, offer a cost-effective solution only when inorganic salt hydrates are used as heat storage materials /1/.

ACKNOWLEDGEMENT

Financial support by the Commission of the European Communities under Contract No. 643-78-7 ESD and by the Federal Ministry of Science and Technology (BMFT) under Grant No. ET 4284 A is acknowledged.

REFERENCES

1. Abhat, A., Aboul-Enein, S. and Malatidis, N.A., "Latent Heat Thermal Energy Storage-Determination of Properties of Storage Media and Development of a New Heat Transfer System, Final Report, EG Contract No. 643-78-7 ESD, IKE, Stuttgart, FRG, July 1980 (To be released)
2. Gmelins Handbuch der anorganischen Chemie, Verlag-Chemie GmbH, 1966
3. Bell, C.R., R. Jäger, F. and Körzen, W., "Systemstudie über die Möglichkeit einer stärkeren Nutzung der Sonnenenergie in der Bundesrepublik Deutschland. In: Sonnenenergie II, Umschau Verlag, Frankfurt, 1977, pp. 49-61

LONG TERM STORAGE

SUMMARY AND OVERVIEW

Prof. E. Hahne
Universität Stuttgart, Federal Republic of Germany

In this chapter seven papers are presented, two on aquifers, two
on storage in the soil, one on storage in hard rock, one on storage in
big water reservoirs and one general paper on the future of long term
storage of solar energy in Europe.
The state of the art with respect to long term storage of solar energy
has hardly been established yet and therefore the variety in the contents
of the papers is rather large.

In his paper Torrenti summarizes the various large scale seasonal
storage systems that have been built and in some cases partly tested up
till now. He compares the various options with respect to energy density,
price expectations etc. and concludes that there are only very few
solutions that show potential application possibilities in collective
large scale projects. The bottleneck seems to be the costprice and
therefore solutions to use the undisturbed soil seem to offer the best
chances.

In their papers Tsang and Iris show the problems involved in long
term thermal energy storage in aquifers: geological formations containing
and conducting water, which can be found at depths ranging from a few
metres to hundreds of metres.
Tsang's paper is focussed on basic studies to understand the fundamental
thermohydrologic processes, on the identification of key parameters and
on sitespecific modelling studies.
In the paper presented by Iris some more attention is paid to experimental
results. He shows that heat storage in aquifers is possible without major
technological problems, though only under particular hydrogeological
conditions, at sufficient aquifer size and with adapted heating systems.

In his paper Wijsman presents the design of a group of about 100
solar houses with seasonal heat storage in the soil. He uses the upper
layer of the soil (up to a depth of approximately 20 m), and suggests

a special insitu-method to insert a heat exchanger consisting of vertical tubes into the undisturbed soil. Only the top of the reservoir has to be furnished with an insulation foam layer. He also presents the results of a parameter sensitivity study of all relevant parameters of the total solar heating and hot water supplying system.

Scholz presents the results of a study to use big water reservoirs for seasonal storage of heat for central solar heating plants or district heating systems. He concludes that especially the construction costs are still higher than economically tolerable.

Jouanna describes a mathematical model of non-saturated soils. This model enables very detailed description of the basic phenomena occurring during practical behaviour of these soils.

Andersson introduces the storage problems involved in using seasonal heat storage in hard rock. He describes the multiple well system and the heat transfer to and from circulating water in a large number of boreholes, which can be considered as a huge heat exchanger.

All these papers together range from extensive theoretical parametric studies on effects of soil properties, operating criteria, sizing problems, system optimization, recovery temperatures, COP's etc.; and in some cases back-up experimental work on field experimental basis is going on.

Some conclusions may be mentioned:
1. There seems to be a remarkably good agreement between some of the models and the experimental results, in spite of the fact that actually very little is known about heat transfer in fluids flowing through heat exchangers in soils and fluid flows in soils and also about the thermophysical properties of soils.
2. There is no unique solution. Geological formation, ecological requirements, location relative to other energy sources - such as an existing or newly planned power station -, geographical location, climatic conditions etc. are important boundary condtions for the type of system and the final site-specific solution.
3. Other sources of energy might be used to charge these seasonal storage systems rather than solar collectors. The inclusion of a heat pump in some of the systems might be advantageous to really use the energy down to low temperatures, even below the water freezing point.

4. The future way of approaching the important subject of seasonal storage
   of solar energy is to go on with and to carry out extensive numerical
   studies, followed by experimental investigations. The size or scale
   of the experiments will depend on the problem under investigation,
   and need not be so large as often pointed out during this symposium.

SEASONAL STORAGE OF SOLAR ENERGY: WHICH
FUTURE FOR EUROPE?

R. TORRENTI
Ecole Nationale Supérieure des Mines de Paris
Centre d'Energétique
Sophia Antipolis
F- 06560 VALBONNE
Tel: (93) 33 05 58

1. INTRODUCTION

Only 20% (in Roma) to 30% (Kobenhavn) of the annual
amount of solar radiation reaching a horizontal surface
is available from October to March [1]. Thus the use of a
seasonal storage in a solar-heating system leads to
- increase (i.e double for instance) the amount of
  energy delivered by a given solar collector area
  in rather sunny european countries (typically France).
- allow the use of solar energy for heating purposes
  in northern countries where the incoming energy during the winter months is too low (typically Sweden)
Starting from these considerations an increasing number of
research and engineering teams have begun to design, from
the mid-seventies on, storage technologies and complete
solar-heating systems allowing the year round operation
of solar collectors [2]. The aim of this paper is:
- to present the storage technologies which are under
  study or have been developped these last years
- to select the main characteristics of some realizations or projects and analyze, through the experimental and computer results, which could be the future of seasonal storage of solar energy in Europe.

2. SEASONAL STORAGE TECHNOLOGIES

One attractive way to store thermal energy is the
use of phase-change materials. Therefore latent-heat storage technologies have been widely developped, so that
many industrial firms all over the world have already begun to commercialize some storage units based on salt hydrates or organic materials. But the cost of these units,
commonly three to five times more expensive than water
storage units, limits their use to short-term storage only.
The two other ways to store thermal energy, namely sensible heat storage and reversible chemical reactions, have
been underway to design seasonal storage technologies.

2.1 Sensible heat storage

Figure 1 shows the different sensible heat storage technologies which can be used on a seasonal basis in a solar heating system. These are rather simple technologies, based on well-know engineering techniques and quite common materials. Therefore many demonstration projects have been already completed today (especially water storage reservoirs [3]), or are the point to (see [4] for aquifer storage) or could be rapidly tested on a full scale basis (refer to [5] for solar ponds).

2.2 Reversible chemical reactions

In the field of chemical heat storage, chemical heat pumps (CHP) are no doubt the most attractive devices which can allow the seasonal storage of solar energy in heating systems especially in the case of individual houses [6]. Though the characteristics required for a CHP are somewhat different from a sorption refrigerator these two devices operate in a same way and the fair experience coming from the latter can be used to develop a CHP. Some CHP-systems are thus quite advanced on a research basis, $H_2SO_4-H_2O$ for instance [7] while in Sweden a recent firm, TÉPIDUS, is on the point to commercialize a system based on Na_2S. [6]

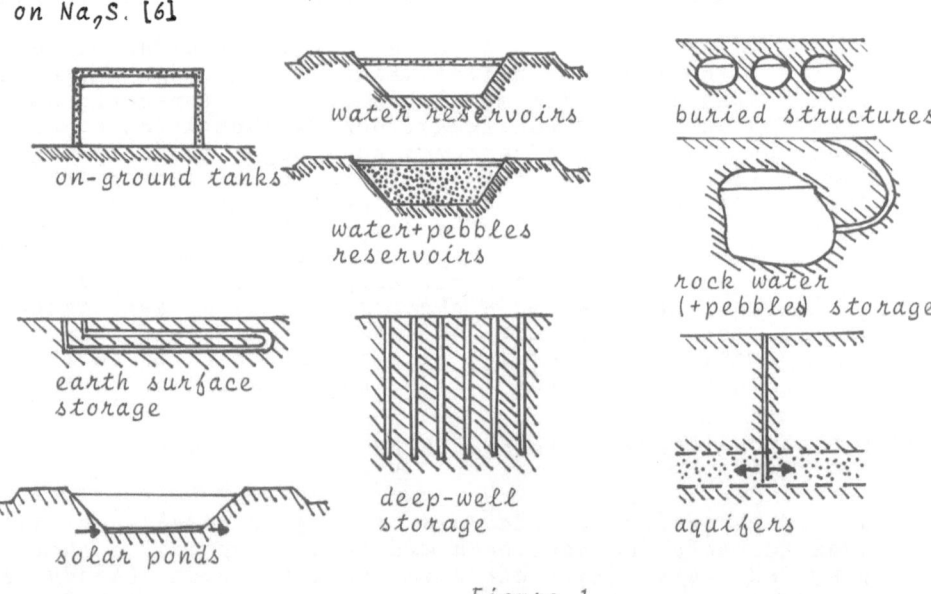

water reservoirs

buried structures

on-ground tanks

water+pebbles reservoirs

rock water (+pebbles) storage

earth surface storage

deep-well storage

aquifers

solar ponds

Figure 1

LOW TEMPERATURE SEASONAL STORAGE OF SOLAR ENERGY

Overview of useable technologies

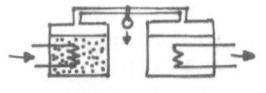

chemical heat pumps

3. FIRST RESULTS

3.1 Sensible heat storage

The first realizations or projects have shown that seasonal storage of solar energy proved feasible for heating purposes as well in the case of aquifer storage [4] than water reservoirs [3] or ground storage [8] The cost of such solar-heating systems reveals to be so that:
- the pay-out period is lower than the life expectancy of the components, allowing to say that these systems are economic.
- the price of a delivered kWh is comparable to that of system equipped with a short-term storage.

These points demand to be analized as first results, derived from a few peculiar studies, based on specific economic hypothesises, and though they do not bear large extrapolations, they can give raise to a certain optimism for the coming projects. Some large scale experiments can thus be justified, involving for instance higher temperature storage (this is the case of aquifer storage: the heliogeothermal systems [4] can be considered as a first step allowing to design higher temperature storage when remaining technological problems are solved).

3.2 Chemical heat pumps

The characteritics of this kind of storage are very attractive: high energy density (500 kWh/m^3 to 1000 kWh/m^3) and constant temperature storage. The first results seem to confirm that these two points can be matched quite easily for a large number of systems. However the future of this solution definitely depends of the investiment cost and on the long-term performances of the storage unit. Much work remains to be achieved on these two points. The experience obtained on latent-heat storage, which involves quite similar problems concerning the heat transfer and reactions reversibility, leads to be very cautious as for the possible use of CHP as seasonal storage. The necessity to operate CHP with 2 heat sources is another parameter which could also limit the future of this technology or at least increase the cost of solar-heating systems based on this kind of storage.

3.3 Technological and systems choice

It is obvious today there is no magical solution among low temperature storage technologies based on sensible heat or reversible chemical reactions. The investiments cost for the different solutions (expressed in $/kWh for instance) are often very comparable. The choice of a given storage technology will in fact mainly depend

on the project to be considered (climate, number of hou--
sings to be connected, ground area and nature of soil available, fixed parameters: type of solar collectors or heat
emitters for instance). Water storage (reservoirs or rock
caverns) or ground storage (deep well or aquifer) will
probably be used in many of future projects. Chemical
heat pumps may allow to enlarge the area of solar heating
with seasonal storage (for individual houses for instance)
but will probably not be the only solution to be considered in the future. In the same way, the answers to some
questions related to the future of seasonal storage will
probably be dependant on the projects to be considered:

- which are the geographical limits of seasonal storage in Europe? (in connection with this question:
  what is the relative interest of long term storage-
  for some weeks-for southern european countries?).
- What is the optimum solar coverage percentage?
  (are 100% solar heating-systems preferable to
  70% to 80%?).
- Could solar ponds be used in Europe as solar collectors and seasonal storage devices?.

4. CONCLUSIONS

Considering the work achieved during this four or
five past years in different european countries it appears
today that seasonal storage for solar heating system:
- IS NOT AN UTOPY ANY LONGER: there is a certain number of storage technologies that have proved feasible and can be considered today in a large number
  of solar-heating projects. Some research and development work has however to be done to get a better
  knowledge of the behaviour of such systems and
  provide reliable technologies.
- IS NOT THE PANACEA OF SOLAR-HEATING: if seasonal
  storage seems to be the only way to solar heating
  in northern countries, it will be probably lead to
  the same economic interest than short term storage
  in more sunny countries, apart from huge-and cer-
  taily rare-projects. Bearing on seasonal storage
  technologies to foreseea large scale use of solar
  energy seems optimistic. This is no doubt more
  dependent on non- technical than on technical obstacles.
Seasonal storage of solar energy for heating purposes is feasible today. Such systems can highly contribute
to the development of solar energy use and needs to be developped further on . This will take time but need the
the confidence of anyone today.

REFERENCES

[1] CEC European atlas of solar radiation. W. Grösschen-Verlag KG, Dortmund, Germany 1980.

[2] Refer for instance to the program of the following meetings
  - 1979 ISES Congress, Atlanta, USA.
  - French-Swedish Symposium on Seasonal storage of solar energy, October 1980, Sophia Antipolis, France.
  - International TNO-Symposium on Thermal storage of solar energy, November 1980, Amsterdam, The Netherlands.

[3] E. ÖFVERHOLM- The swedish seasonal thermal storage program with application to the building sector. International TNO-Symposium, Nov. 1980, Amsterdam.

[4] P. IRIS- Experimental study of heat storage in aquifer. International TNO-Symposium, Nov. 1980, Amsterdam.

[5] H .TABOR- Storage capatibility of solar ponds, International TNO-Symposium, Nov. 1980, Amsterdam.

[6] W. RALDOW- The absorption process for storage of low-temperature heat. International TNO-Symposium, Nov. 1980, Amsterdam.

[7] A.L. MICHAELS- An overview of the USA program for the development of the thermal energy storage for solar energy applications. nternational TNO-Symposium, Nov. 1980, Amsterdam.

[8] A. WIJSMAN- A group of solar houses with seasonal heat storage in the soil. International TNO-Symposium, Nov 1980, Amsterdam.

THEORETICAL STUDIES IN LONG-TERM THERMAL ENERGY STORAGE IN AQUIFERS

CHIN FU TSANG

1. INTRODUCTION

One of the most promising methods for long-term thermal energy storage is the use of underground aquifers. Aquifers are geological formations which contain and conduct water. They may be found at depths ranging from a few meters to hundreds of meters. For many years some of these aquifers have been used for liquid waste disposal and for storing fresh water, oil products, and natural gas. Their use for hot water storage was first suggested in the early 1970's.

In 1978, the Lawrence Berkeley Laborotory (LBL) of the University of California organized and hosted the First International Workshop on Aquifer Thermal Energy Storage. Active workers from nine countries participated in this workshop and their contributions were published in the Workshop Proceedings [1]. Since the Workshop, a periodic Newsletter [2] has kept researchers abreast of the current status of various projects worldwide. Many of these projects were recently reviewed in two survey papers [3 and 4]. Currently, much experimental and theoretical work is being carried out to study the concept of aquifer thermal energy storage. Furthermore, three large-scale "demonstration" experiments were initiated in the United States.

The present paper will describe the LBL theoretical studies in this field. The implications of our results for the implementation of this concept will also be discussed. Our theoretical studies are carried out in two directions:

(1) basic or generic studies to understand the fundamental

thermohydrologic processes and to identify key parameters, and

(2) site-specific modeling studies to understand experimental

observations and to simulate or predict field results.

Earlier work at LBL was mainly along the first direction with the emphasis on detailed modeling for proving the feasibility of the concept. Many of the results have been published already [5 - 7]. In the section following, we shall describe our recent basic studies. Then our site-specific studies will be discussed under two headings, Field Simulation and Study of Alternative Field Designs. A brief conclusion will complete the paper.

2. BASIC STUDIES

Our recent basic studies emphasize the understanding of the energy recovery factor (i.e., the ratio of energy recovered to energy stored) as a function of aquifer properties and storage parameters. The goal is to arrive at optimal choices of aquifer and storage arrangements. Dimensionless parameter groups that will be useful in the planning and design of practical projects are being studied and validated.

So far in our studies we have neglected buoyancy flow. This is the case for low permeability aquifers or for storage of low-temperature water. Hellstrom, Tsang, and Claesson [8] recently studied the problem and from their work came a criterion which may be used to verify the applicability of this assumption. On the other hand, the results obtained for the functional dependence of the recovery factor may still be true for cases of significant buoyancy flow.

Since many computations have to be made for a study of functional dependence, a simple numerical model [9] is used. Besides assuming no buoyant flow, this model also assumes a steady-state fluid flow field in a laterally infinite, uniform aquifer. The recovery factor is calculated for a series of values of aquifer thickness, storage volume, aquifer and aquitard thermal conductivity, caprock thickness, cycle time period, velocity-dependent dispersion, and the number of cycles. In all cases, equal volumes of fluid are injected and produced. As illustrations, some of the results are shown in Figures 1-4.

Figure 1 shows the energy recovery factor as a function of thermal radius, R_{th}, and aquifer thickness, H. Each dashed line traces the recovery factor for a given fluid volume. There is an optimal value of R_{th}/H which yields the maximum recovery factor for each volume. Generally, the recovery factor is a much more sensitive function for small values of R_{th} and H than for large values. Figure 2 shows the recovery factor as a function of volume for a series of values of aquifer thickness, H. The recovery factor increases rapidly at first, then levels off. Figure 3 shows the recovery factor as a function of the aspect ratio, R_{th}/H, for a series of aquitard to aquifer thermal conductivity ratios. As this ratio decreases, the aspect ratio which yields the maximum recovery factor increases. Figure 4 shows energy recovery as a function of the time period of a single injection-storage-production cycle for several different injected volumes. The aquifer thickness for each volume is such that the aspect ratio is optimal. Lines "a" show the results for a cycle with no storage period, i.e., production begins as soon as injection ends. In this figure, injection and production periods are equal. We have found that varying the relative injection and production periods for a given storage period has only a minor effect on the recovery factor. Lines labeled "b" show the results for a cycle with equal injection, storage, and production periods. Lines labeled "c" show the results for the hypothetical cycle that is all storage period, the hot water being instantly injected and later instantly produced. This represents the limit of very short injection and production periods.

Detailed results of this work are described in a paper under preparation [9]. Systematic graphs of the recovery factor as a function of key dimensionless parameters will be included and their use for practical field applications demonstrated. Plans for further calculations include the incorporation of gravity effects.

3. FIELD SIMULATION

A series of experiments were carried out during 1978-1979 by the Auburn University [10]. We performed a modeling study of these experiments and succesfully simulated the observations without adjusting any parameters. The experiments include two injection-storage-recovery cycles. The first six month injection-storage-production cycle involved the storage of 55,000 m^3 of water at about 55°C. The injection took 79.2 days, at the end of which the hot water was stored for 52.5 days. Production was then started at an average rate of 245.6 gpm until the recovered water temperature fell to 32.8°C. At that point 66% of the injection energy was recovered. The second injection-storage-production cycle was carried out in essentially the same manner, using 58,000 m^3 of water at an average temperature of 55.4 °C. When the production temperature had dropped to 33°C, a recovery of 76% of the injected energy was realized.

The first stage of our simulation calculations involved the determination of the hydraulic parameters of the aquifer (the transmissivity and storativity), and the location of a linear hydrologic barrier through well test analysis. Conventional well test type curve analysis techniques require a constant or carefully controlled flow rate. To get around this limitation,

LBL has developed a computer-assisted analysis method, program ANALYZE [11, 12] that can handle a system of several production and injection wells, each flowing at an arbitrarily varying flow rate. This program was applied to the Auburn case, treating the injection period also as a part of the well test data [13].

With parameters thus obtained, the LBL three-dimensional, complex geometry, single-phase model, CCC, was used to make detailed modeling studies. A radially symmetric mesh was assumed. There is one major hydrologic parameter that was not determined by well test analysis. This parameter, the ratio of vertical to horizontal permeability, has to be inferred from field experience and parameter studies. After making a preliminary parameter study, we decided to use a value of 0.10 for this ratio. The same ratio was suggested by the USGS [14].

Results of the simulation include the recovery factor, plots of production temperatures versus time, as well as temperature contour plots and temperature profiles at various times during the injection, storage, and production periods. Both the first and second cycles have been successfully simulated. For the first cycle, the simulated recovery factor of 0.68 agrees well with the observed value of 0.66. For the second cycle the simulated value is 0.78, and the observed value is 0.76. The details of the comparison between simulated and observed energy recovery can be studied in production temperature versus time plots (Figures 5 and 6). For both cycles, the initial simulated and observed temperatures agree ($55^{o}C$). During the early part of the production period, the observed temperatures decreased slightly faster than the simulated temperatures. During the latter part, the simulated temperatures decreased faster than the observed temperatures so that by the end of the production period the simulated and observed temperatures again agree ($33^{o}C$). The discrepancy over the whole range is, at most, 1-2 degrees.

Temperature contour maps of vertical cross-sections of the aquifer at given times (e.g., Figure 7) show the details of buoyancy flow, heat loss through the upper and lower confining layers, and the radial extent of the hot water in the aquifer. Buoyancy flow is important in this rather permeable system. Comparison with temperatures recorded in observation wells throughout the aquifer show that the simulated temperature distribution agrees generally with observed temperatures. However, these discrepancies are much larger than the difference between calculated and observed production temperatures. Apparently there are local variations in the aquifer which tend to average out. Temperatures versus radial distance at given depths and times are also plotted (e.g. Figure 8) and, from these profiles, the effects of thermal conductivity and dispersion on the shape of the thermal front can be studied.

In order to prove the mesh-independence of these results, the first cycle has been modeled again, using first a coarser mesh (doubling the radial step) and then a finer mesh (half the radial step). The coarse mesh recovery factor is 0.65, to be compared with a value of 0.66 using our first mesh. Interestingly, the coarse mesh simulation yields a recovery factor slightly closer to the observed value than does the original simulation, so the increased numerical dispersion may be more closely simulating thermal dispersion due to local heterogeneities in the aquifer. Temperature as a function of radial distance and the production temperature as a function of time also confirm the insensitivity of the results to the mesh chosen.

Based on these results [15], one may conclude that (a) we understand the physical processes involved in the ATES system at the Auburn field site, thus giving us confidence in dealing with confined aquifers of a similar type, and (b) the LBL numerical model "CCC" is a satisfactory code that may be useful for further applications.

4. STUDY OF ALTERNATIVE FIELD DESIGNS

Besides simulation of experimental results, we also perform parameter sensitivity studies as well as modeling to support experimental planning and design. In the case of the Auburn experiments, several parameter variation calculations were made to study results to be expected for different arrangements.

Figure 9 shows the effect of partial penetration of the storage well into the aquifer. With the Auburn field parameters, if the storage and retrieval well penetrates the full thickness of the aquifer, the production temperature (solid line) drops steadily with time from the storage temperature of 55°C and the recovery factor, ε, is 69%. However, if during production the well is withdrawing water from only the upper half of the aquifer, the decrease of production temperature (broken line) during the initial production period is much slower. This may be of significant interest since in most applications production temperature decrease should be minimized over the main part of the production period.

In anticipation of the next series of planned Auburn experiments where water at 90°C will be stored, we performed a series of calculations for this

190

temperature. Figure 10 shows the production temperatures for one such case. As one might expect, the recovery factor is much lower because of the higher buoyancy flow associated with the higher temperature.

Table 1 summarizes the recovery factors using the model "CCC" for four storage temperatures and three values of aquifer permeability. In all the cases, storage volume is assumed to be 55,000 m^3, thickness of the aquifer is 21 m, and permeability of the aquitard is 10^{-5} of that of the aquifer. The Auburn experiment corresponds to the 52 d permeability case. This table represents a substantial amount of computation, including different cycle periods and different penetration arrangements. The variations of the recovery factor are clearly seen. Details of these studies are presented and discussed in a paper under preparation.

5. CONCLUSIONS

Substantial work has been done at LBL and elsewhere in both basic and site-specific studies. The phenomenology of the thermohydraulic flow associated with aquifer thermal energy storage is reasonably well understood. A few questions which are not mentioned in the present paper still remain, such as dispersive phenomena, aquifer regional flow control, well injectivity, and total system efficiency analysis. With the provision that an aquifer is properly selected and carefully characterized, we believe that an aquifer energy storage system can be successfully designed.

ACKNOWLEDGEMENTS

Continued cooperation and asistance from colleagues, T. Buscheck, C. Doughty, G. Hellstrom, D. Mangold, and J. Wang are much appreciated. Work done under the auspices of the Department of Energy, Division of Energy Storage, through the Pacific Northwest Laboratory.

REFERENCES

1. Proceedings of Aquifer Thermal Energy Storage Workshop. May 10-12, 1978. Lawrence Berkeley Laboratory, Berkeley California, LBL-8431.
2. ATES Newsletter, a quarterly review of aquifer thermal energy storage. C. F. Tsang, editor, published by Lawrence Berkeley Labotatory, Berkeley, California 94720.
3. Tsang, C.F. May 27-June 1, 1979. A Review of Current Aquifer Thermal Energy Storage Projects. Invited paper at the International Assembly on Energy Storage, Dubrovnik, Yugoslavia, LBL-9834.

4. Tsang, C.F., Hopkins, D.   February 8-9, 1979. Aquifer Thermal Energy
    Storage - a Survey. Invited paper at the Symposium on Recent Trends on
    Hydrogeology, Berkeley, California, to be published as a Special Paper of
    the Geological Society of America.
5. Tsang, C.F., Lippmann, M.J., Witherspoon, P.A.   1976. Numerical Modeling
    of Cyclic Storage of Hot Water in Aquifers. Symposium on Use of Aquifer
    Systems of Cyclic Storage of Water. Fall Annual Meeting of the American
    Geophysical Union, San Francisco.
6. Tsang, C.F., Lippmann, M.J., Witherspoon, P.A.   January 16-21, 1978.
    Underground Aquifer Storage of Hot Water from Solar Energy Collectors.
    Proceedings of International Solar Energy Congress, New Delhi, India.
7. Tsang, C.F., Buscheck, T., Mangold, D., Lippmann, M.J.   May 10-12, 1978.
    Mathematical Modeling of Thermal Energy Storage in Aquifers. Proceedings
    of Aquifer Thermal Energy Storage Workshop, Lawrence Berkeley Laboratory,
    Berkeley, California, LBL-8431.
8. Hellstrom, G., Tsang, C.F., Claesson, J. Heat Storage in Aquifers: Buo-
    yancy Flow and Thermal Stratification Problems. LBL-11059, to be submitted
    to Journal of Geophysical Research.
9. Doughty, C., Hellstrom, G., Claesson, J., Tsang, C.F. paper in pre-
    paration.
10. Molz, F.J., Parr, A.D., Andersen, F.P. "Thermal Energy Storage in a
    Confined Aquifer - Second Cycle," paper to be submitted to the Journal of
    Water Resources Reseach.
11. Tsang, C.F., McEdwards, D., Narasimhan, T.N., Witherspoon, P.A.   April
    13-15, 1977. "Variable Flow Well Test Analysis by a Computer Assisted
    Matching Procedure," Paper No. SPE-6547, 47th Annual California Regional
    Meeting of SPE of AIME, Bakersfield, California.
12. McEdwards, D., Tsang, C.F.   October 19-21, 1977. "Variable Rate Multiple
    Well Testing Analysis," proceedings of Invitational Well-testing Sym-
    posium, Berkeley, California.
13. Doughty, C., McEdwards, D., Tsang, C.F. "Multiple Well Variable Rate Well
    Test Analysis of Data from the Auburn Univesity Thermal Energy Storage
    Experiment," LBL-10194.
14. Papadapulos, S.S., Larson, S.P.   1978. Aquifer Storage of Heated Water:
    Part II: Numerical simulation of field results; Groundwater, v. 16,
    no. 4.
15. Tsang, C.F., Buscheck, T., Doughty, C.   February 1980. Aquifer Thermal
    Energy Storage - A Numerical Simulation of Auburn University Field Exper-
    iments. Submitted to Journal of Water Resources Research.
16. Buscheck, T., Doughty, C., Tsang, C.F., paper in preparation.

Recovery factor

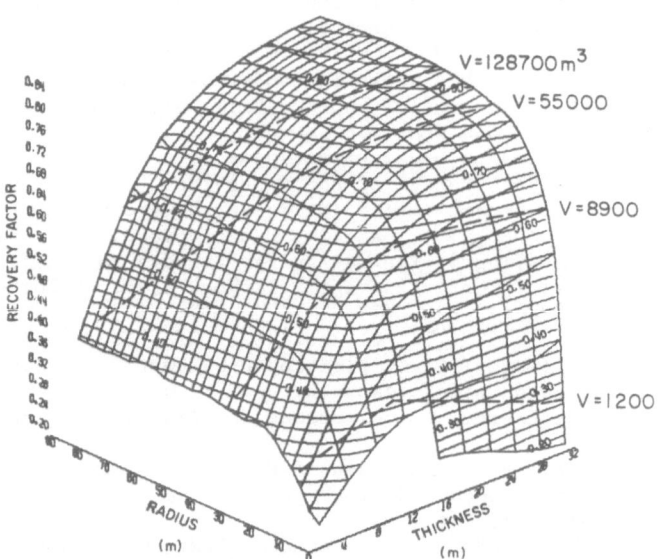

FIG. 1. The recovery factor as a function of aquifer thickness and storage radius. The latter is defined as the radius reached by the storage temperature front at the end of injection period.

Recovery factor vs. $R_{TH}^2 H$

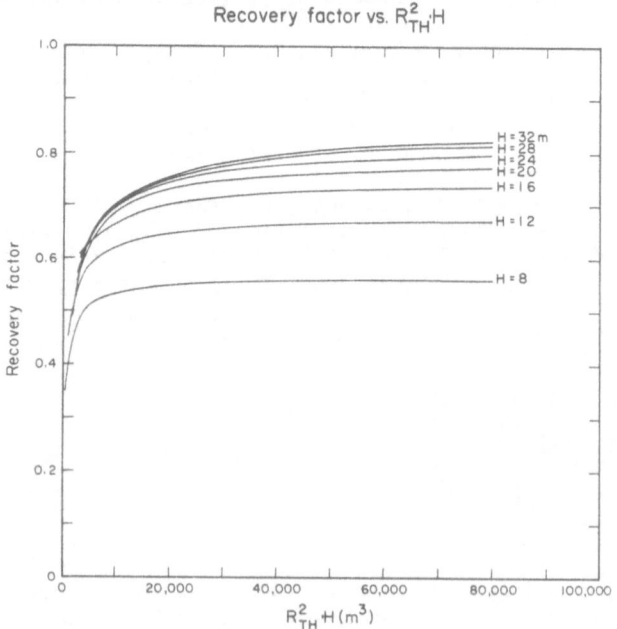

FIG. 2. The recovery factor as a function of storage volume, implicity given by $R_{th}^2 H$ for different values of aquifer thickness H.

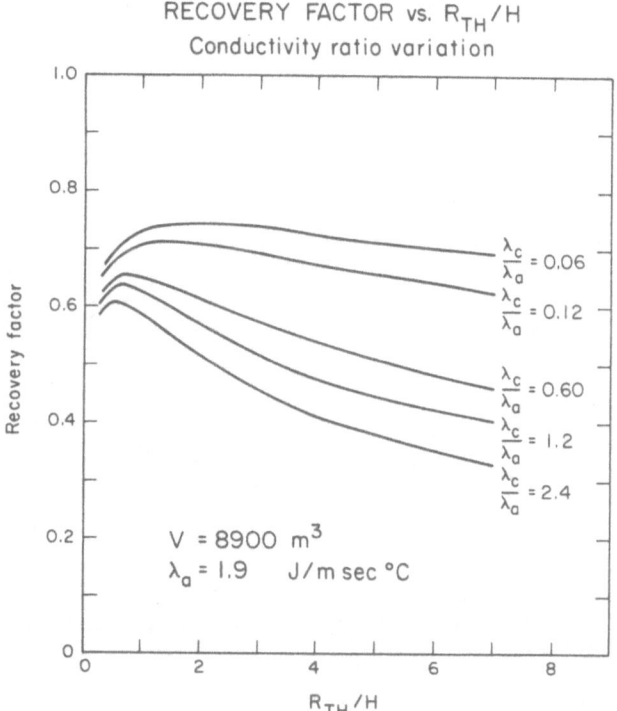

RECOVERY FACTOR vs. R_{TH}/H
Conductivity ratio variation

$\frac{\lambda_c}{\lambda_a} = 0.06$

$\frac{\lambda_c}{\lambda_a} = 0.12$

$\frac{\lambda_c}{\lambda_a} = 0.60$

$\frac{\lambda_c}{\lambda_a} = 1.2$

$\frac{\lambda_c}{\lambda_a} = 2.4$

$V = 8900 \ m^3$

$\lambda_a = 1.9 \quad J/m \ sec \ °C$

FIG. 3. The recovery factor as a function of aspect ratio, storage radius over aquifer thickness (R_{th}/H), for different aquifer and aquitard thermal conductivities (λ_a and λ_c).

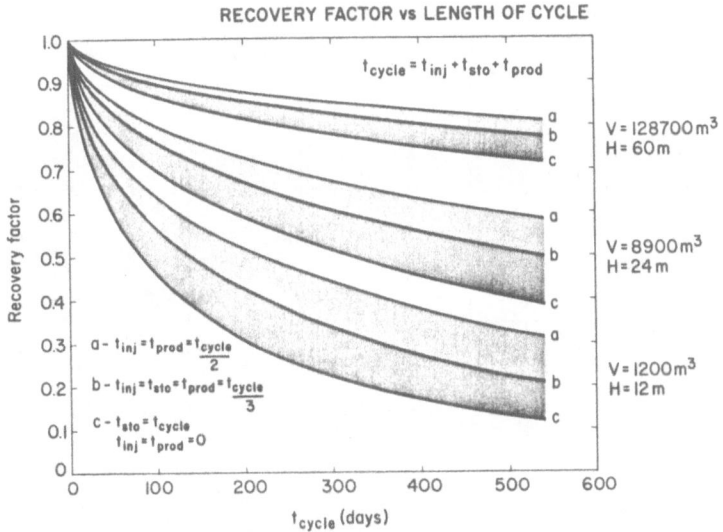

RECOVERY FACTOR vs LENGTH OF CYCLE

$t_{cycle} = t_{inj} + t_{sto} + t_{prod}$

$V = 128700 \ m^3$
$H = 60 \ m$

$V = 8900 \ m^3$
$H = 24 \ m$

$V = 1200 \ m^3$
$H = 12 \ m$

$a - t_{inj} = t_{prod} = \frac{t_{cycle}}{2}$

$b - t_{inj} = t_{sto} = t_{prod} = \frac{t_{cycle}}{3}$

$c - t_{sto} = t_{cycle}$
$t_{inj} = t_{prod} = 0$

FIG. 4. The recovery factor as a function of cycle time periods (t_{cycle}).

194

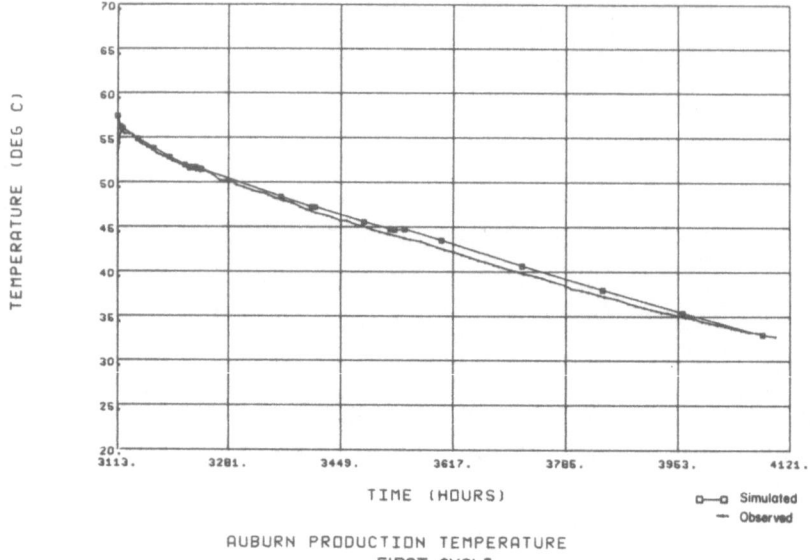

FIG. 5. Observed and simulated production temperatures as a function of time
for the first cycle.

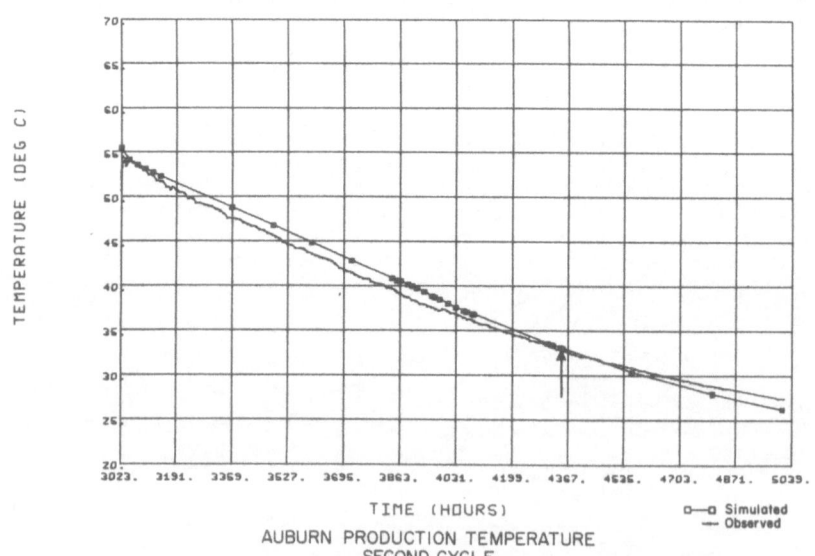

FIG. 6. Observed and simulated production temperatures as a function of time
for the second cycle.

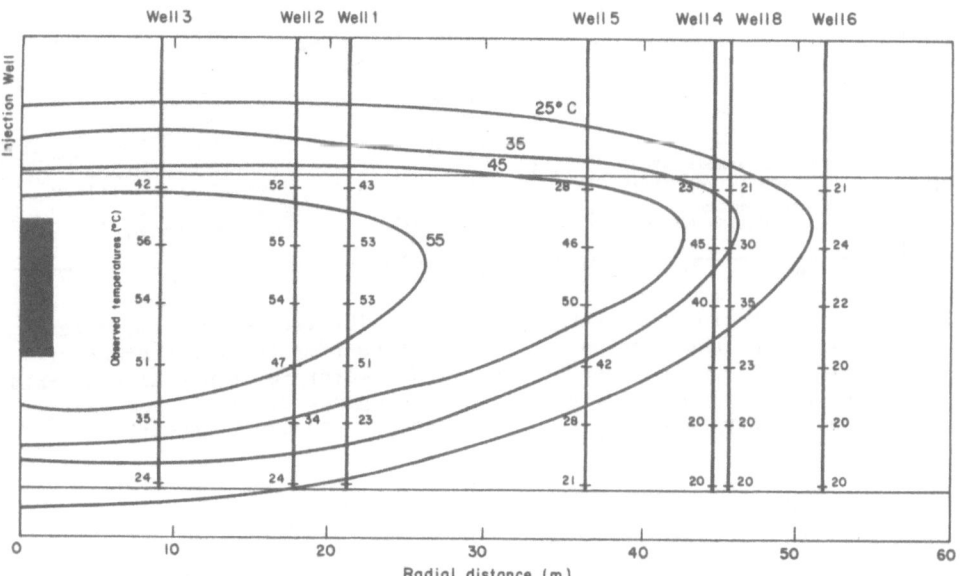

AUBRNO 2
Calculated temperature
t = 1900 hrs.

FIG. 7. Simulated temperature contours in a vertical cross section of the aquifer at the end of the injection period of the first cycle. Observed temperatures are also indicated.

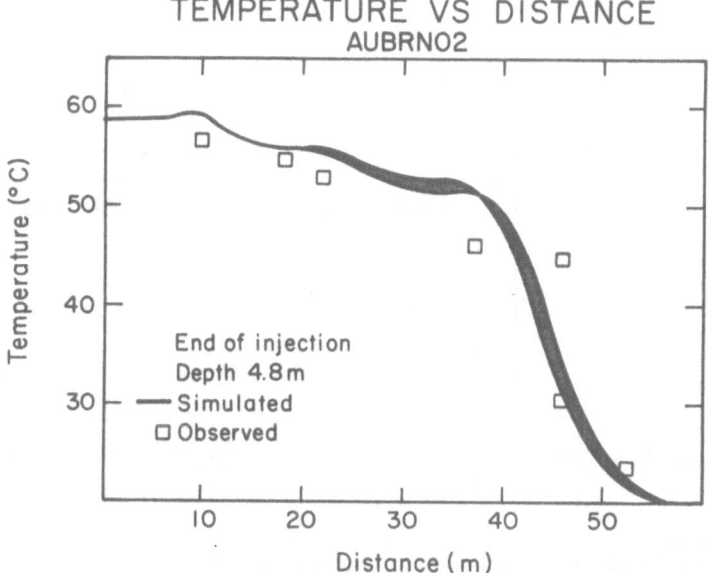

FIG. 8. Temperature versus radial distance at the end of injection period for the first cycle. Shaded curve indicates simulated values, boxes show observed values.

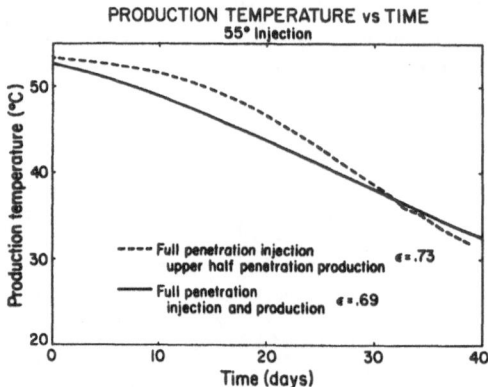

FIG. 9. Production temperature versus time for the storage of 55°C water for two well penetration arrangements (ε is the energy storage-recovery factor).

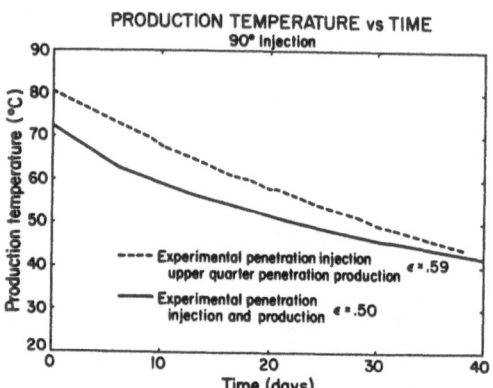

FIG. 10. Production temperature versus time for the storage of 90°C water for two well penetration arrangements. (ε is the energy storage-recovery factor).

K_a \ T_{inj}	36°C	55°C	70°C	90°C
15d	0.69	0.68		0.58
52d	0.67	0.57 (1/2) 0.62	0.46	0.34 (1/2) 0.42 (1/3) 0.44
		(60-60-60) 0.65 (60-120-60) 0.57		(60-60-60) 0.42 (1/2) 0.51
175d		0.31 (1/2) 0.39 (1/3) 0.41	0.24 (1/2) 0.31 (1/3) 0.34	

Unless otherwise indicated:

$(t_{inj}, t_{store}, t_{prod}) = (90-90-90)$ days

$T_o = 20°C$

(1/2) or (1/3) indicates production from the upper 1/2 or upper 1/3 of the aquifer thickness respectively.

TABLE 1. Recovery Factor calculated with numerical model "CCC" for different injection temperatures and aquifer permeabilities.

EXPERIMENTAL STUDY OF HEAT STORAGE IN AQUIFER

P. IRIS*

ABSTRACT

In 1977-78, 20.000 m^3 of water at 33°C were injected during the summer into a shallow phreatic aquifer at 14°C, situated at Campuget, in the south of France, and then partly recovered during the next winter.

The principal aim of this test was to compare numerical calculations on models and field measurements in a large scale experiment. This comparison gave good results and permitted to estimate the hydrogeological conditions under which long term heat storage in aquifer is feasible.

As a result of this research, a new method is proposed, which associates heat pumps, heat storage and low cost solar collectors for economically efficient space heating. This method called "heliogeothermy" will be implemented in Aulnay-sous-Bois, near Paris, in 1981, in the framework of 200 new collective dwellings.

1. INTRODUCTION

In recent years, solar energy research for active systems has been emphasized in France, for the purpose of space heating. In the case of such systems, seasonal storage of heat seems attractive. On the other hand, seasonal heat storage is also needed to recover industrial thermal wastes which may be also used for space heating.

For seasonal storage with large amounts of heat to consider, it seems economically sound to use natural systems instead of engineered stores. The aquifers in the underground appear therefore as the first reasonable targets (1,2,3).

In 1976, the Ministère de l'Environnement et du Cadre de Vie (Plan Construction) decided to support research in this area. The Ecole des Mines

* Centre d'Informatique Géologique, Ecole des Mines de Paris, 35 rue St Honoré, 77305 Fontainebleau, France

C. den Ouden (ed.), Thermal Storage of Solar Energy. All rights reserved.
Copyright © by TNO and Martinus Nijhoff Publishers, The Hague/Boston/London.

together with Electricité de France, Direction des Etudes et Recherches
(Département Application de l'Electricité) proposed to perform a large scale
experiment in an aquifer, in order to test the technical difficulties, the
predictability and the efficiency of such systems.

The aquifer was very shallow for this first experiment, in order to reduce
the costs of the wells; this experiment took place in the framework of 3.800
m^3 of greenhouses heated by two 280 Mcalories/h heat pumps (water-water). These
greenhouses were able to produce the amount of heat during summer and to use
the heat withdrawn from the underground storage during winter, at low cost.

2. THE EXPERIMENT

The aquifer is 9 m thick with its free surface at 2 m under the soil
surface. It lies in alluvial deposits with a high permeable strata at the
top (pebbles between 2 and 8 m) and sand at the bottom. At 11 m, there is
an impervious and thick strata of clay. The natural velocity is low (about
5 cm/day), the natural temperature is about 14°C.

The experiment lasted from July, 4, 1977 to March, 15, 1978, according
to the following chart:

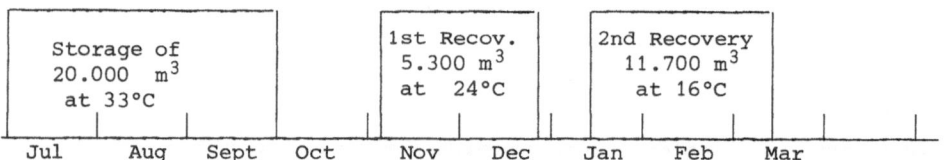

20.000 m^3 of water were stored, and 17.000 recovered; overall, 18.5%
of the heat were recovered (heat calculated on the basis of the natural
temperature of the aquifer). The evolution of the temperature of recovery
is shown on Fig.1 (see next page).

Fig.2 shows the aspect of the thermal storage at the end of the injection
period (October, 1, 1977). This figure was drawn on the basis of different
thermal loggings realised at this period in 9 observation wells distributed
on the area of the storage.

The "hot bubble" is mainly axisymmetric with a little anisotropy in the
direction of the gradient. The lateral thermal front is not sloppy, which
seems to show an important lateral diffusion in the aquifer. According to

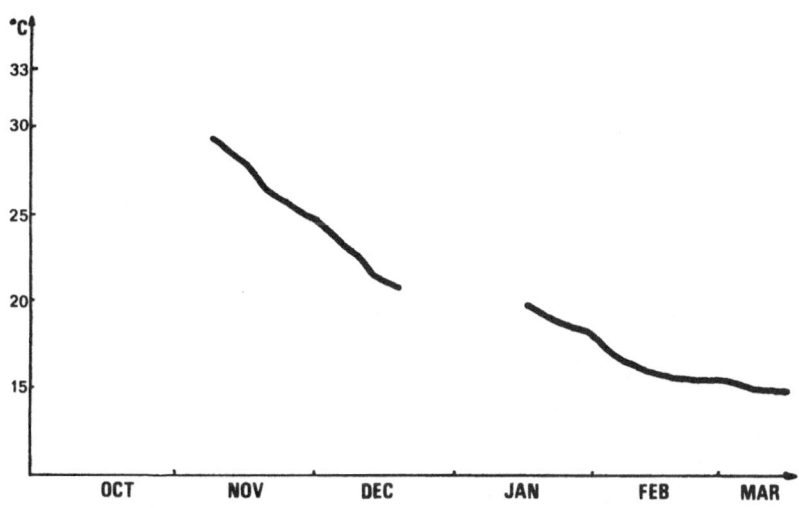

FIG.1 - Evolution of the temperature of recovery

the general shape of the storage, it seems that vertical thermal exchanges by heat conduction above and underneath are also important; at least, most of the heat remains at the top of the aquifer which is due to the fact that the most permeable strata is located at this level; variation of fluid density with temperature can also explain this phenomenon.

On the technical point of view, no clogging was observed during the period of injection in spite of the fact that calcium carbonate was at a saturating concentration in the water at 14°C. Due to the kinetic of calcium carbonate precipitation, which was apparently very slow, no noticeable effect was detected.

On the other hand, the hot water was contaminated by bacteria while circulating in the installation of greenhouses. Because of the favourable temperature conditions in the aquifer (33°C), a strong bacterial contamination of the aquifer was generated in the area of the storage. Another test was performed after the experiment, which consisted in heating the medium at about 30°C with resistors in the wells: no spontaneous bacteria development was observed in the aquifer, and it seems that the contamination during the storage was mostly due to surface bacteria. In any case, moderate chlorination can help to control this problem.

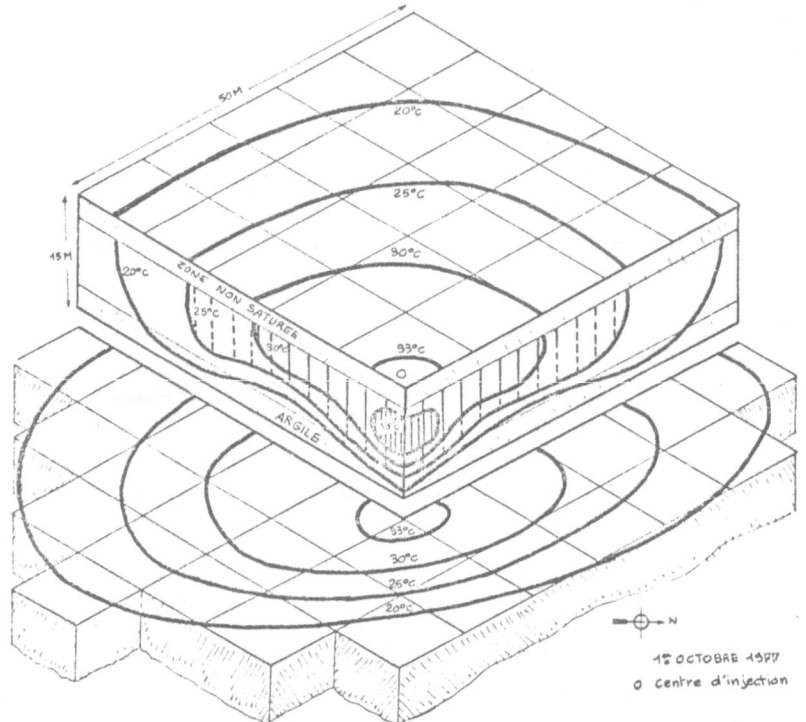

FIG.2 - General aspect of the storage at the end of injection

3. NUMERICAL INTERPRETATION OF THE EXPERIMENT

The purpose of the following interpretation was to understand and explain the curve on Fig.1 and to study how models do fit well with reality (4).

The numerical models developed to simulate heat transfer in aquifers have two governing equations: the mass balance equation (1) which permits to calculate the hydraulic heads field under the given conditions. This equation also permits to calculate the velocity field with the Darcyan relation (1') which is the basic relation in mass transfer in aquifers. On that basis, it is possible to calculate the temperature field with the heat balance equation (2). Generally, the first equation is solved in steady state in order to simplify the calculations. These models are based on the hypothesis that the thermal equilibrium between the solid and the fluid phasis in the medium is instantaneous compared with the time schedule of the simulated experiment.

$$(1) \quad \operatorname{div} \overline{\overline{K}} \ \overrightarrow{\operatorname{grad}} \ H = 0 \quad \text{(steady state)}$$
$$(1') \quad \overrightarrow{V} = -\overline{\overline{K}} \ \overrightarrow{\operatorname{grad}} \ H$$

$$(2) \quad \text{div} \, (\underset{=}{\lambda} \, \overrightarrow{\text{grad}} \, \theta - \gamma_f \, v\theta) = \gamma_m \, \frac{\partial \theta}{\partial t}$$

with $\underset{=}{K}$: tensor of hydraulic conductivity

$\quad$ H : hydraulic head $(H = \frac{p}{\rho g} + z)$ $\qquad$ p: pressure

$\qquad\qquad\qquad\qquad\qquad\qquad\qquad\qquad$ ρ: volumic mass of the fluid

$\quad$ λ : tensor of equivalent thermal conductivity which depends on

$\qquad$ normal heat conductivity and kinematic dispersion:

$$\lambda = \lambda_o + \underset{=}{\overline{\alpha}} \cdot \gamma_f \cdot |\overrightarrow{v}|$$

$\qquad$ λ_o : thermal conductivity $\qquad$ α: tensor of intrinsic dispersivity

$\quad$ θ : temperature of the medium

$\quad$ γ_f: heat capacity of the fluid

$\quad$ γ_m: heat capacity of the medium

$\quad$ $\overrightarrow{V}$: macroscopic velocity of filtration (Darcyan velocity).

According to the aspect of the storage (Fig.2), a radial multilayers model was developed to interpret the experiment (the method of calculation was the linear finite elements in the formulation of Galerkin).

The first thing we did was to measure very precisely all the governing parameters of the equations. Among them, we measured the horizontal and vertical permeabilities of the aquifer (in the pebbles: $K_h = 3.6 \ 10^{-4}$ m/s, $K_v = 3.6 \ 10^{-5}$ m/s), the thermal conductivity of the aquifer (λ_o) and of the zone overlying ($\lambda_o = 2$ W/m°C), and the intrinsic dispersivity which depends on the scale of heterogeneity of the medium. This last parameter is not directly measurable and was obtained by calibration of the model on the thermal loggings realised during the injection period in the observation wells.

Secondly, we tried to justify the validity of the hypotheses contained in that axisymmetric formulation of the problem, which are as follows:

1) the effect of variation of density is supposed neglectible,

2) the effect of natural velocity of the fluid in the aquifer is also supposed neglectible.

This effects have been tested with two special models. Concerning the first point, an axisymmetric monolayer model was developed in which the equation (1) is solved in pressure (not in piezometric head) with the density function of temperature (according to the Boussinesq approximation). On Fig.3 are represented the calculated isotherms in the pebbles strata after 60 days of injection. These curves are vertical, which shows that this effect is neglectible in our case (under other conditions, with high temperature

202

it can be very important with the tilting of the thermal front).

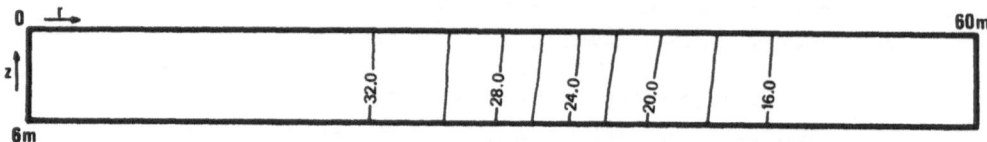

FIG.3 - Isotherms in the medium; calculations with the density effect (60 days of injection)

A monolayer model in cartesian coordinates was also developed in order to take into account the effect of natural velocity in the aquifer. Fig.4 shows the influence of this effect by two simulations, one with the gradient measured at Campuget, the other one without it (no vertical heat diffusion is taken into account). It is not completely neglectible, but it does not explain either the experimental curve of recovery (Fig.1). The energy lost due to that phenomenon represents 6% of the total heat stored.

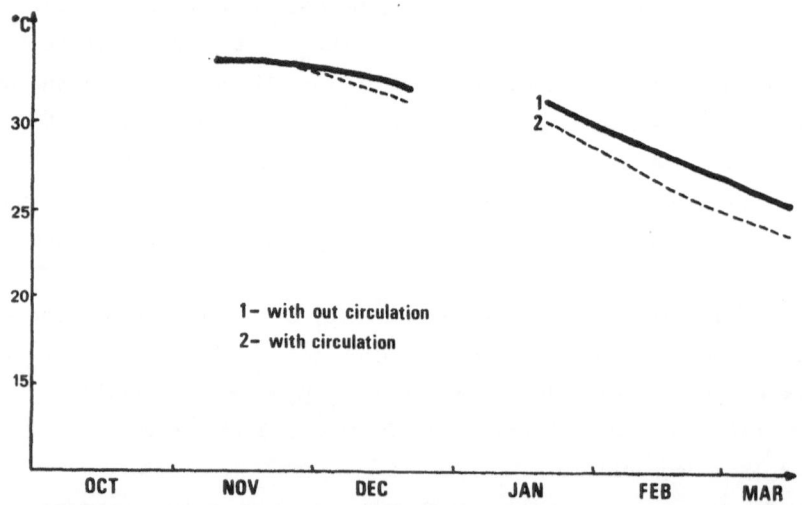

FIG.4 - Influence of the local gradient, recovery temperature with and without the effect of natural circulation in the aquifer

Fig.5 shows the result of the global simulation calculated with the multilayer radial model. The difference between the calculated curve and the measured one is explained by the natural circulation in the aquifer as shown on Fig.4

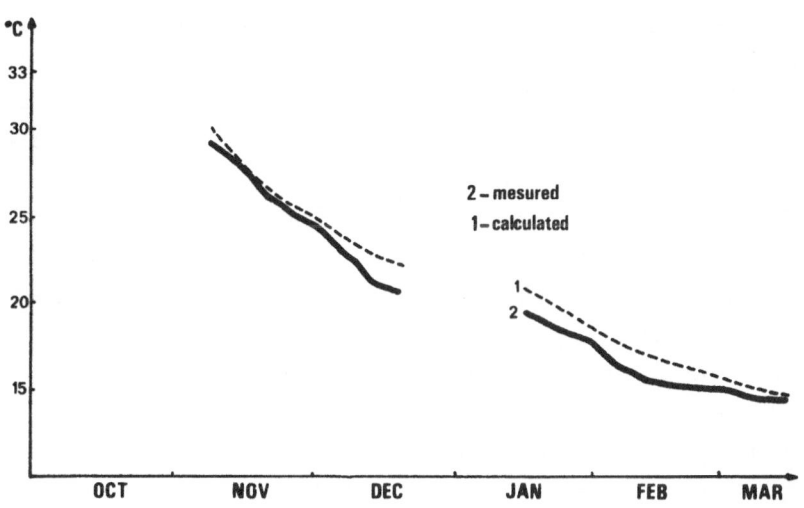

FIG.5 - Global simulation with multilayer axisymmetric model

Different calculations have also shown that 33% of the heat stored was lost by thermal exchange through the soil surface towards the atmosphere because of the low thickness of the zone overlying the aquifer (2 m). Under deeper conditions, better results can be obtained as it was shown in an experiment realised in Alabama (5), where 65% of the heat stored were recovered in a semi-annual cycle: the aquifer was 20 m thick and situated at 40 m depth.

4. RECOMMENDATION FOR HEAT STORAGE IN AQUIFERS

Regarding the results of the experiment of Campuget, it is obvious that particular hydrogeological conditions are necessary to obtain good performances in a long term heat storage in aquifer. The numerical models, the validity of which has been shown with this interpretation, permit to know what are these conditions. Different calculations have been realised in France by Ecole des Mines and the Bureau de Recherches Géologiques et Minières (SAUTY and MENJOZ - 6). It is shown that a sufficient depth is necessary (about 15 m between the top of the aquifer and the soil surface according to the average thermal diffusivity of soils) in order to avoid heat exchange with atmosphere during the seasonal storage; sufficient thickness of the aquifer is also needed, to limit the influence of exchanges at the upper and lower limits of the storage: 10 m thick seems to be the lower acceptable order of

204

magnitude. At least, a sufficient scale is necessary for the storage
(minimum 20.000 m³) to limit the influence of the lateral thermal diffusion
and the influence of the natural circulation in the aquifer which also must
be as low as possible. This last point can be controlled by a radial system
of wells of pumpage and injection.

Under such conditions, good efficiency can be expected for a seasonal
storage as it is shown on Fig.6 in the case of a 10 m thick aquifer where
the same volume as the one injected is pumped back and where, after 5 cycles,
70% of the heat stored is recovered (heat calculated on the basis of the
natural temperature of the aquifer, which is 10°C in that case and storage
temperature 60°C).

A point is that the withdrawn water has a decreasing temperature with
time which is due to heat diffusion in the underground medium. Consequently,
the heating system working with the storage needs to have a very low return
temperature in order to utilize as much heat as possible, at it is shown on
Fig.7: the curve represents the effective useful heat as a function of the
return temperature of the heating system in the case of the 10 m thick aquifer
after 5 cycles of storage-recovery.

According to actual solar collectors prices and efficiency at high temperature, it seems that heat storage in aquifer is better adapted to thermal
industrial wastes at low cost.

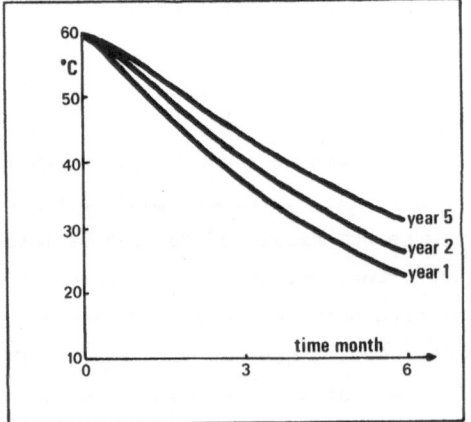

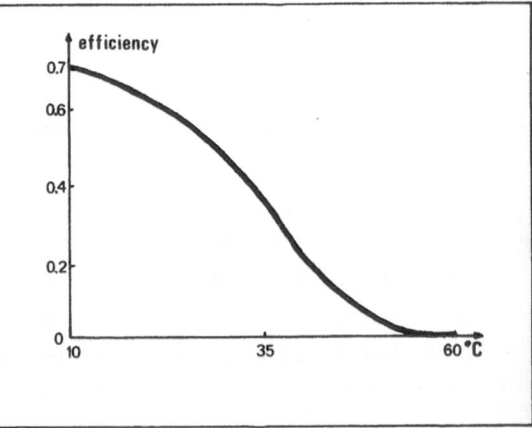

FIG.6 - Temperatures of recovery

FIG.7 - Recovered heat function of the
return temperature of the heating system
(case of a doublet)

At least, on a practical point of view, it is important to store water which is pumped from the same aquifer, in order to avoid or limit the chemical problems and the possibilities of non compatibility between the injection water and the water of the aquifer. Different systems are possible as doublets with two wells, a supply well and a storage well, or with several wells, a storage well at the center of a radial system of supply wells.

5. FURTHER PROJECTS: THE HELIOGEOTHERMAL DOUBLET OF INTERSEASONAL RECHARGE

We have seen that one of the difficulties of solar heat storage is that the temperature of the recovered water drops during the pumping period; it means that high temperature solar collectors are needed, which is an expensive way of valorizing solar energy.

On the other hand, aquifers generally remain at a constant temperature (about 10-15°C) and so, represent a very performing cold source for heat pumps for space heating.

How to exploit such a thermal resource ?

Different systems are possible:

i) withdrawal of the water from the aquifer, and disposal of the water cooled by the heat pump in the drainage network. This solution is only acceptable in areas where the water resource is large, and the users far apart, otherwise the water resource will soon be depleted.

ii) reinjection of the cooled water into a second well, creating a "geothermal doublet". The water resource is conserved, but the aquifer is progressively cooled. The two wells have to be placed far apart (e.g. several hundred meters) or the cool water will soon be recycled, and the efficiency of the system will drop (7).

In any case, a large portion of the aquifer will be cooled (about 7 ha in 10 years for heating 200 collective dwellings in France in an aquifer of 15 m thickness).

Such a system is therefore not widely applicable in a dense urban area.

iii) heliogeothermal doublet of interseasonal recharge. The method that we propose (patent pending by P. IRIS) works as follows:

WINTER   SUMMER

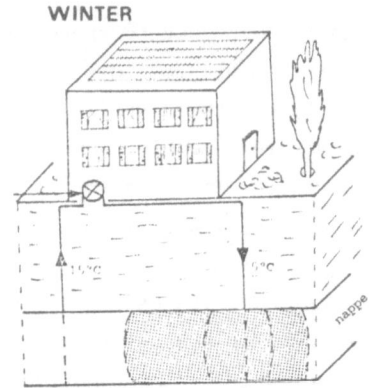   

- In winter, the water is pumped at the natural temperature of the aquifer
(10-15°C according to depth and location), and reinjected after cooling by
the heat pump (5°C).
- In summer, the "cold bubble" is pumped back, heated by solar collectors
working at a very low temperature, and injected at the producing well at the
initial temperature of the aquifer (10-15°C).

This method has the following advantages:
- conservation of both the water resource and the energy resource of the
aquifer indefinitely,
- small distance needed between the two wells, compatible with land ownership
in urban areas,
- behaviour of the energy source at a constant temperature throughout winter,
making it possible to design a simple heating system with heat pumps,
- collection of solar heat at very low temperature (10°C) below the air tempe
rature, using very simple collectors (e.g. array of pipes on a roof without
any glass cover) with a very high efficiency,
- feasibility exists at any scale, from individual houses to large-scale
programs (whereas heat storage is only feasible on a large scale).

This method will be implemented in 1980-81, in France, at Aulnay-sous-
Bois, near Paris, in an aquifer lying at 60 m below ground. This project
was funded jointly by the Commission of the European Communities and the
Plan Construction, Ministère de l'Environnement, and COMES, in France, and
involves the Ecole des Mines de Paris, the Cabinet ANTONELLI and the Housing
Company FFF. 220 housing units will be heated in this way.

Estimated efficiency of the systems shows a primary energy saving of
70% as compared to electric direct heating (50% as compared to oil heating)

at a global cost of the same order of magnitude as usual solutions (oil,
or electric heating).

6. CONCLUSION

The Campuget experiment shows that aquifer heat storage is feasible,
without any great technical difficulties. The numerical models of simulation
of heat transfer in aquifer are well fitted with field measurements and are
able to predict the efficiency of a storage in other hydrogeological conditions.

Good results can be obtained in appropriate hydrogeological conditions,
with low gradient of natural circulation, sufficient thickness of the aquifer
and sufficient depth. Sufficient volumes of storage are also needed. Such
a system seems to be well adapted to thermal industrial wastes storage at
large scale.

The application of this concept to solar energy shows, however, that
instead of using heat storage in the aquifer for space heating, one can
achieve a more economical result in winter by using the water at its natural
temperature in the aquifer by means of heat pumps, and subsequent storing
of chilled water in the aquifer. The following summer, this chilled water
is brought back to the normal temperature of the aquifer and reinjected at
the production well, thus recreating the resource. The heating of the water
can be done using very simple solar collectors. This system permits to
generalize the use of heat pumps on aquifers (even in urban areas).

This project has been given the name of "heliogeothermy" and will be
implemented in France in 1980-81, at Aulnay-sous-Bois, near Paris, for
220 housing units.

REFERENCES

1. DUMON B. 1977. Energie solaire et stockage d'énergie. Masson, Paris.
2. MEYER C, TODD D, 1973. Conserving energy with heat storage wells.
   Environmental Science and Technology, vol.7, n°6, June 1973.
3. MARSILY G de, 1978. Peut-on stocker de l'énergie dans le sol ? Les
   Annales des Mines, Avril 1978.
4. IRIS P, 1980. Contribution à l'étude de la valorisation énergétique des
   aquifères peu profonds. Thèse de Docteur-Ingénieur, Ecole des Mines de
   Paris
5. MOLZ FJ et al, 1978. Thermal energy storage in confined aquifers.
   Experimental results. Water Resources Research, vol.15, n°6, p. 1509-1514.
6. SAUTY JP and MENJOZ A, 1979. Stockage longue durée en nappe phréatique
   de calories à basse température pour l'habitat. Rapport Plan Construction.
7. ANDREWS CB, 1978. The impact of the use of heat pumps on groundwater
   temperature. Groundwater, vol.16, n°6, Nov-Dec 1978.

A GROUP OF SOLAR HOUSES WITH SEASONAL HEAT STORAGE IN THE SOIL

A.J.Th.M. WIJSMAN

INSTITUTE OF APPLIED PHYSICS TNO-TH, DELFT, The Netherlands

ABSTRACT

In the Netherlands a study has been carried out on the feasibility of seasonal storage of solar heat using the soil as storage medium. This study was carried out in the period January 1978 to July 1980 by the Delft Soil Mechanics Laboratory in cooperation with Philips Eindhoven and the Institute of Apllied Physics (TPD-TNO-TH) in Delft.
The work was financed by the CEC and the Dutch National Solar Energy Programme.

In this paper a description of the solar system with seasonal heat storage is given. Results of computercalculations concerning the various heat fluxes in the solar system and results of a parameter sensitivity study are shown.

1. INTRODUCTION

A conventional solar heating system for a house consists of a collector array mounted onto the south facing roof of the house and a short term (ST-) heat storage reservoir in the house. This solar heating system delivers a part of the heat demand for space heating and for the domestic hot water supply.

The function of the ST-heat storage is to store the surplus of collected heat on a clear day for a next cloudy day. Under N.W. European climatological conditions an enlargement of the ST-heat storage reservoir has only a small effect on the solar contribution of the system to the heat demand.

An important increase of the solar contribution can be expected from extension of the system with seasonal or long term (LT-) heat storage: the collected surplus of solar heat in summertime is stored for consumption in wintertime.

For seasonal heat storage one can use large waterreservoirs, but this solution seems to be expensive. A cheaper solution can be heat storage in the upper layer of the soil, in which case the soil itself acts as storage material.

In principle the seasonal heat storage reservoir in the soil is nothing more than a layer of soil (up to a depth of approximately 20 m) with a heat exchanger. The 'reservoir' is not bounded by walls. Only at the top the reservoir is furnished by an insulation foam layer, at the other sides the soil itself acts as insulation.

Because of heat transfer by conduction the seasonal heat storage reservoir is losing heat to the surrounding soil. For a sufficient performance of the seasonal heat storage large dimensions of the reservoir are necessary. So a group of solar houses connected to a central heat storage reservoir is essential for seasonal heat storage.

2. DESCRIPTION OF THE SOLAR SYSTEM WITH SEASONAL HEAT STORAGE

The group of solar houses is situated around the seasonal heat storage reservoir in the soil (see figure 1).

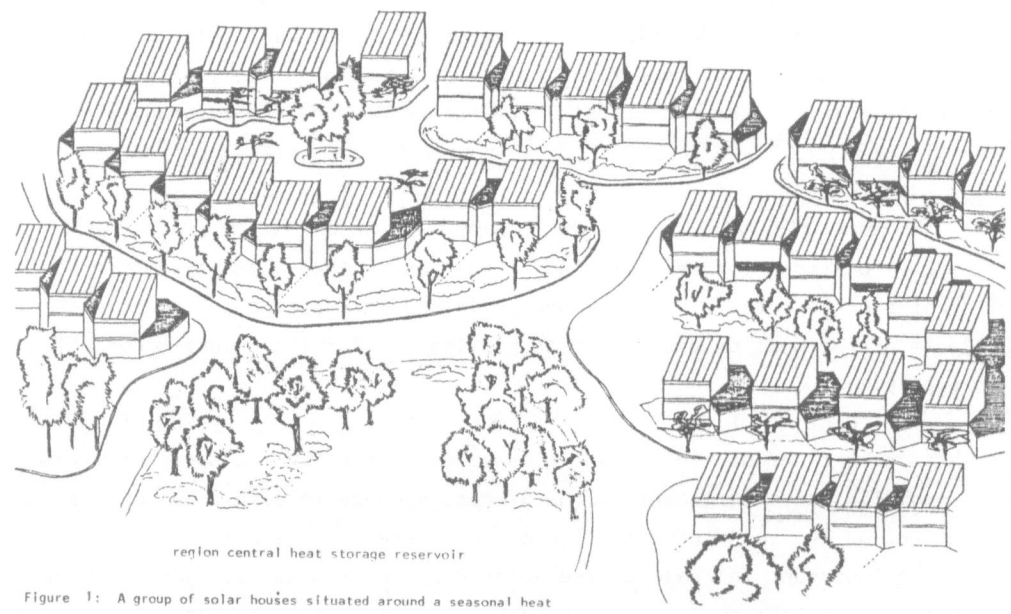

region central heat storage reservoir

Figure 1: A group of solar houses situated around a seasonal heat
storage reservoir in the soil.

This group of houses is divided into a number of subgroups, which are each separately connected by a ringmain with a certain part of the seasonal heat storage reservoir in the soil (see figure 2).

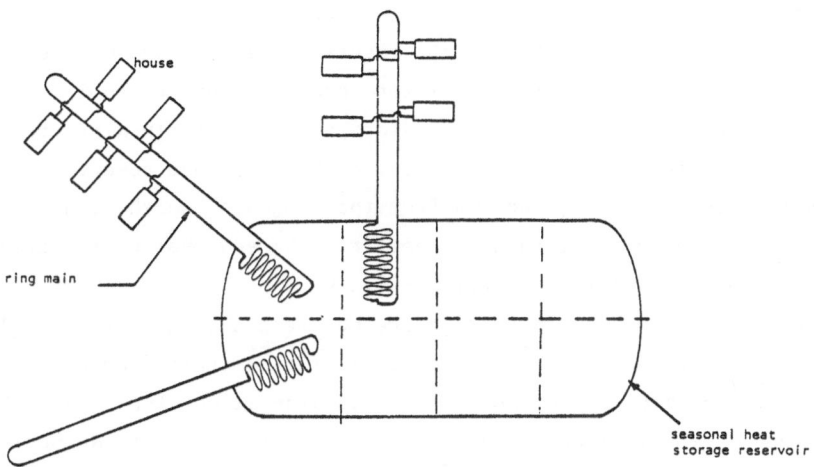

Figure 2:   Location and connection of a group of solar houses to
a seasonal heat storage reservoir in the soil.

Heat exchange between the heat transfer fluid and the soil takes place in
a heat exchanger.

   Each solar house has a solar heating system with a short-term heat
storage reservoir. The solar heating system contributes to the heat demand
for space heating and the heat demand for domestic hot water. In figure 3
a block scheme of the system is given.

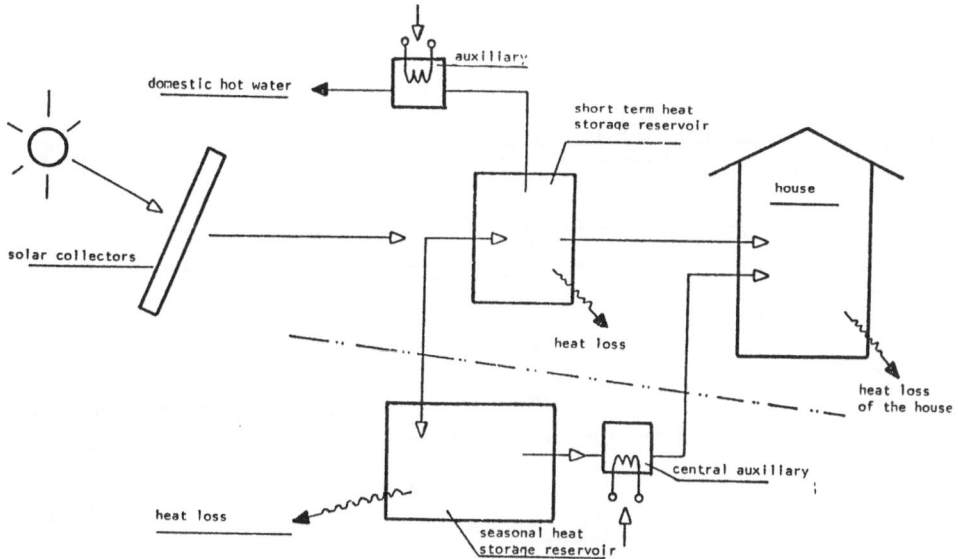

Figure 3:   Scheme of the solar heating system with seasonal heat storage.

212

In the first instance heat storage and heat consumption take place in and from the short term heat storage reservoir. In summertime the surplus of the gained solar heat (the short term heat storage reservoir is full) is transported to the seasonal heat storage reservoir. In wintertime heat is primarily withdrawn from the short term heat storage reservoir, if this is insufficient heat is withdrawn from the ringmain, which is held at the required heating temperature with heat from the seasonal heat storage reservoir and/or from the central auxiliary.

Remark: Basicly the solar collectors can be placed either centrally (around or on the central heat storage reservoir) or distributed (onto the roofs of the houses). Besides system performance considerations the final choice is determined by the available landarea and landprices. In the Netherlands has been chosen for distributed placed solar collectors.

2.1 System in the solar house (see figure 4)

The gain of solar heat:

The solar heat gained by the solar collectors is transferred in a counter flow heat exchanger from the heat transfer medium in the gaincircuit to the heat transfer medium in the storage circuit.

In the storage circuit the gained solar heat is emitted to the short term heat storage reservoir or to the seasonal heat storage reservoir in the soil. This is arranged by two 3-way valves K_1 and K_2. Primarily the gained heat is stored in the short term heat storage reservoir.

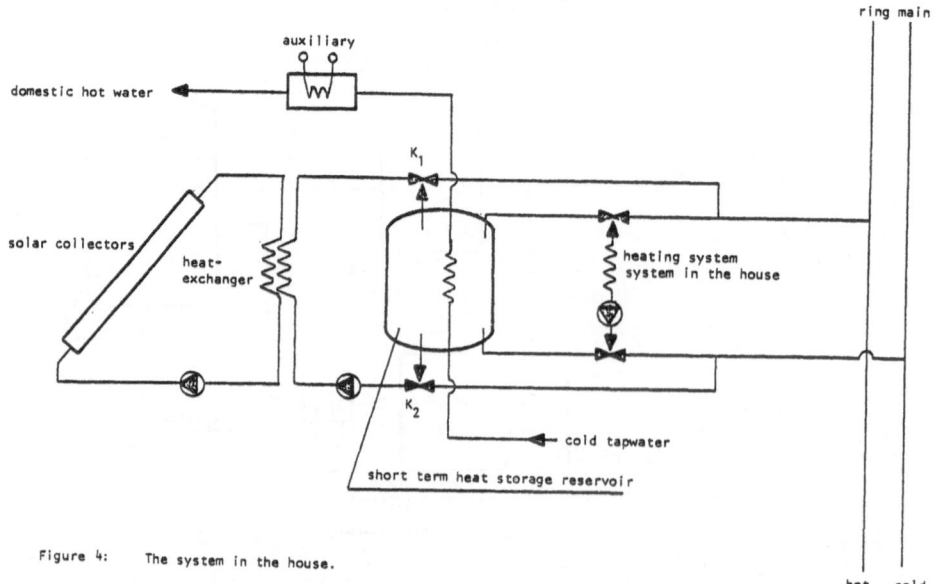

Figure 4: The system in the house.

The consumption of the stored solar heat

The stored solar heat can be used for space heating and for domestic hot water.

Space heating: if there is a heat demand in the house this heat is primarily withdrawn from the short term heat storage reservoir. If the total heat demand cannot be delivered by the short term heat storage reservoir this heat is completely wihtdrawn from the ringmain, which is held at the required temperature with heat from the seasonal heat storage reservoir and if necessary with heat from the central auxiliary.

Domestic hot water: the domestic water is preheated by a heat exchanger in the short term heat storage reservoir. If the tap water does not have the required temperature it is subheated by a conventional auxiliary.

Auxiliary heating for space heating happens centrally and for domestic hot water individually.

2.2 The seasonal heat storage reservoir

The group of solar houses is connected with the central seasonal heat storage reservoir in the soil by a heat distribution pipingnetwork. The gained solar heat is stored in the soil material itself. The seasonal heat storage 'reservoir' has no walls (see figure 5) and is not furnished around with insulation material, the soil itself operates as an insulator.

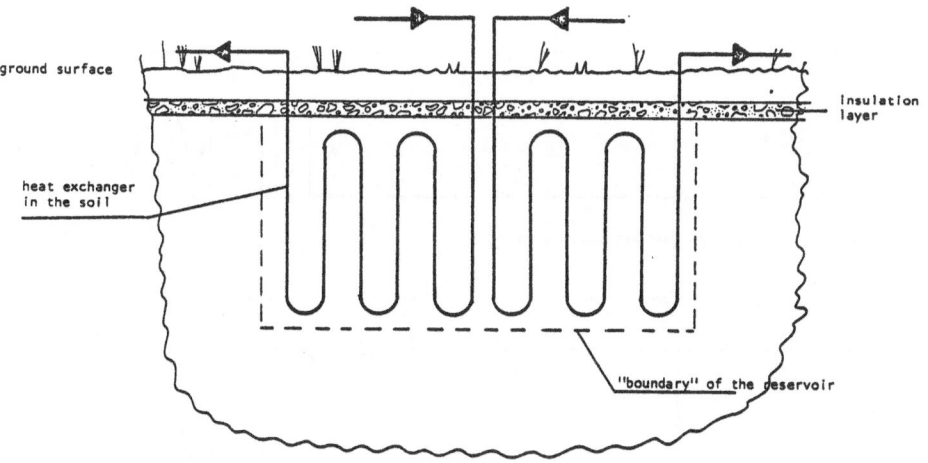

Figure 5: The seasonal heat storage reservoir

Only at the top the 'reservoir' is furnished with an insulation layer at a
depth of about 1 m below ground level; in this way the heat losses at the top
are limited and the biological climate in the top soil layer is maintained.
The heat exchange between the heat transfer medium and the soil happens in a
heat exchanger. The heat exchanger consists of vertical tubes, which should
be installed in the soil by a special insitu-method.

Because of heat transfer by conduction the seasonal heat storage reservoir
is losing heat to the surrounding soil. Apart from heat losses by conduction
there can be additional losses by convection of the ground water in porous
soil. In some cases a vertical wall at the edge of the reservoir is necessary
to limit these heat losses.
Because of temperature stratification in the seasonal heat storage reservoir
in the charging mode the hot transfer fluid enters the heat exchanger in
the centre of the reservoir and leaves the heat exchanger at the edge. In
the discharging mode the fluid direction is opposite.

2.3 The piping network between the houses and the central seasonal heat storage reservoir.

The piping network consists of a number of ringmains, which each separately
connect a subgroup of solar houses with the seasonal heat storage reservoir.
In figure 6 such a ringmain is given for the summermode respectively the
wintermode.

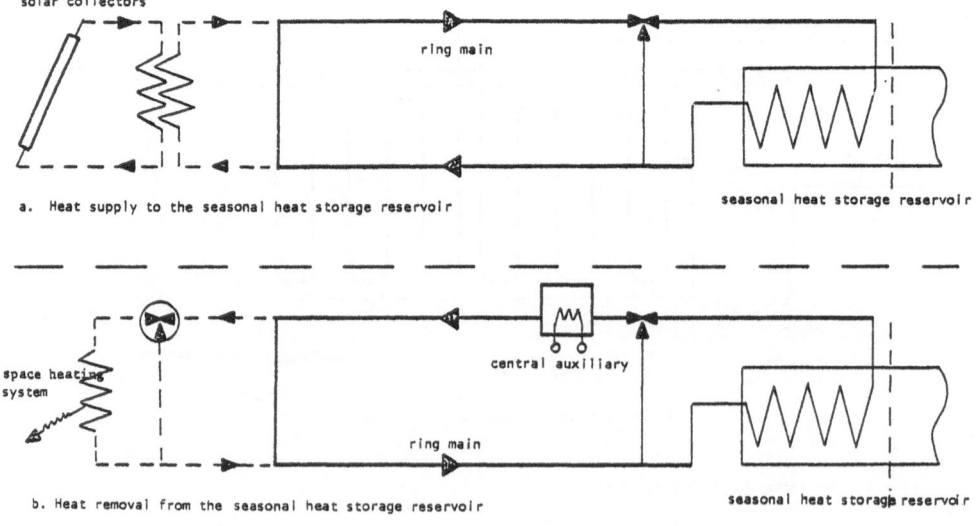

Figure 6: The ringmain between the houses and the seasonal heat storage reservoir.

Summermode: charging of seasonal heat storage reservoir. In the ringmain the transfer fluid is flowing in the direction shown in figure 6a. The heat exchanger in the soil and the solar houses are out of the circuit. If the solar houses are beginning to deliver heat to the ringmain, the ringmain is first heated up to the temperature level of the seasonal heat storage reservoir. If this temperature level is reached the heat exchanger in the soil is taken into the circuit and charging of seasonal heat storage reservoir takes place.

Wintermode: discharging of the seasonal heat storage reservoir. In the ringmain the transfermedium is flowing in the opposite direction as in the summermode (see figure 6b).

The temperature in the supply of the ringmain is continuously kept at temperature T_S, which is determined by the type of heating system in the solar houses. The heat, necessary to keep the supply at that temperature is withdrawn from the seasonal heat storage reservoir. If this is not sufficient the rest heat is delivered by a central auxiliary.

3. SOLAR CONTRIBUTION OF THE SYSTEM WITH SEASONAL HEAT STORAGE

To predict the solar contribution from seasonal heat storage in the soil a computer simulation model for a group of solar houses connected to a central heat storage reservoir in the soil has been developed. As input hourly weather data are used. The model of the seasonal heat storage reservoir is based on heat transfer in the soil by conduction only.

Data of the solar system.

The total system consists of a group of 200 solar houses.

- Each house has at design conditions a maximum heat demand for space heating of 6 kW. The maximum temperature level of the heating system is 35 $^{\circ}$C. The daily domestic hot water (DHW) consumption is 175 l of 50 $^{\circ}$C/10 $^{\circ}$C.

- The solar system of each house consists of 25 m^2 solar collectors mounted onto the south facing roof of the house and a short term (ST-) heat storage reservoir of 1 m^3. The solar collectors are flat plate collectors of the High Performance type (conversion factor η_o = 0.80, heat loss factor U_1 = 2.0 W/m^{2}K; a fictive collector).

- The central heat storage reservoir is in saturated clay soil and has a volume of about 37.000 m^3. The height of the reservoir is 20 m. The top insulation layer has a thickness of 0.50 m. The heat storage capacity is about 600 MJ/K per house. This means 185 m^3 of soil per house, which is compatible with 140 m^3 of water.

- The piping network between the houses and the central heat storage reservoir
consists of 10 ringmains, each with 20 houses. From the lay-out of the
group of houses it is found, that the length of each ringmain is about
2 * 250 m.and the length of the piping between a house and the ringmain
is about 2 * 20 m. The whole pipingnetwork is well insulated: heat loss
is about 0.30 W/K.m$_{piping}$.

With weather data of "a" Dutch reference year the solar contribution of
the above defined solar system has been calculated. Because of the low
start temperature of the central heat storage reservoir it takes a few
years before an equilibrium yearcircle has been reached.

In figure 7 the annual heat fluxes in the system (per house) are given
after a few years of operation. From this figure it follows, that from the
total annual insolation (25960 kWh) about 54% is gained by the solar
collectors (13900 kWh) . The heatlosses in the pipingnetwork and the central
heat storage reservoir are 5200 kWh. So the net solar contribution is 8700
kWh (34% of the solar insolation).

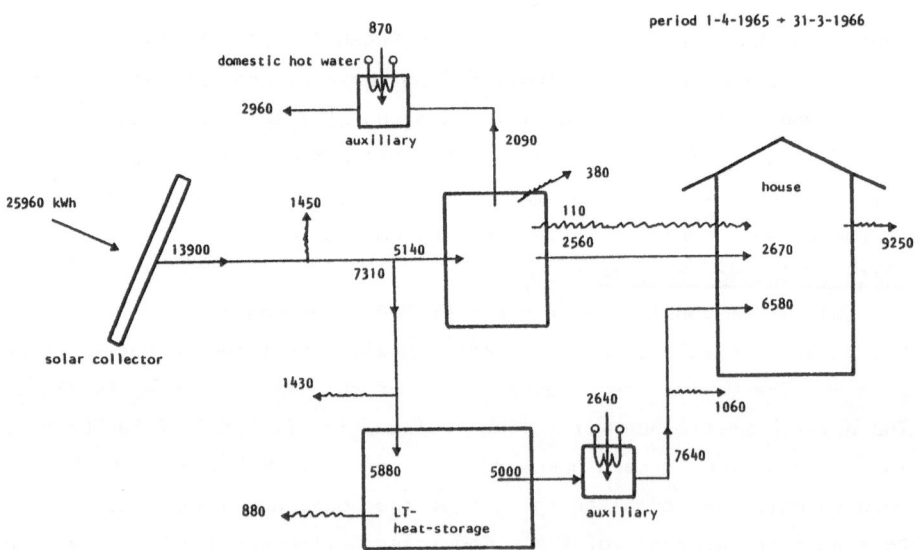

Figure 7: Annual heat fluxes in the solar heating system, which is connected to a
LT-heat storage reservoir.

In the table below it is given how the total heat demand is covered.

	DHW	space heating	total	
total heat demand	2960 kWh	9250 kWh	12210	-
ST	2090	2670	4760	39%
LT	-	3940	3940	32%
Auxiliary	870	2640	3510	29%

The net solar contribution of 8700 kWh is 71% of the total heat demand for domestic hot water and space heating. The solar contribution per m^2 solar collector is about 350 kWh; this is compatible with the annual solar contribution (per m^2) of a solar water heater.

From the computercalculations it was found, that heat supply to the central heat storage reservoir takes place from the beginning of April till the end of October.

In figure 8 the heat supply for space heating during the year is shown: untill the end of January no auxiliary heating is needed.

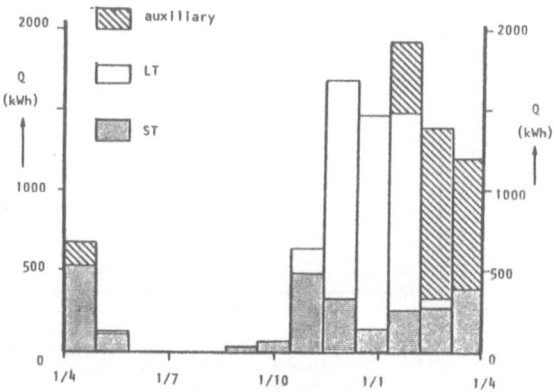

Figure 8:   Heat consumption of the house for space heating.
Heat delivery by respectively ST, LT and auxiliary.

The temperature in the central heat storage is at the end of the winter about 25 °C and at the end of the summer about 60 °C.

218

4. RESULTS PARAMETER SENSITIVITY STUDY.

With the computer simulation programme a parameter sensitivity study has been carried out. At this sensitivity study only the considered parameter is varied in the reference system (the total system defined in chapter 3). Some important results are given in this chapter. In the figures you find on the vertical axe the solar contribution (ST, ST + LT) and on the horizontal axe the considered parameter.

4.1 Heat demand of the house for space heating.

The total solar contribution ST + LT is only slightly influenced by the heat demand of the house. Only the rate between ST- en LT-contribution is changing.

* The arrow points out the reference value.

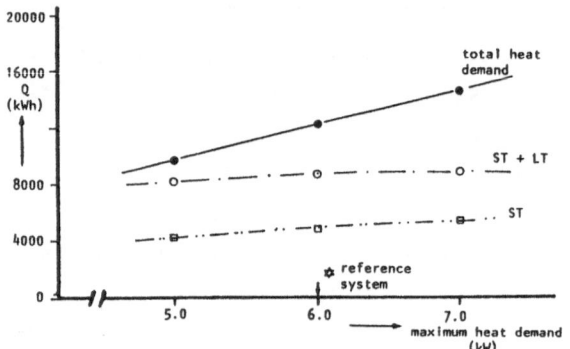

Figure 9: The influence of the maximum heat demand of the house for space heating.

4.2 Temperature level of the space heating system

The total solar contribution (ST + LT) is strongly influenced by the temperature level of the space heating system. Particulary the contribution from LT decreases at higher temperature levels.

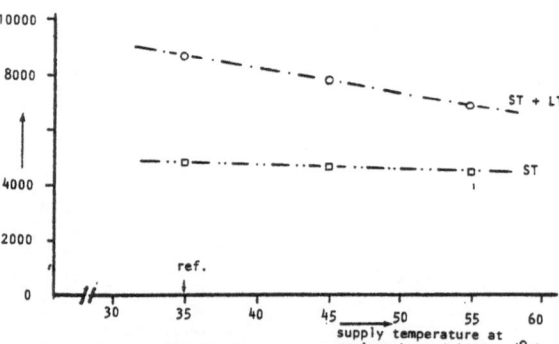

Figure 10: The influence of the temperature level of the space heating system in the house.

4.3 Size of the collector area

The solar contribution from ST is only slightly influenced by the size of the collector area. The solar contribution from LT however increases strongly. At a size of 32 m^2 the reference house has 100% of the space heating load from solar.

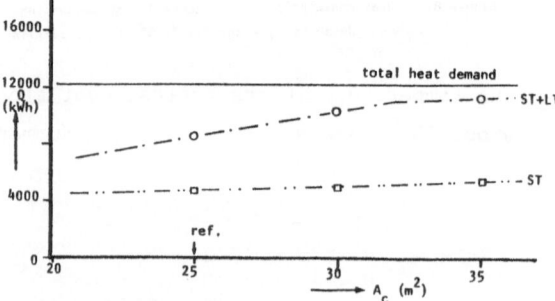

Figure 11: The influence of the size of the collector area A$_c$

4.4 Type of solar collectors.

The influence of the collectortype has been studied for
fictive flat plate solar
collectors: the conversion-
factor η_o = constant = 0.80
and the heat loss factor U_1 =
variable.

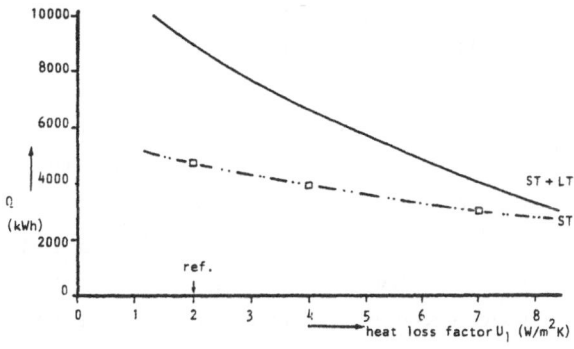

Figure 12: The influence of the solar collector type.
The conversion factor η_o = constant = 0.80.

The total solar contribution
is strongly influenced by the
collectortype.

For an indication:
U_1 = 7 W/m²K : black painted, single glazing collector
U_1 = 4 W/m²K : spectral selective, single glazing collector
U_1 = 2 W/m²K : High Performance collector.
It was found, that for solar collectors with a high heat loss factor U_1
the switch-criterion for heat supply to ST- of LT-heat storage is very
important.

4.5 Insulation of the piping network

As can be seen from figure 12
the rate of insulation of the
piping network is very important.
From district heating systems
it was recently found, that
heat losses of 0.15 W/K.m piping
for high temperature systems
are reasonable. The solar contribution becomes than about
400 kWh/m² solar collector.

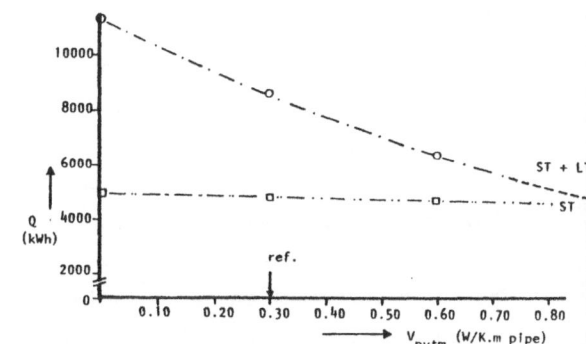

Figure 13: The influence of the insulation of the piping in the
total system.

4.6 Other system parameters

Other systemparameters, which have been studied are the size of the ST-storage reservoir, the switch-criterion for heat supply to ST- of LT-heat-storage, the size of the heatexchanger in the soil, the soil properties, the LT-storage capacity per house, the number of houses etc. The results will not be discussed now.

Remarks on parameter sensitivity study:

At this study only one parameter is varied at the same time. Some parameters have a slight influence, others a strong influence. At the design of a total system most components are different from the reference value. Attention have than to be paid to the fact how all those deviating components influence the solar contribution of the system.

Other important items at the design of the system are the cost/performance of every component and limitations of every component. For instance: the number of solar collectors, mounted onto the roof of a Dutch terraced house, is limited to 25 - 30 m^2.

5. CONCLUSIONS:

Extension of a solar heating system with seasonal heat storage can give in the Dutch climate about a twice as high solar contribution than from Short Term heat storage only. To get this solar contribution High Performance Solar collectors have to be used, the piping network should be as short as possible and well insulated, the heating system should be of a low temperature level.

At the design of a total system a lot of attention should be paid to above mentioned items. Under such conditions a 6 kW-house gets more than 70% of the heat demand for space heating and domestic hot water from solar. The solar contribution per m^2 solar collector is about 350 - 400 kWh, which is compatible with the solar contribution of a solar water heater.

SEASONAL STORAGE OF LOW TEMPERATURE HEAT IN BIG WATER
RESERVOIRS

F. SCHOLZ, KFA JÜLICH

1. INTRODUCTION

Seasonal storage of heat for house heating would be very
attractive as the house heat demand is high in the winter
and very low or zero during the summer. If big thermal storage systems could be safely constructed at acceptable costs
the sources for this low temperature heat could either be
operated all the year round (high efficiency, lower capital
costs) or solar as well as waste heat which mainly or likewise arise during the summer could be used for (peak) heat
demand in the winter. Mainly the following two applications
of seasonal heat storage would be of great interest:

1.1. District heating systems based on combined generation
of heat and power (CHP) could be operated at optimal efficiency all the year round. In the summer and whenever the
heat demand is lower than heat production the surplus energy
could be stored and extracted in the winter when the demand
is higher than the rate of production.

1.2. Central solar heating plants (CSHP) with seasonal storage could be operated theoretically without any other complementary heat source like fossil fired boilers or heat pumps.
They would use in winter time mainly the heat collected
during the summer and stored in the reservoir.

As mentioned above also different sources of waste heat
could be integrated into district heating by long term storage.

In case 1.2 seasonal storage is absolutely necessary as solar heat is mainly available during the summer and needed prevailingly in the winter for space heating. In case 1.1 seasonal storage would decrease the capacity of the CHP-plant and perhaps save some energy. A much lower percentage of the annual heat demand has to be stored compared to case 1.2 as the CHP-plant is operated (and needed also from the electrical energy side) during the winter, whereas the heat source of case 1.2, the sun, is less or hardly available during the winter.

Water, though old-fashioned, still proved to be a suitable heat storage medium expecially in the low temperature range sufficient for space heating for various reasons:

Water is at the same time the medium for storage, transport and heat transfer.

Intermediate heat exchangers and temperature differences for heat transfer are avoided.

Water has a relatively high specific heat and offers a good storage capacity in combination with a reasonable temperature spread which is also appreciated for transport reasons.

For its use within the house it is important that water is not toxic, not inflammable and not corrosive.

Water is easily available, relatively cheap and easy to handle.

Like all seasonal storage systems big water reservoirs must be extremely cheap as they can be used only once a year. /1/

2. POSSIBLE DESIGNS OF HOT WATER RESERVOIRS

Most existing water reservoirs are operated at elevated pressures, as the storage temperature often exceeds 1oo $^{\circ}$C. Mainly cylindrical or spherical steel tanks are in use. As it is not absolutely necessary to have upper heating temperatures in excess of 1oo $^{\circ}$C there are promising attempts to construct water reservoirs at ambient pressure.

2.1. Big pressureless steel tanks

In the case of pressureless steel tanks the cost per unity
of stored energy is far below that of a pressurized vessel
though the temperature spread is only a fraction of that
available in pressure vessels of e.g. 1o to 2o bar /2/ (also
referred to in /14/). Big pressureless steel tanks are in
operation in some places in Scandinavian countries /3/ and
under construction also in Germany. Fig. 1 shows the heat
reservoir of Uppsala in Sweden engineered by Tore J. Hedbäck
AB. It contains nearly 3o.ooo m^3 of water. The water level
may vary by 4 m. A floating diffusor injects or withdraws
hot water close to the water surface level and a fixed device
close to the bottom handles the same amount of cold water in
opposite direction /4/. A thermocline develops between the
upper warm and the lower cold tank content. These accumulators are operated for short term storage only, as their costs
are too high for seasonal storage.

To lower the costs there are proposals in Sweden to
construct large uninsulated rock caverns /5/. A pilot plant
may be constructed in the near future.

2.2. Storage lakes

In Germany big artificial water lakes originally proposed
by G. Schöll /6/ have been studied. To extremely reduce the
costs it was thought feasible that the excavated soil could
be used to form dams, thus creating a very cheap additional
storage volume above surrounding ground level. Plastic foils
were assumed to be sufficient for sealing the pit against
water losses. A floating and sealing plastic foam insulation
should reduce the heat losses at the surface whereas the soil
itself should act as a thermal insulant.

Studies at KFA Jülich /7/ and at Stadtwerke Mannheim /8/
showed, however, that at least partly plastic materials were
not suitable and that a minimum thermal insulation towards
the soil is required. Also further technological problems
led to the conclusion that big artificial thermal storage
lakes are very difficult to realize or that their construc-

tion costs would at least be prohibitive.

A small prototype (64o m^3) of such an earth pit reservoir has been constructed in Studsvik, Sweden. Fig. 2 shows that storage facility together with an office blook heated by the water from that reservoir. Solar collectors are mounted on a rotating lid.

2.3. Advanced steel membrane reservoirs

In order to overcome the various difficulties of a big laketype reservoir the Mannheim team proposed a steel membrane water reservoir, which also goes back to G. Schöll. Cylindrical tanks as used in Uppsala and widely in the oil industry cannot be constructed for very large volumes. An upper limit may be about 15o.ooo m^3 restricted by the circumferential stresses in the cylindrical wall close to the bottom of such vessels. The biggest oil tanks built in Germany have volumes of 1oo.ooo m^3. Demands for seasonal storage in district heating systems can easily be far beyond 1o^6 m^3. Therefore G. Schöll proposed steel membrane tanks which have the form of a flat shell /11/. The side walls are not vertical but are given the shape of a standing quarter of an ellypsis. Stresses then occur only in the direction of this ellyptical contour. For heights of about 12 m the thickness of these membrane side walls is claimed to be about 7 mm only. The weight forces of the water above the ellyptical contour must be taken by pillars or a supporting ring connected to the upper edge of that shell-shaped vessel. Therefore reservoirs of very great diameters or even with rectangular ground plans are feasible. A feasibility study for this advanced design is underway with an experienced manufacturer in Germany.

Fig. 3 gives an artists impression of the Mannheim proposal /8/. The main advantage compared to the artificial lake is the fact that the vessel has a vapour-tight steel skin against the water. Thus very effective conventional and cheap insulating materials can be used. The bottom of the vessel is situated below normal ground level and above

groundwater level. The excavated soil forms a dam which acts
as a safety barrier in the case of important leaks of the tank.
The upper ceiling, insulated also on the outside, rests on
pillars and is welded to the side walls.

3. THERMODYNAMIC REMARKS

Thermal losses of seasonal storage systems should be very
low because of the long storage periods involved. Losses to
the environment not only reduce the storage capacity but
also the stored exergy by reducing the temperature of the
water. Further exergy losses occur by the inevitable heat
transport from the upper hot to the lower cold water across
the thermocline. The outer losses can be reduced by improving
the thermal insulation, which is of course limited by costs.
The inner exergy losses cannot be influenced by simple measures. They can be reduced by distributing the hot water with
minimum turbulent mixing on the upper surface at the beginning of the loading process. Experiments showed that this
problem can ce overcome. The development of the thermocline
(vertical temperature distribution) can be predicted by
relatively simple computer models /12/, /13/. Fig. 4 shows
a comparison between measured and calculated vertical temperature profiles during a loading process of an experimental
storage container. Similar curves for the partly loaded experimental reservoir calculated with different values for the
effective thermal conductivity are given in Fig. 5. The
thickness of the thermocline grows with time and can be in
the order of 8 to 1o m after some 3.000 hours. Therefore it
reduces the volume efficiency, too. As a consequence the
storage height should be as big as possible. But it has been
said in chapter 2 that relatively cheap reservoirs are
rather flat. This is also unfavourable for the outer heat
losses as the ratio of volume to surface is very poor for
flat plates. As a consequence we propose to split up the
big reservoir in different smaller ones /1/, /14/. Thus at
least the ratio of horizontal cross section to height is
improved. If one loads these reservoirs in series, feeding

the water from the bottom of magazine 1 to the top of
reservoir 2 one can transfer the thermocline into reservoir
2. Then number 1 is loaded completely and further inner
exergy losses are avoided. The split up reservoir version
provides great advantages, with regard to safty and safe
supply but the costs will increase, too.

As for the load changing strategies an "optimized" sceme
has been developed by a German utility /15/ for a decentralized solar heating system. Fig. 6 shows this sceme, which is
a sound compromise between a severly stratified reservoir
and simple means to achieve this effect. An auxiliary electrical heating system offers the possibility to use also
cheep week-end electrical energy for storage or to increase
the upper temperature for the heating system when the storage
temperature is lower than needed by the users.

4. ECONOMICAL CONSIDERATIONS

In the Juelich project an industrially backed study was
performed for a new residential district heated area for
14o.ooo people /7/. In case 1 it was assumed that a 5 million
m^3 storage lake close to the consumer together with a cogeneration power station 2o km away should provide the annual
heat demand. The CHP-plant should deliver heat for about
7.ooo hours a year at constant level while the storage lake
should act as a buffer thus providing heat at higher demands
to the grid or storing heat at low demand periods. Also the
maintenace or outage times of the CHP-plant should be bridged
by the lake. In this case about 24 % of the annual heat
demand was to be stored in the lake and the base load of the
CPH-plant resulted to be about 43 % of the peak demand.

In case II a fossil fired heating plant was to take over
the peak supply. The CPH-plant delivered max. 58 % of the
peak load. In this case only 9 % of the annual heat was to
be supplied by the heating plant.

The result was that case II (without storage) led to lower
annual heating costs. Assuming equal costs the storage volume would have to be constructed at costs of 4 to 7 DM/m^3.

Feasible costs of the reservoirs were evaluated at 25 to
4o DM/m^3. From the state of to day's-knowledge the costs would
be at least two or three times as much.

Consequently the plans to build an experimental facility
at Juelich were cancelled. Experimental and theoretical work
concerning the hydraulic and thermal behaviour of stratified
water reservoirs together with first step material tests were
carried on. The results of those activities were available
to the subsequent projects, e.g. Mannheim.

5. SUMMARY AND CONCLUSIONS

The prevailing unfavourable results of the different
projects and studies with regard to big seasonal thermal
storage systems lead us to the conclusion that this technology will not be introduced at large scale within the near
future, mainly for economical reasons. Although, in our
studies, the boundary conditions have been considered preferably for district heating systems based on CPH the conditions for centralized solar heating plant seem to be similar.
The solar situation may even be worse in case of a monovalent
solar system. The fraction of annual heat demand which must
be stored is about 2.5 to 3 times higher compared to the CPH
system as solar heat is preferably available in the summer
and house heating energy is mainly needed during winter time.
The CPH-plant, however, is in operation at highest duty in
the winter and in reduced form or even at zero power (e.g.
for maintenance) in the summer.

At least under German conditions big short term thermal
storage systems in district heating networks based on CPH,
both with available technology, seem to be a better option
and offer perhaps the same potential in saving (fossil)
energy compared to seasonal storage which does not exist and
seems to be unattractive from economic reasons.

Such reservoirs are successfully in operation in Sweden
since several years (the biggest in Uppsala 3o.ooo m^3) and
are now under construction or projected also in Germany.
They will perhaps save more energy indirectly as they are

able to reduce or avoid some drawbacks or difficulties which today are stumble stones on the way to a large introduction of the energetically most favourable CPH technology /16/.

The utmost desirable use of solar energy for house heating based on collectors threatens to fail if one insists on seasonal storage. On will have to make compromises. A good proposal may be to use collected solar energy for the summer base load of a district heating system. In many cases the network itself acts as a buffer or additional reservoirs may be built or used to bridge bad weather periods during the summer or during transition times. In this way a lot of fossil fuel can be saved which otherwise is simply burnt in a central heating plant with low efficiency under part load conditions. Even in smaller district heating systems with backpressure CPH-plants it may be more economical to switch off the CPH-plant during the summer and to use solar heat, as electrical energy can be easily provided in the summer by existing base load power stations, e.g. by brown coal or nuclear fueled or even hydropower stations with better efficiencies.

Another favourable possibility to go solar is the indirect way of using solar energy by heat pumps. Here natural seasonal storage systems (with zero construction costs) like ambient air, soil, ground- or running water are used as a source for heat synchronously with the demand in the winter. This technology may be an attractive way for small villages or single houses too far away from being supplied by a district heating system or even from gas pipelines.

REFERENCES

1. Scholz F, Langzeitspeicherung in Verbindung mit Kraft-
   Wärme-Kopplung. Heizung Lüftung Haustechnik (HLH) 31 (1980)
   Nr. 1o p. 374-381
2. Bundesministerium für Forschung und Technologie, Einsatz-
   möglichkeiten neuer Energiesysteme; Programmstudie "Sekun-
   därenergiesysteme" Teil V, Fernwärme, Bonn 1975 p. 337,
   Weiss Druck + Verlag, Monschau-Imgenbroich
3. Västeras Stads Värme Kraftwerk AB, Wärmespeicherung-Tech-
   nik und Wirtschaftlichkeit-Konventionelle Speicherung.
   Fernwärme international-FWI 8 (1979) p. 2o-23
4. Martensson K, Ergebnisse von Temperatur-Messungen im Wär-
   mespeicher. Report of Uppsala Kraftvärme AB 1980 15 pages
5. Jonsson B, Bjurström S, Hultin S-A, Energieeinsparung
   durch Nutzung unterirdischer Räume. Proceedings of the
   WEC 1980, Munich, Vol. 1B, p. 489-5o9
6. Schöll G, Warmwasser-Großspeicher. VDI-Berichte 233 (1974)
   p. 33-35
7. Scholz F, Warmwasserseen als Langzeitwärmespeicher
   VDI-Berichte 288 (1977) p. 15-26
8. Geipel W, Große Warmwasserspeicher in der Fernwärmever-
   sorgung, Wunschdenken oder Realität?
   Fernwärme international-FWI 9 (1980) Heft 4 p. 319-327
9. Margen P, Roseen R, Central Solar Heat Stations and the
   Studsvik Demonstration Plant.
   The 13th Intersociety Energy Conversion Engineering
   Conference, USA 1978
1o. Kittel A, Bauten für die Lagerung von Mineralöl und Gas.
    Energie 29 (1977) Nr. 4 S. 94-96
11. Schöll G, Die wirtschaftliche Nutzung der Sonnenenergie.
    Akademie für Solar- und Speichertechnik Wolfschlugen.
    Gemeinnützige Stiftung 1978. 31 pages
12. Leyers HJ, Scholz F, Tholen A, Theoretische und experi-
    mentelle Ermittlung der Temperaturverteilung in geschich-
    teten Warmwasserspeichern. VDI-Berichte 288 (1977)
    p. 27-32
13. Adä W, Bach H, Striebel D, Modell für einen Langzeitspei-
    cher. Statusbericht Sonnenenergie I Bundesministerium
    für Forschung und Technologie, Bonn 1980 VDI-Verlag
    p. 333-344
14. Scholz F, Warmwasserspeicher in Fernwärmesystemen mit
    Kraft-Wärme-Kopplung. Brennst.-Wärme-Kraft 31 (1979)
    Nr. 1o p. 397-4o2
15. Dietrich B, Langzeitspeicher für Wohngebäude. Statusbe-
    richt Sonnenenergie I Bundesministerium für Forschung
    und Technologie, Bonn 1980 VDI-Verlag p. 311-331
16. Scholz F, Sind Wärmepumpen eine energetische Alternative
    zur Kraft-Wärme-Kopplung? Energie 32 (1980) Nr. 4
    p. 149-153

230

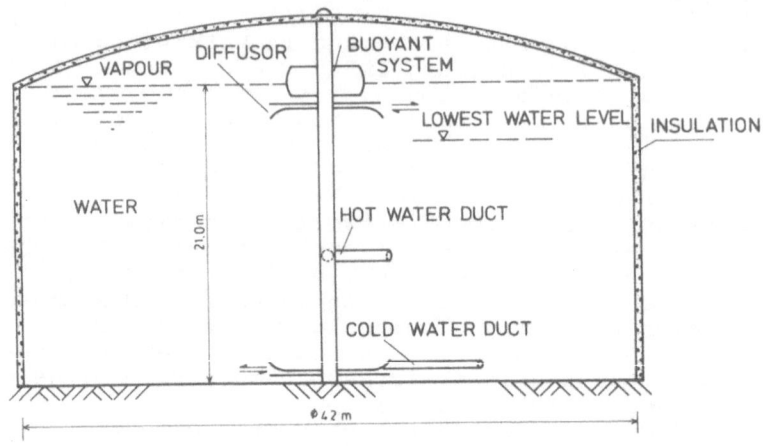

.FIGURE 1   Steel tank reservoir Uppsala

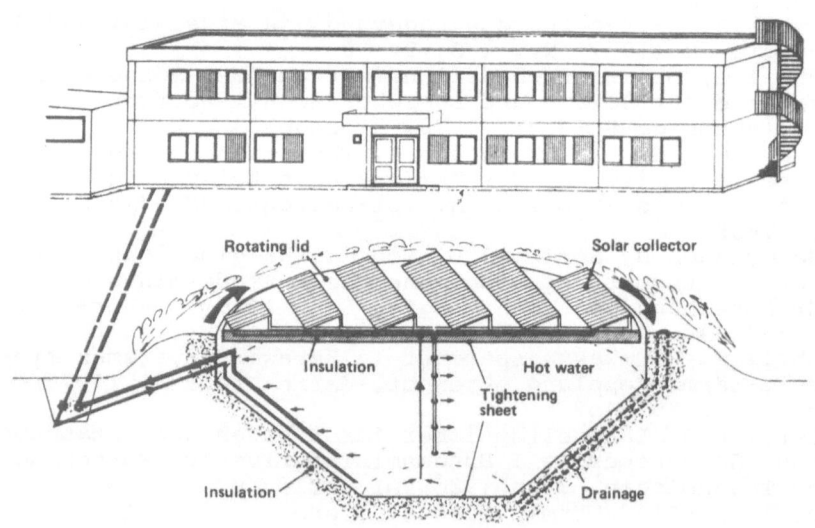

FIGURE 2   Earth pit reservoir Studsvik

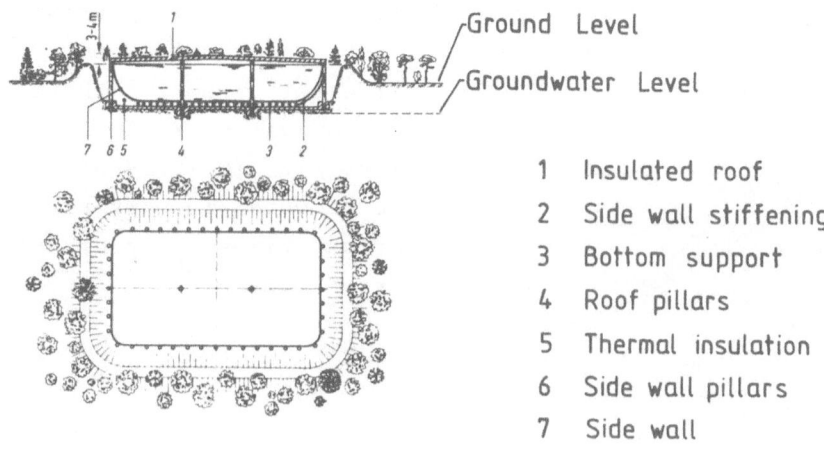

FIGURE 3 Advanced steel membrane reservoir (Mannheim)

1 Insulated roof
2 Side wall stiffening
3 Bottom support
4 Roof pillars
5 Thermal insulation
6 Side wall pillars
7 Side wall

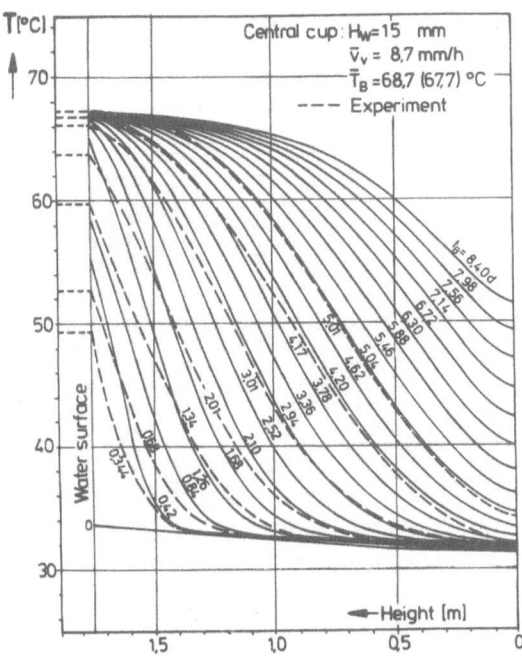

FIGURE 4 Vertical temperature distribution during loading, of the RWE container, calculated (drawn lines) and measured curces /12/

232

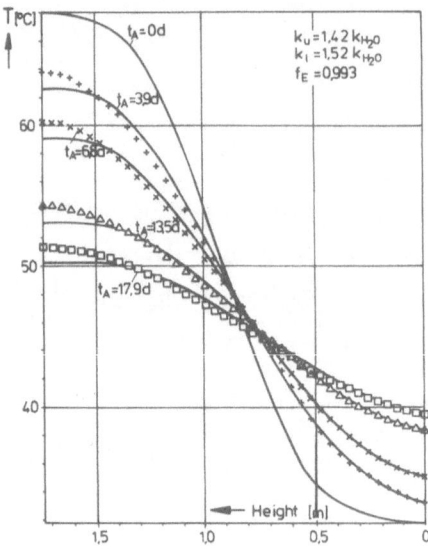

FIGURE 5 Development of the thermocline after partial loading of the tank for different effective thermal conductivities k of the water /12/ (Symbols are experimental results after various times t_A in days after loading)

1 Heat exchanger in collector loop
2 " " " heating loop
3 Low temperature heating system
4 Selector valve
5 Mixing valve
6 Loading branch for auxiliary heating

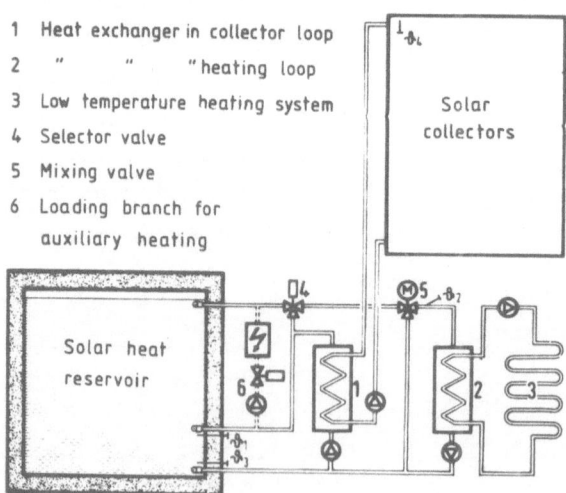

FIGURE 6 Optimized System coupling of a solar house heating plant /15/

THEORETICAL MODEL OF HEAT AND MASS TRANSFER IN NON-SATURATED SOILS WITH
PHASE CHANGE

J.C. Benet and P. Jouanna
Civil Engineering laboratory - University Montpellier 2
Place E. Bataillon - 34060 MONTPELLIER Cedex FRANCE

1. INTRODUCTION

The purpose of this work was to elaborate a mathematical model describing the evolution of non-saturated soil used as a heat storage medium.
The high temperatures encountered in this application lead to important
elementary phenomena as far as heat and mass transfer are concerned but which
are often negligible at ambient temperatures.

A non-saturated soil may be considered as a triphase medium made up
of a solid phase assumed to be invariable and macroscopicaly homogeneous
and isotropic, a liquid phase assumed to be pure, and a gas phase consisting of a mixture of air and water vapour. The thermodynamics of irreversible and linear processes (T.I.P.) model [1,2] is particularly suitable
for describing such a system placed in a non-equilibrium situation. However,
although this model, when based only on the writing of a single GIBBS
equation following the mean motion, is suitable for mixture of gas and
solutions, it does not enable correct representation of non-saturated porous
media for two main reasons :

a) the only constitutive equations which can be decuced from the source
of entropy using a single GIBBS equation written following the mean motion
of all the components concern the equations for the diffusion of these
comporents in relation to the mean motion. However, diffusion velocities
thus defined are not accessible using the experiment which, in a non-saturated
porous medium, gives access essentially to :

- the velocities of the liquid phase and the gas phase in relation
to the solid phase, i.e. filtration velocities [3, p. 248] ;

- the diffusion velocities of air water vapour in relation to the
mean motion of the gas phase [4, p. 238] .

b) in the T.I.P. model temperature is introduced by the GIBBS equation;
consequently a model based on a single GIBBS equation does not allow

234

introduction of different temperatures for the various phases and therefore
prohibits taking into account the heat transfer between phases which is
nevertheless of prime importance in heat storage and geothermal applications.

Models of non-saturated soil based on a single GIBBS equation are
given in references [5] , [6] and [7] . The writing of two GIBBS equations
is proposed by MARLE [8] for a saturated medium. Finally, GUELIN [9]
extends this writing to a non-saturated medium making a distinction between
a solid phase and the liquid + gas mixture.

In order to meet conditions above we propose extending the model [9]
by writing three GIBBS equations - one for each phase: solid, liquid and
gas. The phenomena which then become accessible are as follows:

- distinction between the filtration velocities of the liquid phase
  and the gas phase, this separation being necessary in a number of
  applications involving temperature.
- phase changes which are sometimes of prime importance when conditions
  are far removed from ambient temperatures e.g. in the case of water
  vapour exchanges in a non-saturated soil.
- heat transfert between solid, liquid and gas also important when
  conditions are far removed from temperature equilibrium between
  phases. To avoid overcomplication of the presentation of the model,
  we assume here that liquid and gas temperatures are identical.

We will not, in the present article, go into theoretical considerations
of microscopic physics which enable the writing of a GIBBS equation following
the mean motion of each phase [10] , [11] . Only the compatibility of such
an approach with the writing of experimental laws is envisaged.

2. MASS,MOMENTUM AND ENERGY BALANCES

The balance equations for mass, momentum and energy, written for the
solid phase,liquid phase and gas phase respectively, are derived from the
general model of heterogeneous media given by TRUESDELL and TOUPIN [12] .

2.1. <u>Parameters for the description of a non-saturated porous medium;
state variables.</u> ρ_i, v_i^k , u_i , s_i , h_i represent the apparent mass density,
velocity,specific internal energy,specific entropy and enthalpy respectively
of the phase or component i ; where : i =1 for the solid phase, i=2 for
the mixture : liquid+gas,i=e for liquid water,i=a for air,i=v for water
vapour . The same variables without indices refer to the mean motion and

are defined as follows:

$$\rho = \sum_{i=1,e,a,v} \rho_i \tag{1}$$

$$\rho v_i^k = \sum_{i=1,e,a,v} \rho_i . v_i^k \tag{2}$$

$$\rho . \Omega = \sum_{i=1,e,a,v} \rho_i . \Omega_i \tag{3}$$

where Ω represents specific internal energy, entropy or enthalpy.

The variables with index g concern the partial mean motion of the gas phase and are defined as follows:

$$\rho_g = \rho_a + \rho_v \tag{4}$$

$$\rho_g . v_g^k = \rho_a . v_a^k + \rho_v . v_v^k \tag{5}$$

$$\rho_g . \Omega_g = \rho_a . \Omega_a + \rho_v . \Omega_v \tag{6}$$

The diffusion flux of phase i or of constituent i in relation to the total mean motion is defined as:

$$J_i^k = \rho_i (v_i^k - v^k) \ , \ i=1,e,a,v \ ; \ \sum_{i=1,e,a,v} J_i^k = 0 \tag{7}$$

The diffusion flux of air or water vapour in relation to the mean motion of the gas phase is defined as:

$$J_i^{'k} = \rho_i (v_i^k - v_g^k) \ , i=a,v \ ; \ J_a^{'k} = - J_v^{'k} \tag{8}$$

The temperature of the solid phase is indicated as T_1 and the temperature of the liquid phase+gas is referred to as T_2 . The state variables selected for defining the state of the system are:

$$\rho_1 \ , \ \rho_e \ , \rho_a \ , \ \rho_v \ , \ T_1 \ , \ T_2$$

ρ_1 being the apparent mass density of the solid phase, and assumed to be uniform and invariable.

2.2. Mass balance equations . The mass balance equations are written as follows [12,p.472] :

$$\frac{\partial}{\partial t} \rho_e = - (\rho_e . v_e^k)_{,k} - J \qquad \text{(liquid water)} \tag{9}$$

$$\frac{\partial}{\partial t} \rho_v = - (\rho_v . v_g^k)_{,k} - J_{v,k}^{'k} + J \quad \text{(water vapour)} \tag{10}$$

$$\frac{\partial}{\partial t} \rho_a = - (\rho_a . v_g^k)_{,k} + J_{v,k}^{'k} \qquad (\text{air}) \tag{11}$$

$$\frac{\partial}{\partial t} \rho_g = - (\rho_g . v_g^k)_{,k} \qquad (\text{gas phase}) \tag{12}$$

$$\frac{\partial}{\partial t} \rho = - (\rho . v^k)_{,k} \qquad (\text{all constituents}) \tag{13}$$

J represents the phase change velocity, i.e. the mass of liquid water transformed into vapour per unit of time and volume of porous medium.

2.3. Momentum balance. The symbol $\overset{\shortmid}{\phi}$ is used below to represent the derivative of the quantity ϕ according to the motion of a phase or a constituent. The symbol $\overset{\centerdot}{\phi}$ is used to represent the derivative of the quantity ϕ following the mean motion . These two material derivatives are given by $[12, \text{p.470}]$:

$$\overset{\shortmid}{\phi} = \frac{\partial}{\partial t}\phi + \phi_{,k} \cdot v_i^k \tag{14}$$

$$\overset{\centerdot}{\phi} = \frac{\partial}{\partial t}\phi + \phi_{,k} \cdot v^k \tag{15}$$

For each constituents , i= 1,e,a,v ; the momentum balance is given as $[12, \text{ p. 567}]$:

$$\rho_i \cdot \gamma_i^k = \rho_i \cdot (\overset{\shortmid}{v_i^k}) = - P_{i,m}^{km} + \rho_i \cdot f_i^k - \lambda_i^k , \quad i= 1,e,a,v \tag{16}$$

γ_i^k is the acceleration of constituent i ; γ_1^k and v_1^k for the solid being nil; f_i^k is the extrinsic load per unit of mass of constituent i; λ_i^k is the supply of momentum of constituent i; P_i^{km} is the partial pressure tensor of constituent i .

For the mean motion and for the gas phase, the momentum balances are formaly written as follows:

$$\rho \cdot \gamma^k = \rho \cdot \overset{\centerdot}{v}^k = - P_{,m}^{km} + \rho \cdot f^k \tag{17}$$

$$\rho_g \cdot \gamma_g^k = \rho_g \cdot \overset{\shortmid}{v}_g^k = - P_{g,m}^{km} + \rho_g \cdot f_g^k - \lambda_g^k \tag{18}$$

where:

$$\rho \cdot f^k = \sum_{i=1,e,a,v} \rho_i \cdot f_i^k \tag{19}$$

$$\rho_g \cdot f_g^k = \rho_a \cdot f_a^k + \rho_v \cdot f_v^k \tag{20}$$

Writing equations (16) at a given point, adding them for i=1,e,a,v and comparing to (17) written at the same point, and taking equation (2) into account, we obtain:

$$P^{km} = \sum_{i=1,e,a,v} (P_i^{km} + \frac{J^k \cdot J^m}{\rho_i}) \tag{21}$$

$$\sum_{i = 1,e,a,v} \lambda_i^k + J(\frac{J_e^k}{\rho_e} - \frac{J_v^k}{\rho_v}) =0 \tag{22}$$

Likewise , writing equation (16) at a given point for i=1,e, adding to equation (18), comparing to (17) and taking (2) and (5) into account, gives:

$$P_g^{km} = \sum_{i=a,v} (P_i^{km} + \frac{J_i^{'k}.J_i^{'m}}{\rho_i}) \tag{23}$$

$$\lambda_g^k = \lambda_a^k + \lambda_v^k - J \cdot \frac{J_v^{'k}}{\rho_v} \tag{24}$$

The partial pressure of the phase or the constituent i is defined by [11, p.667] :

$$p_i = \frac{1}{3} P_i^{kk} \;,\; i = 1,e,a,v,g \tag{25}$$

We define:

$$\Pi_e^{km} = P_e^{km} - p_e \cdot \delta^{km} \tag{26}$$

$$\Pi_i^{km} = P_i^{km} - (p_i + \frac{1}{3} \frac{J_i^{'k2}}{\rho_i}).\delta^{km} \;;\; i=a,v \tag{27}$$

$$\Pi_{Ig}^{km} = \Pi_a^{km} + \Pi_v^{km} \tag{28}$$

Where δ^{km} is the Kronecker tensor.

2.3. **Internal energy balance.** From the internal energy balance and following the mean motion [12,p.614] , the derivation at a given point of the total internal energy gives :

$$\frac{\partial}{\partial t} \rho.u = - \sum_{i=e,a,v} (\rho_i.u_i.v_i^k)_{,k} - \sum_{i=1,e,a,v} J_{qi,k}^k - \sum_{i=e,a,v} P_i^{km}.v_{i,m}^k$$

$$+ \sum_{i=1,e,a,v} \lambda_i^k \cdot \frac{J_i^k}{\rho_i} - \frac{1}{2} J((\frac{J_v^k}{\rho_v})^2 - (\frac{J_e^k}{\rho_e})^2) \tag{29}$$

where J_{qi}^k is the conduction heat flux in the constituent i . Taking into account equations (7),(8),(13),(14),(24),(25),(26),(27),(28) the internal energy balance (29) becomes :

$$\frac{\partial}{\partial t} \rho.u = - (\rho_e.u_e.v_e^k + \rho_g.u_g.v_g^k)_{,k} - \sum_{i=1,e,a,v} J_{qi,k}^k - \Pi_e^{km}. v_{e,m}^k$$

$$- \Pi_{Ig}^{km} \cdot v_{g,m}^k - p_e \cdot v_{e,k}^k - p_g \cdot v_{g,k}^k - \frac{1}{2} J (v_g^{k2} - v_e^{k2} + (\frac{J_v^k}{\rho_v})^2)$$

$$+ \lambda_e^k \cdot v_e^k + \lambda_g^k \cdot v_g^k - (J_v^{'k} (u_v - u_a))_{,k} - P_v^{km} (\frac{J_v^{'k}}{\rho_v})_{,m}$$

$$+ P_a^{km} (\frac{J_v^{'k}}{\rho_a})_{,m} + J_v^{'k} (\frac{\lambda_v^k}{\rho_v} - \frac{\lambda_a^k}{\rho_a}) \tag{30}$$

This energy balance depends only on the velocities v_e^k , v_g^k and the diffusion flux $J_v'^k$.

The internal energy balance for the solid phase can be written thus:

$$\frac{\partial}{\partial t} \rho_1 \cdot u_1 = - J_{q1,k}^k + \hat{u}_1 \tag{31}$$

where $\hat{u}_1$ is the energy received by the solid phase per unit volume of porous medium and unit time.

3. ENTROPY BALANCE.

Taking into account the remark made in the introduction, the GIBBS equation [2,p.23] can be written for phase i per unit of volume , following the motion of this phase :

Solid phase (assumed invariable) :

$$T_1 \left(\frac{\rho_1 \cdot s_1}{\rho} \right)_1^{\cdot} = \left(\frac{\rho_1 \cdot u_1}{\rho} \right)_1^{\cdot} \tag{32}$$

Liquid phase :

$$T_2 \left(\frac{\rho_e \cdot s_e}{\rho} \right)_e^{'} = \left(\frac{\rho_e \cdot u_e}{\rho} \right)_e^{'} + p_e \left(\frac{1}{\rho} \right)_e^{'} - \mu_e \left(\frac{\rho_e}{\rho} \right)_e^{'} \tag{33}$$

Gas phase :

$$T_2 \left(\frac{\rho_g \cdot s_g}{\rho} \right)_g^{'} = \left(\frac{\rho_g \cdot u_g}{\rho} \right)_g^{'} + p_g \left(\frac{1}{\rho} \right)_g^{'} - \mu_a \left(\frac{\rho_a}{\rho} \right)_g^{'} - \mu_v \left(\frac{\rho_v}{\rho} \right)_g^{'} \tag{34}$$

Where μ_e, μ_a, μ_v are the specific chemical potential of liquid water, air and water vapour .

Taking into account the following integral equations:

$$-T_2 \cdot \rho_e \cdot s_e + \rho_e \cdot u_e + p_e - \rho_e \cdot \mu_e = 0 \tag{35}$$

$$-T_2 \cdot \rho_g \cdot s_g + \rho_g \cdot u_g + p_g - \rho_a \cdot \mu_a - \rho_v \cdot \mu_v = 0 \tag{36}$$

and the internal energy balance (30) and (31), writing equations (32) ,(33) and (34) at one point, using (15), adding these equations and using equation (3), the entropy balance at one point is written:

$$\frac{\partial}{\partial t} \rho \cdot s = - \frac{1}{T_1} \cdot J_{q1,k}^k + \frac{\hat{u}_1}{T_1} - \frac{1}{T_2} \sum_{i=e,a,v} J_{qi,k}^k - \frac{\hat{u}_1}{T_2}$$

$$- (\rho_e \cdot s_e \cdot v_e^k + \rho_g \cdot s_g \cdot v_g^k)_{,k} - \frac{\Pi_e^{km}}{T_2} \cdot v_{e,m}^k - \frac{\Pi_{Ig}^{km}}{T_2} \cdot v_{g,m}^k$$

$$+ \frac{1}{T_2} (\lambda_e^k \cdot v_e^k + \lambda_g^k \cdot v_g^k) + \frac{J}{T_2} ((\mu_e - \mu_v) - \frac{1}{2} (v_g^2 - v_e^2 + (\frac{J_v^{'k}}{\rho_v})^2))$$

$$- J_{v,k}^{'k} (\frac{\mu_a - \mu_v}{T_2}) + \frac{1}{T_2} (-(J_v^{'k}(u_v - u_a)))_{,k} - P_v^{km} (\frac{J_v^{'k}}{\rho_v})_{,m}$$

$$+ P_a^{km} (\frac{J_v^{'k}}{\rho_a})_{,m} + J_v^{'k} (\frac{\lambda_v^k}{\rho_v} - \frac{\lambda_a^k}{\rho_a})) \tag{37}$$

Taking into account the following thermodynamics relations [2,p.26]:

$$T_2 (\frac{\mu_i}{T_2})_{,k} = ((\mu_i)_{T_2})_{,k} - \frac{h_i}{T_2} T_{2,k} \quad , i = a, v \tag{38}$$

where the index T_2 shows that the gradient must be evaluated at constant temperature and where h_i is the enthalpy of constituent i, given by:

$$h_i = u_i + \frac{P_i}{\rho_i} \quad ; i = a, v \tag{39}$$

the entropy production can be written:

$$S = - \frac{\Pi_e^{km}}{T_2} v_{e,m}^k - \frac{\Pi_{Ig}^{km}}{T_2} v_{g,m}^k + \frac{1}{T_2} (\lambda_e^k \cdot v_e^k + \lambda_g^k \cdot v_g^k) - J_{q1}^k \frac{T_{1,k}}{T_1^2}$$

$$- J_{q2}^k \frac{T_{2,k}}{T_2^2} + \frac{J_v^{'k}}{T_2} (((\mu_a - \mu_v)_{T_2})_{,k} + \gamma_a^k - \gamma_v^k + f_v^k - f_a^k)$$

$$+ \frac{J}{T_2} ((\mu_e - \mu_v) + \frac{1}{2} (v_e^2 - v_g^2) - \frac{1}{2} (\frac{J_v^{'k}}{\rho_v})^2) + \hat{u}_1 (\frac{1}{T_1} - \frac{1}{T_2}) \tag{40}$$

where:

$$J_{q2}^{'k} = \sum_{i=e,a,v} J_{qi}^k + J_v^{'m} ((\frac{P_v^{mk} - \delta^{mk} \cdot P_v}{\rho_v}) - (\frac{P_a^{mk} - \delta^{mk} \cdot P_a}{\rho_a})) \tag{41}$$

This source reveals the contribution of the various elementary phenomena to the production of entropy :

- dissipation of energy by friction inside the phases in motion;
- dissipation of energy by friction between phases;
- thermal conduction in the phases;
- diffusion of water vapour in the gas phase,
- change of phase of water;
- energy transfer between the solid phase and the other phases.

4. PRESSURES AND SPECIFIC CHEMICAL POTENTIALS IN A NON-SATURATED POROUS MEDIUM .

4.1. Phenomenological pressure and pore pressures

Phenomelogical pressures p_e and p_g, defined by (25) are not measurable. The meaningful quantities for experimentation are the pressures in the pores of phases and constituents :

p_e^* : water pressure in the pores

p_g^* : gas phase pressure in the pores

p_a^* and p_v^* : partial pressure of air and water vapour in the pores,

where:
$$p_g^* = p_a^* + p_v^* \tag{45}$$

Taking into account the perfect gas assumption :

$$p_a^* = \frac{RT}{n_1 . M_a} \cdot \rho_a \tag{46}$$

$$p_v^* = \frac{RT}{n_1 M_e} \cdot \rho_v \tag{47}$$

Where M_a is the fictitious molar mass of air, M_e the molar mass of water, R the perfect gas constant and n_1 is the volume taken by the gas phase per unit volume of porous medium :

$$n_1 = \frac{\rho_g}{\rho_g^*} = (1 - \frac{\rho_1}{\rho_s} - \frac{\rho_e}{\rho_w}) \tag{48}$$

ρ_g^* is the specific mass of the gas phase in the pores, ρ_s is the specific mass of the solid subtance and ρ_w is the specific mass of water.

The pressures p_e^* and p_g^* are related to the capillary pressure p_c and to the capillary suction ψ, expressed in water column, by the equations [3, p. 57]

$$p_c = p_g^* - p_e^* \tag{49}$$

$$\psi = - \frac{p_c}{\rho_w . g} \tag{50}$$

where g is the acceleration due to gravity.

4.2. Specific chemical potentials

Assuming the gas phase to behave as an ideal mixture of perfect gases, the specific chemical potentials of air and water vapour can be written [2, p.204] :

$$\mu_v = (RT_2 \ln p_v^* + \eta_v(T_2)) / M_e \tag{51}$$

$$\mu_a = (RT_2 \ln p_a^* + \eta_a(T_2)) / M_a \tag{52}$$

where $\eta_v(T_2)$ and $\eta_a(T_2)$ depend only on temperature T_2.

The following equations, used below, can be deduced :

$$((\mu_a - \mu_v)_{T_2})_{,k} = RT_2 \left(\frac{1}{M_a} \ln p_a^* - \frac{1}{M_e} \ln p_v^*\right)_{,k} \tag{53}$$

If, in addition, the total pressure of the gas phase is uniform :

$$((\mu_a - \mu_v)_{T_2})_{,\check{k}}^{p_g^*} = - \frac{RT_2 \, M_g}{p_v^* \, M_a M_e} \frac{p_g^*}{(p_g^* - p_v^*)} \, p_{v,k}^* \tag{54}$$

where M_g is the fictitious molar mass of the gas phase :

$$M_g = M_a \left(1 - \frac{p_v^*}{p_g^*}\right) + M_e \frac{p_v^*}{p_g^*} \tag{55}$$

The chemical potential of liquid water in a "normal" state (i.e. where the effect of the surface phases can be ignored) can be written [13, p.122] :

$$\mu_e^o = \frac{\mu_e(T_2)}{M_e} + \frac{p_e^*}{\rho_w} \tag{56}$$

where $\mu_e(T_2)$ depends only on temperature.

At very low water contents, the thermodynamic properties of the liquid water in the porous medium differ from the properties of "normal" water; this results in a rise in the equilibrium pressure of the water vapour in the porous medium (p_{vs}) in relation to the equilibrium pressure of water vapour in the presence of "normal" water (p_{vs}^o), and a deviation in the chemical potential of the water in the porous medium in relation to the chemical potential of water in a "normal" state :

$$\Delta\mu_e = \mu_e^o - \mu_e \tag{57}$$

$\Delta \mu_e$ can be evaluated at constant temperature.

The equilibrium of water vapour with "normal" water leads to [13,p.35] :

$$(\mu_e^o)_{T_2} = (\mu_v(p_v^* = p_{vs}^o))_{T_2} = (RT_2 \ln p_{vs}^o + \eta(T_2))/M_e \tag{58}$$

The equilibrium of water vapour with the water of the porous medium leads to :

$$(\mu_e)_{T_2} = (\mu_v(p_v^* = p_{vs}))_{T_2} = (RT_2 . \ln p_{vs} + \eta(T_2))/M_e \tag{59}$$

i.e. : $(\Delta\mu_e)_{T_2} = \dfrac{RT_2}{M_e} \ln \dfrac{p^o_{vs}}{p_{vs}}$ $\qquad$ (60)

which leads to, using (56),(57) and (60) :

$(d\mu_e)_{T_2} = \dfrac{dp^*_e}{\rho_w} + \dfrac{RT_2}{M_e} \ln \dfrac{p_{vs}{}^o}{p_{vs}}$ $\qquad$ (61)

4.3. Equations between phenomenological pressures (p_e and p_g) and phase pressures in the pores (p^*_e and p^*_g)

For the liquid phase, the GIBBS-DUHEM equation [13, p. 28] is written :

$dp_e = \qquad \rho_e d\mu_e + \rho_e s_e dT_2 = \rho_e (d\mu_e)_{T_2}$ $\qquad$ (62)

according to (61) :

$dp_e = \dfrac{\rho_e}{\rho_w} dp^*_e + \dfrac{\rho_e}{M_e} RT_2\, d(\ln \dfrac{p_{vs}}{p_{vs}{}^o})$ $\qquad$ (63)

For the gas phase, the GIBBS-DUHEM equation is written :

$dp_g = \rho_a d\mu_a + \rho_v d\mu_v + \rho_g s_g dT_2 = \rho_a (d\mu_a)_{T_2} + \rho_v (d\mu_a)_{T_2}$ $\qquad$ (64)

from (51), (52), (46) and (47) :

$dp_g = n_1\, dp^*_g = \dfrac{\rho_g}{\rho^*_g}\, dp^*_g$ $\qquad$ (65)

This result can be interpreted in a simple physical manner, considering a plane section of the porous medium of unit surface. The ratio $\dfrac{\rho_g}{\rho^*_g}$ represents the fraction of this unit surface contained in the gas phase : p_g appears as the internal force in the gas phase per unit of geometrical surface.

When $p_{vs} = p^o_{vs}$, according to (63) p_e can be interpreted logically in the same way as p_g. The second term of the right-hand side of (63) represents the contribution of the surface phases to the internal force of the liquid phase when $p_{vs} < p^o_{vs}$.

5. PHENOMELOGICAL EQUATIONS

5.1. General case

As the phases occupy disjointed portions of space, elementary phenomena occurring in a given phase are independant of thermodynamic forces present in the neighbouring phases.

Taking this remark into account and using the method of the linear thermodynamics of irreversible processes, the CURIE principle [2,p.31] the ONSAGER relations [2,p.35] and assuming the medium to be isotropic; the phenomenological equations can be written :

$$\Pi_e^{km} = -\eta_e \cdot v_{e,k}^k \cdot \delta^{km} - 2 \nu_e \cdot v_{e,m}^k \tag{66}$$

$$\Pi_{Ig}^{km} = - \eta_g \cdot v_{g,k}^k \cdot \delta^{km} - 2 \nu_g \cdot v_{g,m}^k \tag{67}$$

where η_i and ν_i are the LAME coefficients of phase i.

$$J_{q1}^k = \frac{L_{11}}{T_1^2} \cdot T_{1,k} \tag{68}$$

$$\begin{pmatrix} J_{q2}^{'k} \\ \lambda_e^k \\ \lambda_g^k \\ J_v^{'k} \end{pmatrix} = \begin{pmatrix} L_{22} & L_{2e} & L_{2g} & L_{2d} \\ L_{2e} & L_{ee} & L_{eg} & 0 \\ L_{2g} & L_{eg} & L_{gg} & L_{gd} \\ L_{2d} & 0 & L_{gd} & L_{dd} \end{pmatrix} \times \begin{pmatrix} -\dfrac{T_{2,k}}{T_2^2} \\ \dfrac{v_e^k}{T_2} \\ \dfrac{v_g^k}{T_2} \\ F_D \end{pmatrix}$$

(69)
(70)
(71)

where L_{ij} are the phenomenological coefficients and F_D is the thermodynamic force of diffusion of the water vapour.

$$F_D = \frac{1}{T_2} ((\mu_a - \mu_v)_{T_2} ,_k + \gamma_a^k - \gamma_v^k + f_v^k - f_a^k) \tag{72}$$

The change of phase occurs at the liquid-gas separation surface; it is therefore not coupled with the energy transfer occuring at the solid-liquid separation surface :

$$J = L_{rr} \frac{1}{T_2} ((\mu_e - \mu_v) + \frac{1}{2} (v_e^{k^2} - v_g^{k^2} - (\frac{J_v^{'k}}{\rho_v})^2)) \tag{73}$$

$$\hat{u}_1 = L_{qq} (\frac{1}{T_1} - \frac{1}{T_2}) \tag{74}$$

In the absence of transfer of matter between the solid phase and the other phases, the supply of momentum λ_1^k is a pure friction force which inverts itself when the velocities of the mobile phases invert themselves. According to (22) and (24) :

$$\lambda_e^k + \lambda_g^k = J (v_e^k - v_g^k) - \lambda_1^k \tag{76}$$

displaying the property that $\lambda_e^k + \lambda_g^k$ inverts itself with the inversion of v_e^k and v_g^k .

Adding (70) and (71) gives:

$$\lambda_e^k + \lambda_g^k = - (L_{2e} + L_{2g}) \frac{T_{2,k}}{T_2^2} + (L_{ee} + L_{eg}) \frac{v_e^k}{T_2}$$

$$+ (L_{eg} + L_{gg}) \frac{v_g^k}{T_2} + \frac{L_{gd}}{T_2} F_D \tag{77}$$

Inversion of velocities v_e^k and v_g^k does not necessarily imply that $T_{2,k}$ and F_D invert themselves . Consequently:

$$L_{2e} + L_{2g} = 0 \tag{78}$$

$$L_{gd} = 0 \tag{79}$$

The vectorial phenomenological equations are written thus:

$$J_{q2}'^k = - \frac{L_{22}}{T_2^2} T_{2,k} + L_{2e} (v_e^k - v_g^k) + L_{2d} \frac{F_D}{T_2} \tag{80}$$

$$\lambda_e^k = - \frac{L_{2e}}{T_2^2} T_{2,k} + \frac{L_{ee}}{T_2} v_e^k + \frac{L_{eg}}{T_2} v_g^k \tag{81}$$

$$\lambda_g^k = + \frac{L_{2e}}{T_2^2} T_{2,k} + \frac{L_{eg}}{T_2} v_e^k + \frac{L_{gg}}{T_2} v_g^k \tag{82}$$

$$J_v'^k = - \frac{L_{2d}}{T_2^2} T_{2,k} + L_{dd} \cdot F_D \tag{83}$$

5.2. Simplification for the usual applications in porous media.

In a number of applications:

a) the velocities and accelerations are low;

b) liquid phase and gas phase can be considered as perfect fluids [14,p.125] :

$$P_i^{km} = P_i \cdot \delta^{km} \quad ; \quad i=e,a,v,g \tag{84}$$

c) the only extrinsic loads are forces due to gravity:

$$f_i^k = g^k \quad ; \quad i=1,e,a,v,g \tag{85}$$

The momentum balance equations (14) and (18) are then written:

$$-P_{i,k} + \rho_i \cdot g^k - \lambda_i^k = 0 \quad ; \quad i= e,g \tag{86}$$

The flux $J_{q2}'^k$ defined by (41) becomes a pure conduction heat flux:

$$J_{q2}'^k = J_{q2}^k = \sum_{i= e,a,v} J_{qi}^k \tag{87}$$

According to (29),(31) and (84) , the internal energy balance for the liquid and gas phase is, after the above simplifications , written as follows:

$$\sum_{i=e,a,v} \frac{\partial}{\partial t} \rho_i \cdot u_i = - \sum_{i=e,a,v} ((\rho_i \cdot u_i + p_i)v_i^k)_{,k} - J_{q2,k}^k - \hat{u}_1 \qquad \text{(87 bis)}$$

It is justified to neglect the velocities and accelerations compared to specific potential chemical differences . The phenomenological equations are therefore written as follows using (53),(84),(85), and (86):

$$J_{q1}^k = - \frac{L_{11}}{T_1^2} T_{1,k} \qquad (88)$$

$$J_{q2}^k = - \frac{L_{22}}{T_2^2} T_{2,k} + L_{2d} \cdot R \left(\frac{1}{M_a} \ln p_a^* - \frac{1}{M_e} \ln p_v^* \right)_{,k} + \frac{L_{2e}}{T_2}(v_e^k - v_g^k) \qquad (89)$$

$$J_v^{'k} = - \frac{L_{2d}}{T_2^2} T_{2,k} + L_{dd} \cdot R \left(\frac{1}{M_a} \ln p_a^* - \frac{1}{M_e} \ln p_v^* \right)_{,k} \qquad (90)$$

$$v_e^k = - \frac{T_2}{L_{ee}} (P_{e,k} - \rho_e \cdot g^k) - \frac{L_{eg}}{L_{ee}} v_g^k + \frac{L_{2e}}{L_{ee}} \frac{T_{2,k}}{T_2} \qquad (91)$$

$$v_g^k = - \frac{T_2}{L_{gg}} (P_{g,k} - \rho_g \cdot g^k) - \frac{L_{eg}}{L_{gg}} v_e^k - \frac{L_{2e}}{L_{gg}} \frac{T_{2,k}}{T_2} \qquad (92)$$

$$J = \frac{L_{rr}}{T_2} (\mu_e - \mu_v) \qquad (93)$$

$$\hat{u} = L_{qq} \left(\frac{1}{T_1} - \frac{1}{T_2} \right) \qquad (94)$$

Taking (63) and (65) into account , the relations (91) and (92) can be written:

$$v_e^k = - \frac{T_2 \cdot \rho_e}{L_{ee} \cdot \rho_w} (P_{e,k}^* - \rho_w \cdot g^k) - \frac{T_2 \cdot \rho_e \cdot R.T_2}{L_{ee} \cdot M_e}(\ln \frac{P_{vs}}{p_{vs}^\circ})_{,k}$$

$$\qquad - \frac{L_{eg}}{L_{ee}} v_g^k + \frac{L_{2e}}{L_{ee}} \cdot \frac{T_{2,k}}{T_2} \qquad (95)$$

$$v_g^k = - \frac{T_2 \cdot \rho_g}{L_{gg} \cdot \rho_g^*} (P_{g,k}^* - \rho_g^* \cdot g^k) - \frac{L_{eg}}{L_{gg}} v_e^k - \frac{L_{2e}}{L_{gg}} \frac{T_{2,k}}{T_2} \qquad (96)$$

It can be shown [15] that the phenomenological equation of phase change can be stated as follows:

$$J = \frac{\partial}{\partial t} \rho_{vs} + (\rho_{vs} \cdot v_{vs}^k)_{,k} - L_{rr} \frac{R}{M_e} \ln \frac{P_v^*}{P_{vs}} \qquad (97)$$

5.3. Comparison between the phenomenological equations and known experimental results

The validation of phenomenological equations presented in this study resides in their agreement with known experimental laws. As there are practically no experimental result in the case of high velocities and accelerations, comparison can only be made in the important case of low velocities and accelerations (as given in chapter 5.2).

5.3.1 Phenomenological equations of diffusion and thermal conduction

The experimental results concerning diffusion of water vapour essentially concern the case where the total pressure of the gas phase is constant and uniform. In such a case, according to (54), the phenomological equation of diffusion is written :

$$J'^k_v = - \frac{L_{2d}}{T_2^2} T_{2,k} - \frac{L_{dd} \cdot R \cdot M_g}{p_v^* \cdot M_a \cdot M_e} \frac{p_g^*}{p_g^* - p_v^*} p_{v,k}^* \tag{98}$$

The structure of this equation is in complete agreement with known results [16, p. 451] .

Experimental results [17, p.63] show that the first term of (98) is generally negligible when compared to the second term even in the case of high temperature gradients. This implies that the coefficient L_{2d} has to be low when compared to L_{dd}. Combining (89) and (90), the phenomenological equation of conduction in the mixture liquid + gas can be written :

$$J^k_{q2} = (- L_{22} + \frac{L_{2d}^2}{L_{dd}}) \frac{T_{2,k}}{T_2^2} + \frac{L_{2d}}{L_{dd}} J'^k_v + \frac{L_{2e}}{T_2} (v^k_e - v^k_g) \tag{99}$$

In the case of low velocities, this equation is reduced to the first term and fits FOURIER's law.

5.3.2 Phenomenological equation of filtration of phases

When the first two terms of (95) and (96) are considered, it can be seen that they are in complete agreement with generalised DARCY's law [3, p. 248]. This is an important result of this study showing that this law can be deduced from the entropy supply analysis using three GIBBS-DUHEM equations . As far as we know, no experimental result exists on the coupling terms of (95) and (96).

5.3.3 Phenomenological equations of phase change and energy transfer between the solid phase and other phases

An experimental study [15] enabled us to verify that the structure of the phase change equation (97) is correct. In addition, this study

made it possible to measure coefficient L_{rr} and showed that this coefficient
is highly dependent on the water content of porous medium.

Finally, equation (94) conforms to the experimental law governing
heat transfer between bodies at different temperatures [16, p. 470].

6. CONCLUSION

This study shows that the methods of thermodynamics of linear irreversible
processes make it possible to build a model of non-saturated porous media
which agrees with known experimental results. The keystone of such a
model is the writing of the GIBBS equation for each phase according to
its own motion. It must be stressed that this procedure leads to phenomenological relations of firstly filtration of liquid phase and, secondly,
filtration of gas phase, obeying generalised DARCY's law in the case of
low velocities and accelerations.

Conformity of the phenomenological equations drawn from the entropy
source with known experimental laws justifies extension of the T. I. P. model to
new elementary phenomena in soils : changes of phase, chemical reactions,
diffusion of constituents. The essential advantage of this approach is
the separation of various elementary phenomena and the obtention of the
structure of the corresponding phenomenological equations. This structure
is a guide for experiments and, as has been done with phase change [15],
enables measurement of the phenomenological coefficients, avoiding any
empirical attempt insofar as the structure of the law is concerned.

In the case of low velocities and accelerations, the mathematical
model made up of :
a) the mass balance equations (9),(10) and (11) ;
b) the internal energy balance equations (31) and (87 bis);
c) the phenomenological equations (88), (89), (90), (94), (95), (96)
and (97)
makes possible application in a number of varied fields such as heat storage
in soil, high enthalpy geothermal projects, agronomy, etc...

This model is flexible and can be simplified for a given application,
or can be extended for new applications by the introduction of other elementary phenomena such as phase change of the solid, chemical reactions,
diffusion of constituents...

REFERENCES

1. Prigogine I. 1968 . Introduction à la thermodynamique des phénomènes irréversibles.Paris,Dunod.
2. De Groot S.R.,Mazur P. 1969. Non-equilibrium thermodynamics . Amsterdam, North-Holland Publishing company.
3. Scheidegger A.E. 1974. The physics of flow through porous media. University of Toronto press . Third edition.
4. Baver L.D.,Gardner W.H.,Gardner.W.R.1972.Soil physics.New York,John Wiley & sons . Fourth edition.
5. Taylor S.A., Cary J.W. 1964. Linear equations for the simultaneous flow of matter and energy in a continuous soil system .Soil Sci. Soc. Am. Proc., vol. 28,pp. 167-172.
6. Groenevelt P.H.,Bolt G.H. 1969. Non-equilibrium thermodynamics of the soil-water system .J. of hydrology,n°7,pp.358-388.
7. Baijal S.K,Sriramulu A.N. 1975. Some aspects of non-equilibrium thermodynamics in flow through porous media.Indian J. of technology ,vol.13,pp.480-483
8. Marle C. 1965. Application de la thermodynamique des processus irréversibles a l'écoulement d'un fluide à travers un milieu poreux. Bulletin RILEM, nouvelle serie n°29,pp.107-117.
9. Guelin p. 1970.Thermodynamique des filtrations polyphasiques .Thèse de Doc. és Sciences Physiques.Faculté des Sciences de l'Université de Grenoble.
10.Mazur P.,Prigogine I. 1951. Sur l'hydrodynamique des mélanges liquides He^3 et He^4. Physica XVII , n°7,pp.680-693.
11.Prigogine I.,Mazur P. 1951 . Sur deux formulations de l'hydrodynamique et le problème de l'hélium liquide II . Physica XVII,n°7,pp.661-679.
12.Truesdell C., Toupin C. 1960. The classical field theories . Handbuch der physics , band III/1 .Berlin Springer-Verlag.
13.Guggenheim E.A. 1965. Thermodynamique. Paris,Dunod.
14.Germain P. 1973. Mecanique des milieux continus.Paris,Masson et Cie.
15.Benet J.C. , Jouanna P. 1980. Relation phénoménologique de changement de phase de l'eau dans un milieu poreux.Vérification expérimentale et mesure du coefficient phénomenologique .7è Symposium international de l'A.I.R.H. Toulouse,France,ENSEEIHT.
16.Eckert E.R.G. 1959. Heat and mass transfer , Series in mechanical engineering New York, Mc Graw Hill.
17.De Vries D.A., Kruger A.J. 1967. On the value of the diffusion coefficient of water vapour in air. Colloques internationaux du CNRS n° 160. Paris, édition du CNRS.

SEASONAL STORAGE IN HARD ROCK - MULTIPLE WELL SYSTEM

Sören Andersson, AIB, Stockholm
Anders Eriksson, AIB, Stockholm

Introduction

Heat can be stored in hard rock. The heat is transferred to
and from the rock by water. The water circulates in a great
number of boreholes. The system can be can be considered as
a huge heat exchanger, fig. 1.

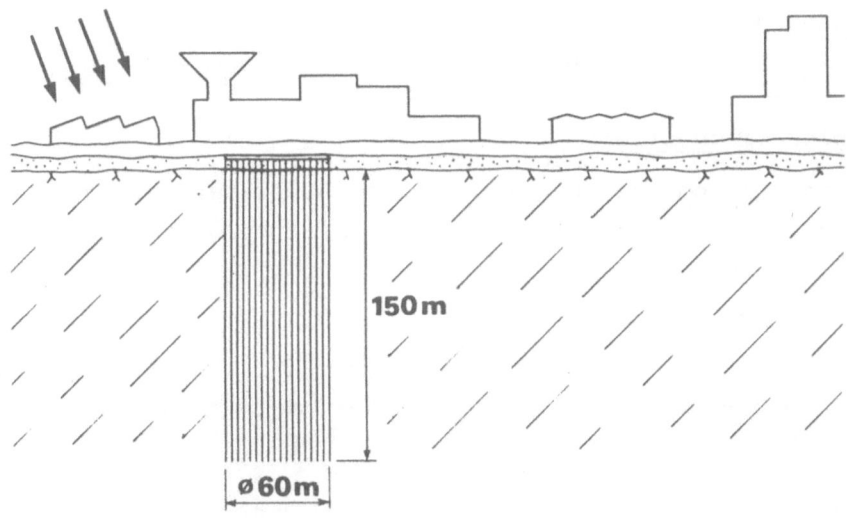

Fig. 1 Heat storage in hard rock. Multiple
well system

The storage is constructed principally without any artificial
thermal insulation. The volume of the storage must be large
to enable acceptable heat recovery and temperature recovery
factors.

A minimum size of the storage volume will probably be
200 000 m³. This volume corresponds to a storage capacity
of about 4 GWh. A working temperature difference of 40°C
is assumed.

The storage probably must, due to economic reasons, be performed with unlined boreholes. Depending on charging and discharging strategies, a certain leakage between the boreholes
is harmless or even positive. Leakage out from the storage
must be prevented or minimized by means of hydraulic control
or grouting.

There is a good availibility in Sweden of hard bedrock,
suitable for construction of multiple well storage systems.
Normally, the bedrock is covered with earth layers, mostly
less than 10-15 m thick.

Principles of function

The storage is charged by circulating warm water through the
boreholes. The circulation system can be open or closed,
fig. 2a and 2b.

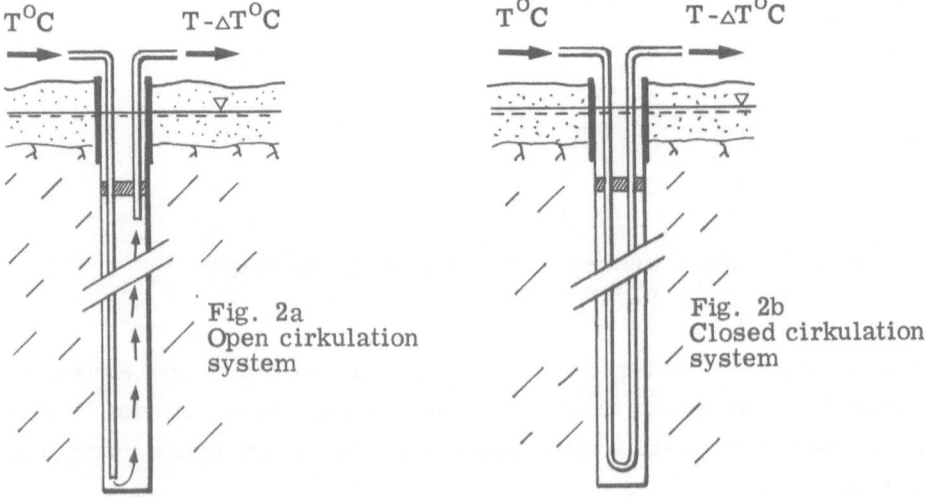

Fig. 2a
Open cirkulation
system

Fig. 2b
Closed cirkulation
system

A closed system according to fig. 2b implies an efficient
hydraulic control. In open circulation systems, fig. 2a,
there will normally be a risk for leakage caused by the over-
pressure in the boreholes. Open circulation systems also imply
a risk for chemical precipitation on exchanger surfaces etc.

Different strategies can be applied for the charging and discharging operations. All parts of the storage volume can be
charged or discharged at the same time, i.e. all the wells
are throughflowed by similar flow at the same temperature
level. A more efficient strategy implies that the storage is
charged with beginning in the centre and then outwards in a
radial direction. Discharging will then be done by a reversed
operation.

Different temperature zones can be formed within the storage
by connecting the boreholes by groups, fig. 3.

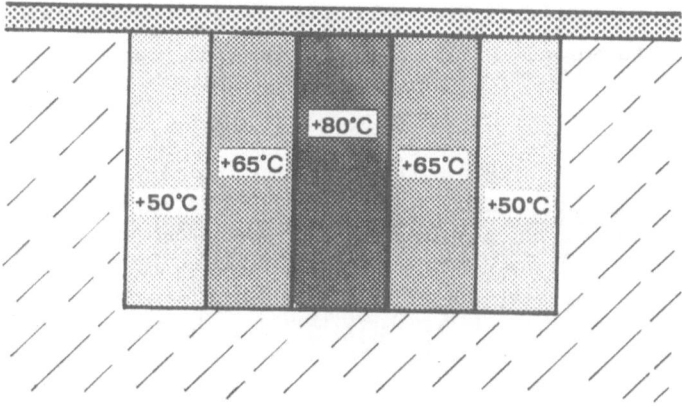

Fig. 3   Temperature zones in a cylindrical multiple
         well storage. Principle sketch

252

The function of the storage is based on heat conduction through
the rock. Heat conduction is a strongly time-dependent process.
Hence, the multiple well storage system must be considered
slow compared with for example insulated water reservoirs. The
peak load capacity of the storage can be improved by an
arrangement of boreholes with small interspaces.

Most of the operation strategies will probably imply the use
use of heat pumps but this question is not yet studied in
detail.

Thermal behaviour

The thermal process in a multiple well storage and in the
surrounding rock is complicated. There is a multidimensional,
dynamical temperature field, which is governed by the injection
and extraction of heat by the water circulating in the well
system. In addition to the annual cycles, there is a transient
thermal built-up during the first cycles.

The thermal behaviour of a multiple well storage has been
studied by Claesson, J [1] for a reference case according to
fig. 4.

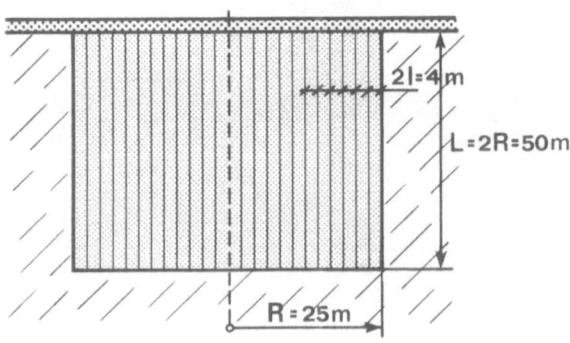

Fig. 4   Reference case. Cylindrical multiple well
          storage in granite

r the reference case the following operational data are
osen.

orage_cycle: injection (3 months) + rest (3 months) +
            + extraction (3 months) + rest (3 months)
jection_and_extraction_rate: Q_o = 1,0 MW
x._injection_temperature: T_i = 90°C
n._extraction_temperature: T_e = 50°C

e fluid (water) temperature T_f is not allowed to exceed
e value T_i. If T_f during injection reaches T_i, the injection
ll continue, not with the constant rate Q_o, but with T_f = T_i.
ere is a similar limit for extraction. When T_f drops to T_e,
e extraction continues with T_f = T_e as long as this temper-
ure is maintained. The behaviour of the storage during the
rst three years is illustrated in fig. 5. Heat has been
jected at the doubled rate, 2 MW, during the first injection
riod.

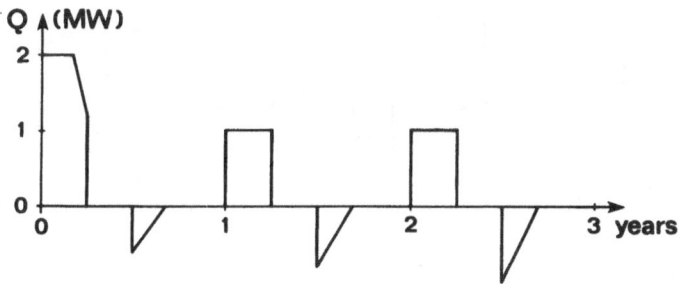

Fig. 5
Thermal behaviour
during the first
three years for
the reference case

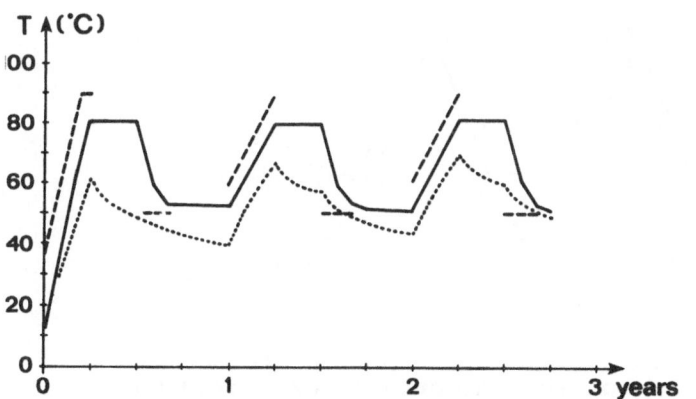

――――― Temperature in the centre of the storage
- - - - -   "   of the injected or extracted water
. . . . .   "   on the surface of the storage at a depth L/2

The extraction period starts after six months. Heat will be
extracted at a constant rate, 1 MW, until the water temperature
has fallen to 50°C.

The injected and extracted amounts of energy are proportional
to the areas given by the Q-curve (upper diagram, fig. 5).
The heat recovery, i.e. the quotient between the extracted and
injected energy, increases slowly with the number of cycles.

A variant of the reference case is shown in fig. 6. The minimum
extraction temperature is now decreased from 50°C to 20°C.

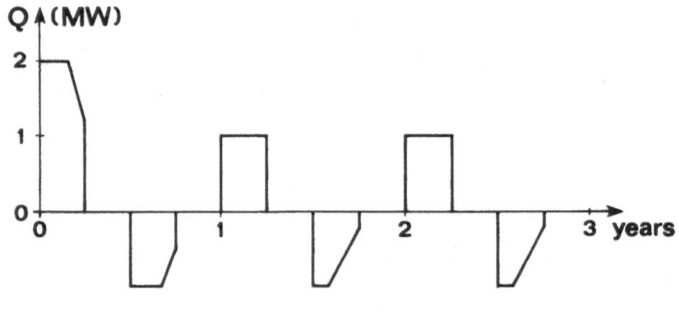

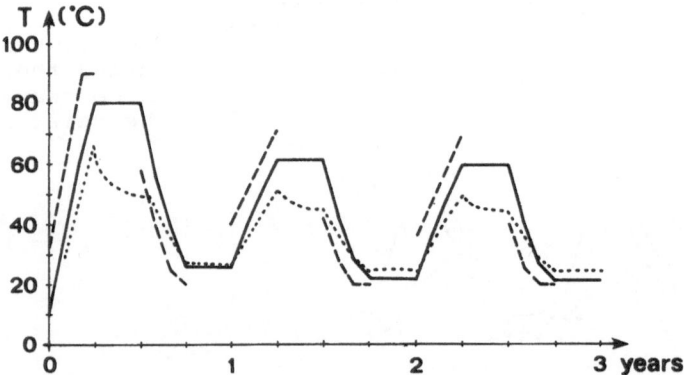

Fig. 6
Thermal behaviour
during the first
three years for
the reference case
when the minimum
extraction temper-
ature is diminishe
to 20°C

——————— Temperature in the centre of the storage
- - - - - - " of the injected or extracted water
· · · · · · · " on the surface of the storage at a depth L/2

Parametric studies

The influence of different storage parameters has been simulated by Claesson, J [1]. The reference case, mentioned before, has been used as a base for studies with a number of varied parameters.

The parametric studies have been related to the storage performance in terms of the heat recovery factor and the temperature recovery factor.

The heat recovery factor tells how much of the injected energy that is regained and the temperature recovery factor gives the quality of the regained heat, i.e. it compares the temperature of the extracted heat with that of the injected. The temperature recovery factor is defined by the use of the surrounding temperature T_O = 10°C.

In fig. 7, 8 and 10 the recovery factors are shown as a function of the following storage parameters

- size of the storage
- thermal conductivity of the rock
- distance between the wells
- lowest acceptable extraction temperature.

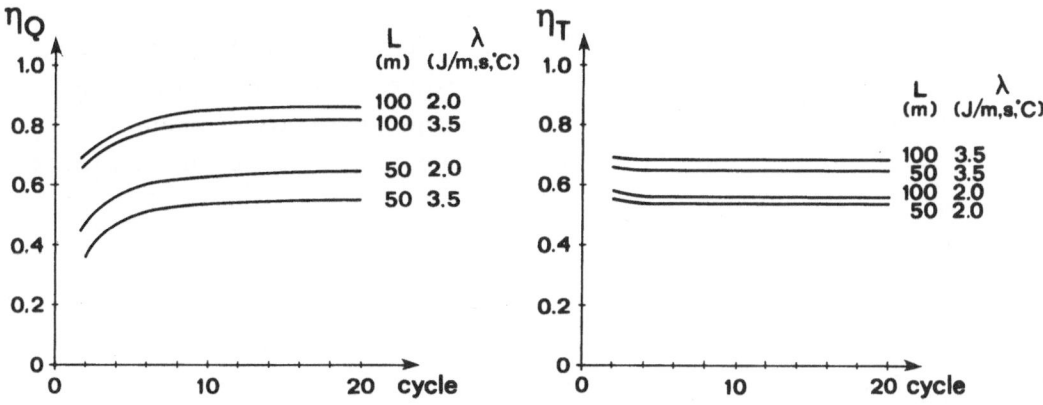

Fig. 7   Heat recovery and temperature recovery factors as functions of the storage size L and heat conductivity λ of the rock. Remaining data according to the reference case fig. 4. T_e = 50°C

For the size L = 100 m, i.e. a volume eight times the storage
volume in the reference case, the injection rate Q_0 is increased
to 8 MW during the first injection period.

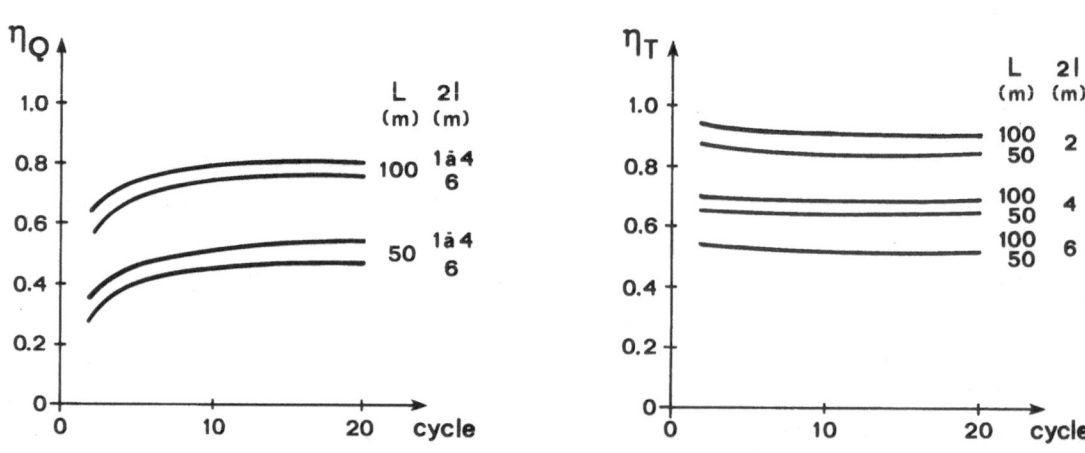

Fig. 8   Heat recovery and temperature recovery factors
          as functions of the storage size L and the
          distance between the wells, 2 l. Remaining data
          according to the reference case.

The distance between the wells is a very important parameter
as the construction cost of the storage very much depends on
the number of wells. However, also the specific storage
capacity depends on the well distance as can be seen in fig. 9.

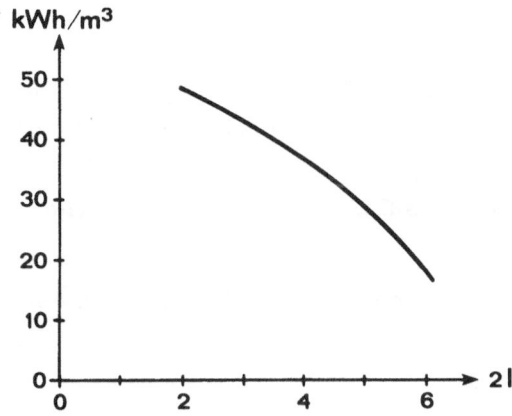

Fig. 9   Storage capacity as a function of the distance 2 l
          between the wells. Storage cycle: injection (5 months)
          + rest (2 months) + extraction (3 months) + rest
          (2 months)

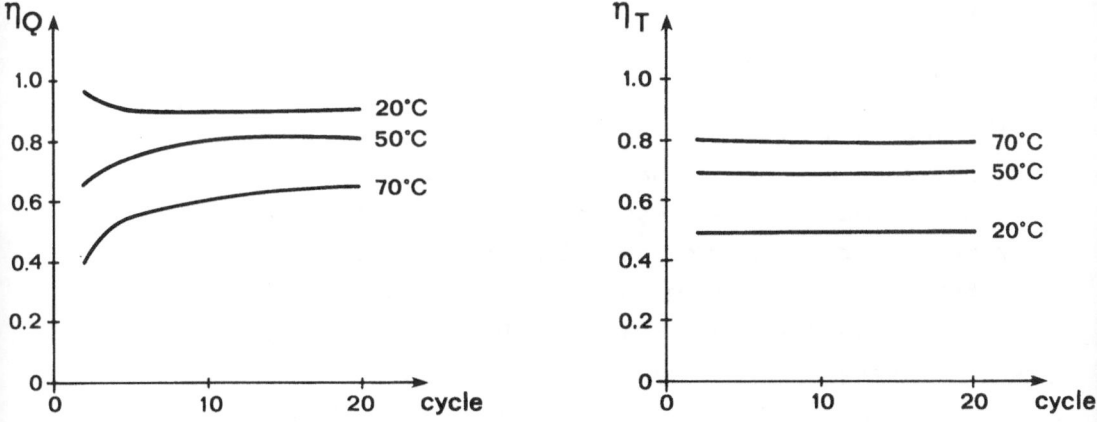

Fig. 10   Heat recovery and temperature recovery factors as
          functions of the lowest acceptable extraction
          temperature T_e. L = 100 m, Q_O = 8 MW. Remaining
          data according to the reference case.

Other parameters influencing the construction cost and the efficiency of the storage are
- the shape of the storage
- the heat capacity of the rock
- the duration of the different periods of the storage cycle
- the thermal insulation, if any, on the top of the storage.

Parameter studies have been made also by Johansson, B, and Nordell, B, [2].

Cost estimate

Preliminary construction cost estimates of a multiple well heat storage have been made by Johansson, B, and Nordell, B, [2]. The estimates, which include the costs for pipe installations, heat pumps, etc, indicate that the multiple well storage is competitive with other types of seasonal heat storages, for example rock caverns.

Further development

Further development of the storage concept, field experiments and development of mathematical models for different operating strategies are performed in collaboration between AIB, Consulting Engineers, Stockholm, the Department of Mathematical Physics, Lund University, and the Department of Hydraulics, Luleå University.

References

[1] Claesson, J, Johansson, M, 1979. Continous Heat Source Model for Ground Heat Storage, Lund University, Department of Mathematical Physics.
[2] Johansson, B, Nordell, B. Berglager - en anläggning för säsonglagring av värme. TULEA 1980:14, Luleå University, April 1980.

CHEMICAL STORAGE
OF THERMAL ENERGY

SUMMARY AND OVERVIEW

Prof. B.J. Brinkworth
University College Cardiff, United Kingdom

In this chapter five papers are presented, two on energy storage in solid absorbing materials, two on chemical heat pumps and the thermodynamical reactions involved and one on solid-solid phase transitions.

Compared to previous chapters - in which the majority of papers discussed engineering problems - this chapter offers more science than technology. The work on chemical storage of thermal energy is still under the hands of scientists and technologists are just beginning to make their contributions.

In his paper Bougard discusses three classes of chemical reactions, dealing firstly with reactions between a solid phase and a gaseous phase, indicating that among these reactions the water adsorption on silicagel will probably play an important part for energy storage below 100 $^{\circ}$C. Secondly he indicates reactions between condensed pahses formed by miscible liquids and the third class concerns reacting systems with three or more solid phases.

Verdonschot shows a thermal storage system based on the heat of adsorption of water in hygroscopic materials which he has studied as a component of an air-based solar space heating system with an integrated heat recovery system. In this system design study he examined the interaction between the storage device and the other components of the total solar installation.

Vialaron gives special attention to chemical heat pumps and provides an overview of French researchwork on chemical solar energy storage. He discusses some economical aspects of chemical storage, such as capital cost, costs of chemicals and sizing problems, and concludes that the best possibilities for chemical storage systems are to be found in systems that need energy storage for one or two weeks.

Fittipaldi discusses a solid-solid phase transition in organo-metallic chemical compounds. These materials often have to be stabilized in polymeric matrices and show, as far as the thermo-physical properties are concerned - remarkable similarities with salt hydrates (see chapter 3).

Raldow presents an advanced Swedish chemical heat pump project for seasonal storage of solar thermal energy. He describes the principle of adsorption storage and the materials used as working pairs, the heat sources and heat sinks utilized and discusses some possible applications.

In general the potential advantages of chemical storage systems are:
- high energy density,
- storage at ambient temperatures,
- possibilities for storage over an indefinitely long period,
- transportability of stored heat.

There are still obstacles to use chemical storage systems, such as:
- they are hardly available, except for a few laboratory prototypes,
- the complexity of the systems, compared with systems using sensible or latent storages,
- higher costs, although we have not enough information to fully justify that.

Some of the problem areas that need to be investigated more in depth are:
- A clear definition and classification of all promising available types of chemical reactions is needed.
- In the field of thermo-chemistry work has to be done on reaction kinetics.
- There is a tremendous shortage of information on thermo-physical properties of the majority of reactants under consideration. Properties of interest are specific heat capacities, thermal conductivities and heat of reactions etc.
- More knowledge on stability of these materials and the reversibility of the reactions in question is needed. Do they undergo some kind of degradation, since these systems must have lifetimes of 15-25 years?
- Some of the processes involve catalysis and if so, there is the question of the regeneration of the catalyst, whether it has a sufficient lifetime etc.

- Many of the processes take place in containers; the various corrosion and mechanical problems in question have to be investigated.
- Some extra work on heat transfer and fluid mechanics has to be done.
- The system design has to be looked at, in particular there is the interesting question of the interaction between the storage device and the other components. We need to be able to calculate the pumping power loss, and we need good designs for keeping the pumping power low, particularly for air systems. Do we need to develop new collectors to take advantage of the special characteristics of the storage system? Do we need to adapt the heat distribution systems, so that they can operate in more favourable temperature ranges?
These are all questions to consider in parallel with the most important one:
- The question of economic evaluation of solar installations with these advanced heat storage systems. We need to do a performance and cost analysis for installed systems for various possible applications such as house heating, industrial process heating etc.

Concluding we can say that there is still a lot of work to be done before we can answer the above-mentioned questions about the applicability of chemical storage systems as a component of solar installations.

THERMOCHEMICAL ENERGY STORAGE AND CHEMICAL HEAT PUMPS

PROF. J. BOUGARD AND R. JADOT

FACULTÉ POLYTECHNIQUE DE MONS, MONS, BELGIUM.

1. INTRODUCTION

The intrinsic diurnal and seasonal variability of solar energy rises the problem of storing heat.

In temperate climate as this of western Europe, the main problem is to store energy during several months. For instance, figure 1 shows the distribution during the year of the contribution of the solar roof and of the thermal load of a mean belgian dwelling. The excess of solar heat stored in summer is enough to offset the deficit of winter solar heat if it is possible to store heat from October to March.

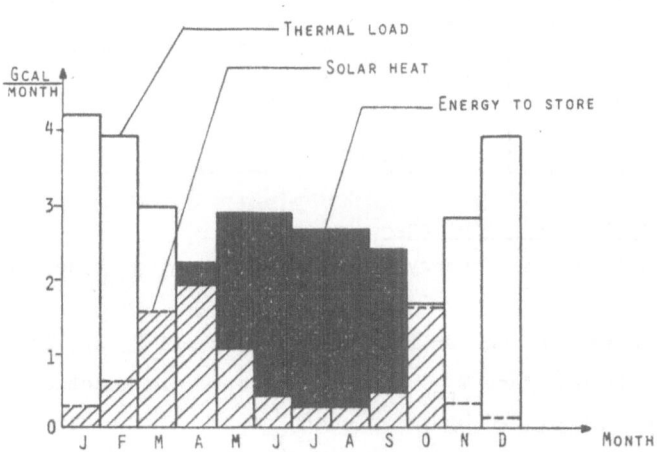

FIG. 1

The classical heat storage systems, in form of sensible or latent heat, are characterized by the fact that the amount of stored heat is an increasing function of the temperature at which the components are kept. Heat losses through imperfect insulation decreases the amount of stored energy and the thermodynamic quality of this one, so that, after several months, the available heat is only a small fraction of the stored energy.

In chemical systems, energy is stored in form of two or more chemical compounds able to react exothermically on request but kept at ambient temperature. Insulation becomes thus a minor problem. The life of the storage is, in principle, infinite.

The main differences between these three storage modes are summarized in table 1.

Table 1

Some properties of the different storage modes

Sensible heat	Heat of fusion	Chemical reaction

– Weak capacity
  Water ($\Delta T=20°C$) :
    0.01 Gcal/m^3
  Rock bed ($\Delta T=20°C$) :
    0.012 Gcal/m^3
– Restitution at
  variable temperature
– Necessary insulation
– Large energy losses
  in long term storage
– Easy working up
	– Weak capacity
	  Glauber's salt :
	    0.09 Gcal/m^3
	  Paraffins :
	    0.04 Gcal/m^3
	– Restitution at cons-
	  tant temperature
	– Necessary insulation
	– Energy losses in
	  long term storage
	– Fairly difficult
	  working up
		– Large capacity
		  0.2–0.5 Gcal/m^3
		
		– Restitution at cons-
		  tant or variable t°
		– No insulation
		– Low energy losses
		
		?

1.1. Principle of chemical storage

In chemical systems, energy is stored in chemical bonds during reversible reactions.

In the storing or regeneration mode, reactants R_i are converted in products P_i at temperature T_{reg} using an endothermic reaction , the heat of reaction being provided by solar energy.

The products are stored at ambient temperature. In order to prevent the natural thermodynamic tendency for the **exothermic** reaction, the products are kept separated or without catalysts.

The mixing of products or the introduction of catalysts makes possible the restitution of energy on request.

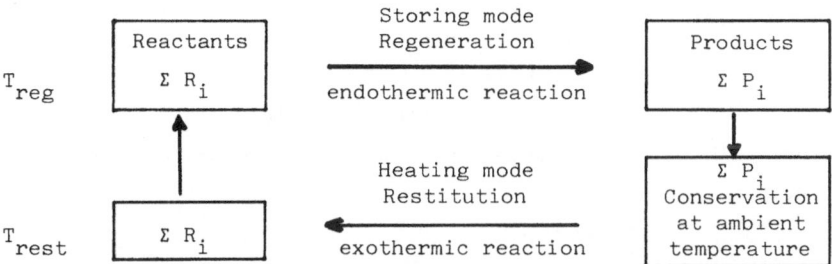

1.2. Thermodynamic conditions

In order to achieve the regeneration of products, affinity of the endothermic reaction must be positive at the regeneration temperature while it must be negative at the restitution temperature. The affinity of the reaction is given by :

$$a = -\Delta H + T\Delta S \qquad (1)$$

where ΔH and ΔS are the variations of enthalpy and entropy during the reaction. This relation gives the minimum regeneration temperature T^* at which $a = 0$

$$T^* = \frac{\Delta H}{\Delta S} \qquad (2)$$

This relation points out the limitation of the capacity of a chemical storage. For instance, using flat plate solar collectors, the maximum temperature obtained is below hundred degrees. The storage capacity is given by : $\Delta H = T^*\Delta S \qquad (3)$

Therefore, for a given regeneration temperature, the greater will be ΔS, the greater will be the storage capacity.

1.3. Classification

The affinity of a reaction is a function of many variables : nature of the components, temperature, pressure and composition of each phase.

However, in practical applications the number of variables really influencing the affinity value is smaller. The chemical reactions used for energy storage can be then classified on basis of this number of variables.

1.3.1. Reactions between immiscible condensed phases and a gazeous phase

The affinity is given by :

$$a = f(T, p, \text{composition of the gaseous phase}) \quad (4)$$

For a reaction such as :

$$AB(S \text{ or } L) \rightarrow A(S \text{ or } L) + B(g)$$

which occurs with a large value of the entropy change ΔS, the equilibrium constant $K(T)$ is given by :

$$K(T) = p \cdot y_B \quad\quad (5)$$

where y_B is the mole fraction of B in the gaseous phase. Heat evolved during the reaction is :

$$Q = \Delta H_r(n_B - n_B^o) \quad\quad (6)$$

where ΔH_r is the heat of reaction, n_B and n_B^o the numbers of mole of component B at equilibrium and in the initial conditions.

1.3.2. Reactions between miscible liquids or solids

Because of the small value of the molar volumes in the condensed phase, the affinity does not depend of the pressure :

$$a = f (T, \text{composition}) \quad (7)$$

For a reaction such as :

$$A(L) + B(L) \rightarrow C(L)$$

we have the following relations :

$$K(T) = \frac{x_C}{x_A x_B} \quad\quad (8) \text{ (ideal solution)}$$

(x = mole fraction in condensed phase)

$$Q = \Delta H_r (n_C - n_C^o) \quad\quad (9)$$

(with the same notations than in equation (6))

1.3.3. Reaction between immiscible solids

Example : $A + B \rightarrow C + D$

In this case, the affinity is only a function of the temperature :

$$a = f(T) \quad\quad (10)$$

The thermal behaviour of such a chemical system is the same that in a phase transition or a congruent melting. If T_{eq} is the temperature at which the affinity is zero, at $T < T_{eq}$ the reaction does not occurs while at $T > T_{eq}$ the reaction is fully performed.

2. CONDENSED PHASE – VAPOUR PHASE REACTIONS

This type of reaction has a very large entropy change and thus a large storage capacity per mass of the reactants. However, if the gaseous phase must be stored for reasons of cost or toxicity, the volume of the system will be large and the volumetric storage capacity will become very low, except if the pressure is high.

If the gaseous component is water vapour, it is then possible to use the water vapour in the atmospheric air in an open storage system.

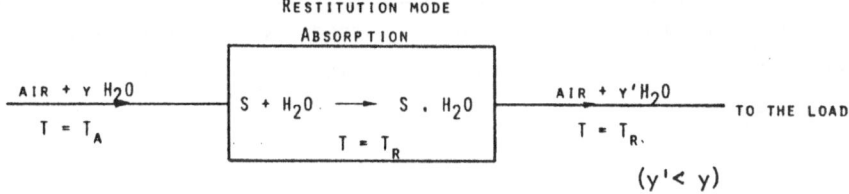

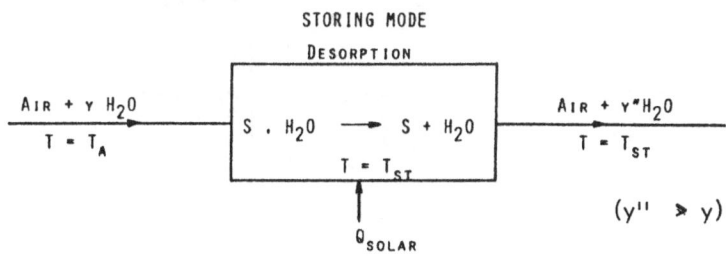

FIG. 2

Y = AMOUNT OF WATER IN AIR IN KG OF H_2O/KG AIR

2.1. Reaction between a solid phase and water vapour – open system

The chemical reactions available for such systems are :

2.1.1. Hydration of inorganic salts. The reaction is :

$$A + kH_2O \rightarrow A.kH_2O$$

(A is an inorganic salt)

Table 2 gives some examples. The volumetric storage capacity of these systems is interesting. But after some cycles, the solid is so altered (slurry formation) that the system becomes unserviceable. A possible solution is to support the active salt by an inert porous matrix leading to a decrease of the volumetric storage capacity.

| | Table 2 | |
Hydrate	T_{reg} (°C)	$Q(Gcal/m^3)$
$KAl(SO_4)_2 \cdot 12H_2O$	80–85	0.25
$Na_2B_4O_7 \cdot 10H_2O$	70–80	0.21
$Na_3PO_4 \cdot 12H_2O$	70–80	0.40
$MgCl_2 \cdot 6H_2O$	80–90	0.13

2.1.2. <u>Water adsorption on industrial adsorbents</u>. The most important water adsorbents used in industrial processes are activated alumina (A.A), molecular sieve (M.S) and silicagel (S). Their adsorption isotherms are shown on figure 3.

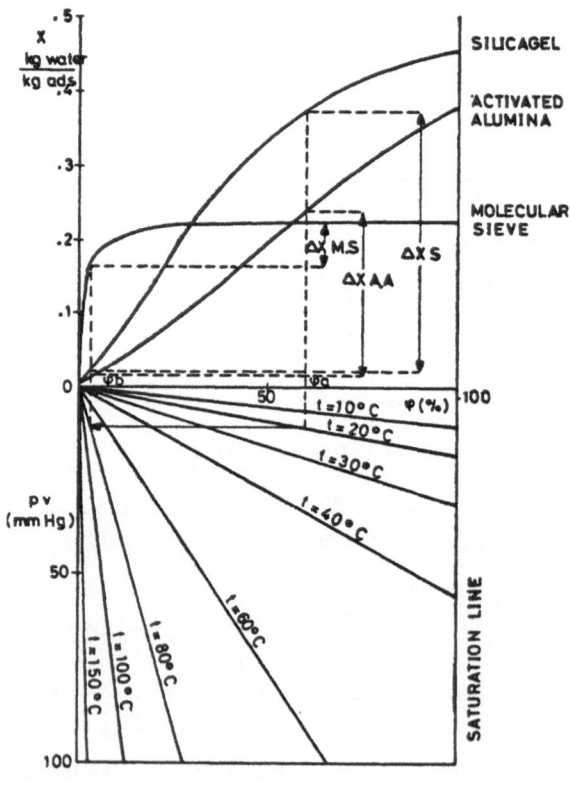

Fig. 3

The working principle is represented on figure 2. During the drying of
the adsorbent (storing time), the relative humidity of the inlet air
changes from about φ_a = 60% at the inlet temperature (Ta) to a much smaller
value φ_b at the regeneration temperature (3% at 80°C). For silicagel and
activated alumina, the amount of water remaining after regeneration at 80°C
(X_{reg}) in the adsorbent is very low, but for molecular sieve the X_{reg}
value remains still very high. Heat delivered by kg of adsorbent during
the restitution period is approximatively given by :

$$Q = \Delta H_{ads} \cdot (X_{rest} - X_{reg}) = \Delta H_{ads} \Delta X \qquad (11)$$

where ΔH_{ads} is the heat of adsorption per kg of water and X_{rest} the value
of X in the restitution conditions. When the regeneration takes place at
80°C, ΔX is very lower for molecular sieve than for activated alumina or
silicagel. Molecular sieve will become interesting only for regeneration
temperature greater than 150°C.

Table 3 gives the main storage properties for the three adsorbents.

Table 3

Adsorbent	T^*_{reg} (°C)	$T_{reg} = T^*_{reg}$ p_v = 10mmHg	$T_{reg} = 80°C$ p_v = 10mmHg
		Q(Kcal/kg of adsorbent)	Q(Kcal/kg of adsorbent)
Molecular sieve	220	240	60
Activated alumina	150	180	145
Silicagel	150	260	240

From this table and also from the figure 3, silicagel seems to be the
most interesting storage agent. However, X_{rest} or ΔX decreases rapidly
when the relative humidity of the air at the restitution temperature
decreases. In order to maintain the storage capacity of silicagel systems
it is necessary to decrease the restitution temperature with as result a
decrease of the quality of the heat provided to the dwelling or to increase
the water content in the inlet air using a saturator. This last possibility is the only one acceptable for heating purposes. The energy for
saturating the inlet air may be provided by solar collectors working at
20-25°C and thus with a good efficiency.

Along the year, such a system is thus in thermal contact with four sources :
the solar collectors working at high temperature and the ambient air,

during storing time (spring and summer), the solar collectors working at low temperature and the ambient air, during heating time.

This storage system is in fact an open intermittent thermochemical heat pump. The system may be closed as shown on figure 4.

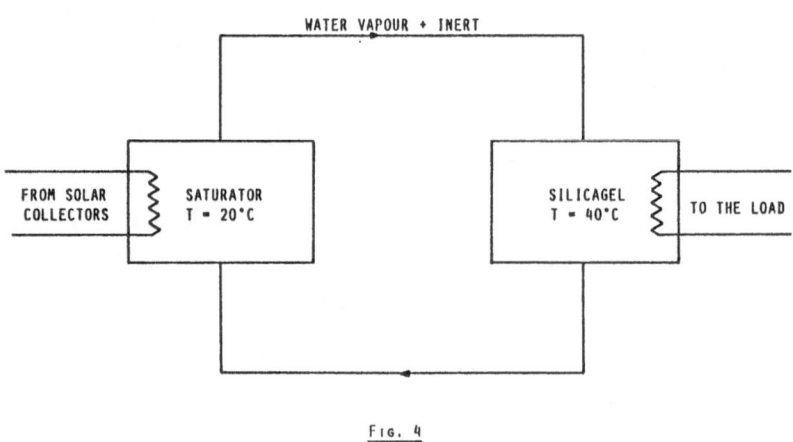

WATER VAPOUR + INERT

FROM SOLAR COLLECTORS — SATURATOR T = 20°C — SILICAGEL T = 40°C — TO THE LOAD

FIG. 4

2.2. Thermochemical heat pumps. Closed system.

A thermochemical heat pump (CHP) consists of two subsystems between which a gaseous compound is transfered using reversible reactions.

During storing time, an exothermic chemical reaction occurs in the tank I, maintained at temperature T_4 by solar collectors

$$AB(s) \rightarrow A(s) + B(g)$$

producing a gas B at the equilibrium pressure $p_I(T_4)$ while in the tank II, at the ambient temperature T_3, occurs an endothermic reaction : gas B is condensed (figure 5a)

$$B(g) \rightarrow B(l)$$

or absorbed (figure 5b)

$$B(g) + C(s) \rightarrow CB(s)$$

under an equilibrium pressure $p_{II}(T_3)$.

During restitution time the back reactions will take place, as shown on figure 5c, using a heat source at low temperature T_1, and the load at temperature T_2.

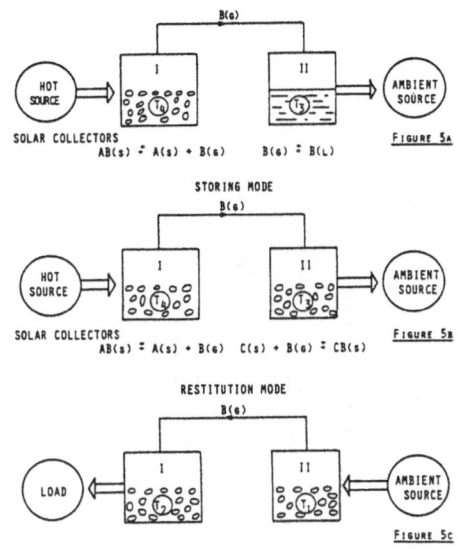

FIGURE 5A

STORING MODE

FIGURE 5B

RESTITUTION MODE

FIGURE 5C

It is clear that the following conditions :

$$\begin{cases} p_I(T_1) > p_{II}(T_2) & (12) \\ p_I(T_4) > p_{II}(T_3) & (13) \end{cases}$$

shown in a diagram $(\ln p, \frac{1}{T})$ on figure 6, must be satisfied for the transfer of the gas B.

The difference between the heating and cooling modes consists of a permutation between the utilisation and the uncostly source .
In the heating mode, the utilisation is the heating of the dwelling provided by the tank I, the uncostly source being at low temperature (5-20°C). In the cooling mode, the utilisation is the cooling of the dwelling provided by the endothermic reaction in the tank II, the uncostly source being at higher temperature (30°C or more).
Some chemical heat pumps are actually studied in some laboratories in the world.

These CHP may be classified in two groups corresponding respectively to the use of figure 5a (reaction-condensation) and figure 5b (reaction-reaction).

274

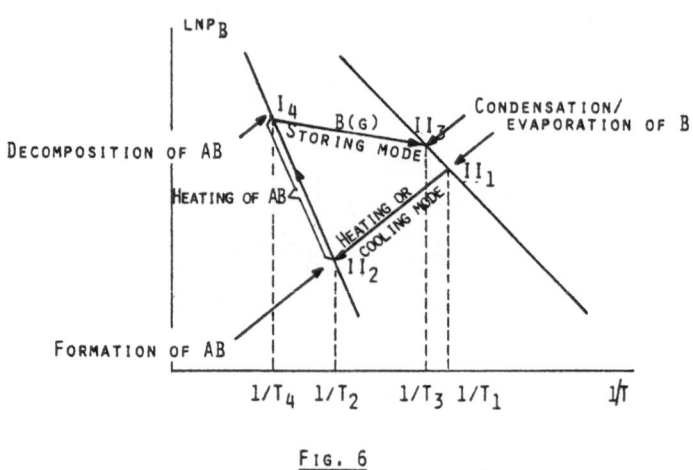

FIG. 6

2.2.1. CHP of group I.

Water/adsorbent. The solid phase is silicagel or molecular sieve. However, the pressure difference between the vessels is very low, only some mmHg and consequently the mass transfer rate is not efficient. Table 4 gives the equilibrium pressures of an hydrate and fully wetted silicagel.

Table 4

t/°C	Restitution		Storage	
Solid	$T_1 = 5$	$T_2 = 40$	$T_3 = 20$	$T_4 = 80$
$MgCl_2 \cdot 6H_2O$	1	16,4	3.55	205
H_2O	6.5	55	17.5	355
Silicagel	3	33	9	292

The transfer of water from the water tank at 5°C (uncostly source) to the load at 40°C is impossible in the two cases of table 4. The temperature of water must be at least 20-25°C. That can be obtained with the solar collectors working at low temperature in winter (see fig. 4).

<u>Methanol inorganic salt</u>. Methanol reacts with some inorganic
salts such as $CaCl_2$ following :

$$CaCl_2 + 2CH_3OH \rightleftarrows CaCl_2 \cdot 2CH_3OH$$

The heat delivered by this reaction is 0,13 Gcal/m^3. The advantage, in
comparison with water system, for a CHP utilisation is the greater value
of the vapour pressure which allows to have a cold source at lower temperature. The vapour pressures are shown on figure 7.

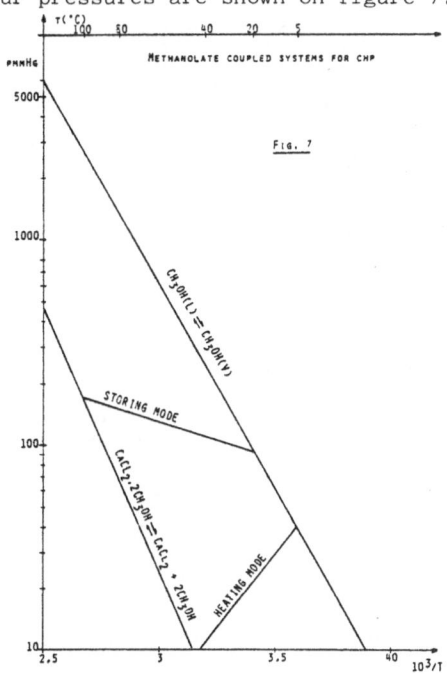

2.2.2. <u>CHP of group II</u>.

<u>Ammoniated salts</u>. The reaction between an inorganic salt and
ammonia gas may be written :

$$Salt.(x-z)NH_3 + z\ NH_3 \rightleftarrows Salt.xNH_3$$

In the majority of cases,the value of z is 4 and the heat evolved is about
10 kcal per mole of NH_3. The storage capacity for such a system is of the
order of 0.2 Gcal/m^3 of bulk material. Figure 8 shows the equilibrium
pressure versus the temperature for some inorganic salts chosen for their
cheapness and meeting the conditions on the equilibrium pressures
(12) and (13). Two possible pairs of ammoniated salts for CHP are shown
in table 5.

Table 5

High temperature ammoniacate	Low temperature ammoniacate
$CaCl_2.8NH_3 \rightleftarrows CaCl_2.4NH_3+4NH_3$	$NH_4Cl+3NH_3 \rightleftarrows NH_4Cl.3NH_3$
$MnCl_2.6NH_3 \rightleftarrows MnCl_2.2NH_3+4NH_3$	$CaCl_2.4NH_3+4NH_3 \rightleftarrows CaCl_2.8NH_3$

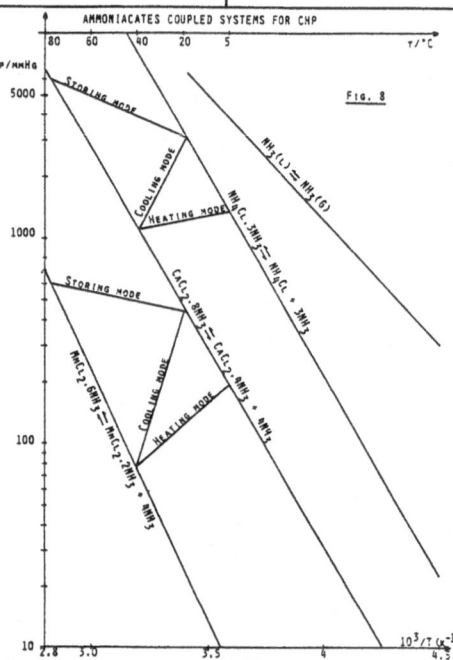

Fig. 8

As shown on table 6, the pressure differences between coupled ammoniacates are very greater than in the case of water systems and give a good mass transfer without fan.

Table 6

Couple 2 systems	$\Delta p_{5 \rightarrow 40°C}$ (mmHg)	$\Delta p_{80 \rightarrow 20°C}$ (mmHg)
$NH_4Cl - CaCl_2$	220	3000
$CaCl_2 - MnCl_2$	120	120

The advantages of ammoniacate systems are :

1) cheapness of the salt ($CaCl_2$ costs about 0.1 $/kg);
2) mass transfer without fan;
3) **large energy storage capacity;**
4) easy regulation by action on the opening of valves on the ammonia transfer pipe.

However, some problems are actually still to be solved :

1) preparation of the fully deshydrated salt especially in the case of $CaCl_2$ and $MnCl_2$;

2) heat transfer between a gas and a solid phase;

3) large volume variation during ammonia absorption or desorption;

4) corrosion problems.

Hydrogen/metal or alloy. Adsorption of hydrogen on two metals or alloys is a candidate for being used in a CHP.

The system actually studied are :

$$2.13FeTiH_{0.1} + H_2 \leftrightarrows 2.13FeTiH_{1.04} \qquad \Delta H = -3.36 \text{ kcal/mole } H_2$$

$$MgNi + 2H_2 \leftrightarrows MgNiH_4 \qquad \Delta H = -15.4 \text{ kcal/mole } H_2$$

$$LaNi_5 + 3.35H_a \leftrightarrows LaNi_5H_{6.7} \qquad \Delta H = -7.26 \text{ kcal/mole } H_2$$

2.3. Conclusions

For storing solar heat in open systems, reaction between water vapour and silicagel seems the most interesting one : a high value of the storage capacity (about three or four times the sensible heat of water), storage in a single tank without insulation in form of dry silicagel. Unfortunately, the low water content in air in the winter time reduces the quality of the delivered heat. This systems are suitable for moderate climatic conditions.

For using in thermochemical heat pumps, reaction between a solid or liquid and a vapour phase other than water, seems more convenient. Among these, the reactions between inorganic salts and ammonia gas are the most suitable for heat storage at temperature below 100°C. The interest of these systems is also their possible utilization as cooling systems.

3. REVERSIBLE REACTION IN LIQUID PHASE

The principle is to realize in a reactor a chemical reaction such as

$$A + B \to C$$

the components A and B being stored separately in two tanks at the ambient temperature, thus without necessary insulation. This procedure is very easy since the two liquid reactants may be transfered in small reactors for delivering heat where and when it is needed. In the storing mode, the back reaction occurs at the regeneration temperature. The principle of energy storage are shown on figure 9.

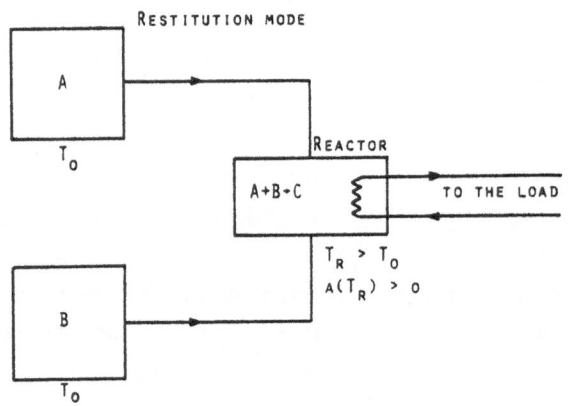

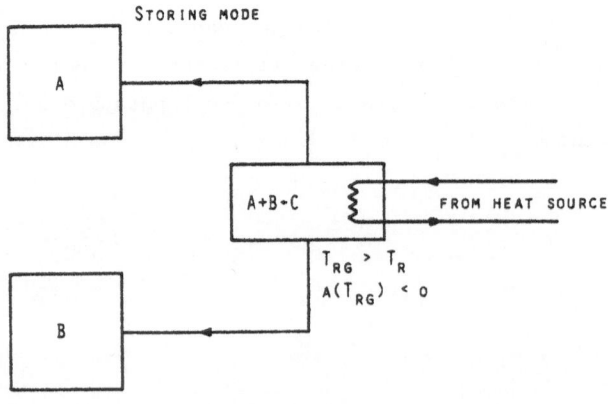

FIG. 9

The affinity of the exothermic reaction (A+B→C) must evidently be positive at the restitution temperature. Because of the Van't Hoff law, the affinity is larger and thus positive too at the ambient temperature. It is thus necessary to separate the reactants A and B for storing after the decomposition of C at the regeneration temperature or to impede the back reaction by catalyst. The heat delivered by the exothermic reaction is given by

$$Q = \Delta H_r (n_c - n_c^o) \qquad \text{(equation 9)}$$

If the reactants A and B are separated the value of n_c^o is clearly equal to zero. For such a system, the heat supplied by the running reaction causes a heating of the reactor and a decrease of n_c value. Thus, it is necessary to maintain the reactor at a temperature as low as suitable for heating the dwelling.

The other relevant equations of these systems are :

$$K(T) = \frac{x_C}{x_A x_B} \qquad \text{(equation 8)}$$

$$\frac{d\ln K(T)}{dT} = \frac{\Delta H_r}{RT^2} \qquad (14)$$

The criteria of success for such reactions for energy storage are thus :
1) a great value of $K(T)$ for the exothermic reactions;
2) a great variation of the equilibrium constant with the temperature in order to realize the decomposition reaction at moderate temperature. That supposes a great value of ΔH_r.

Table 7 shows the influence of parameters $K(T)$ and ΔH_r on the storage capacity Q. The values of Q correspond to the reaction, A+B→C, with separation of reactants A and B ($n_c^o = 0$); a mean molecular weight of 150 and a density of 0.8 are supposed.

Table 7

K(40°C)	n_c	Q Gcal/m³		
		$\Delta H_r = 10$	$\Delta H_r = 20$	$\Delta H_r = 30$
1	0.292	0.015	0.03	0.06
10	0.698	0.035	0.07	0.105
20	0.781	0.040	0.08	0.12
30	0.820	0.042	0.082	0.126

(ΔH_r in kcal/mole)

To be competitive with the more chemical ways of storage such as sensible heat or latent heat, these systems, using generally organic components, less time stable and of higher cost, must have at least a storage capacity of about 0.12 Gcal/m^3 which corresponds to a heat of reaction of 30 kcal/mole and an equilibrium constant of 20 or more.

3.1. Addition to carbonyl groups

Water is the most common compound that can be added to carbonyl groups. For example, reaction between acetaldehyde and water is :

$$H_2O + CH_3 - \overset{\overset{O}{\|}}{C} - H \rightarrow CH_3 - \overset{\overset{OH}{|}}{\underset{OH}{C}} - H$$

Thermodynamical data for dissociation of the hydrates of various aldehydes and ketones are shown in figure 10.

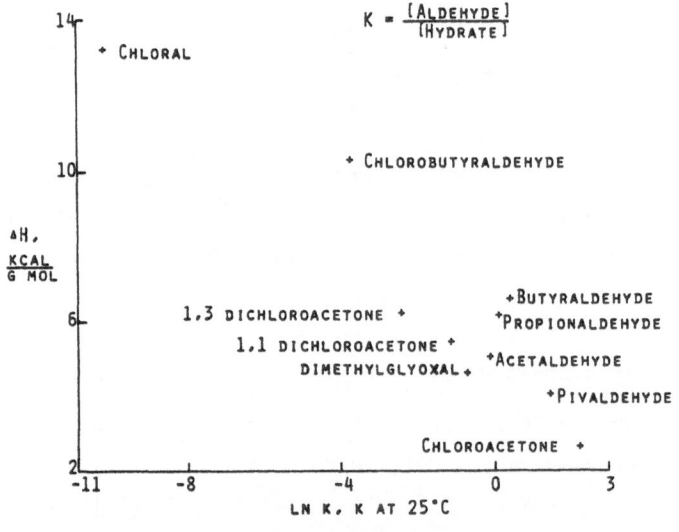

Fig. 10

PLOT OF LN K AT 25°C VS. ΔH OF REACTION FOR THE DISSOCIATION OF THE HYDRATE OF VARIOUS ALDEHYDES AND KETONES.

For an equimolar mixtures of the chloral-water system one can estimate the storage capacity to 0.12 Gcal/m^3 assuming the "best" K value that can be obtained. This value is subject to some uncertainties because of the lack of data on mixing effects. The reactions with water are very interesting because it is not necessary to store it.

A number of compounds besides water can be added to carbonyl groups. The compounds include bisulfites and alcohols. Data for these reactions are even less complete than for hydration reactions.

3.2. The Diels-Alder reactions

A second set of candidates, is the Diels-Alder reactions in which two molecules of reactants give one molecule of product.
The principle of this reaction is the cyclisation of two organic molecules, one with a double bond and other with two conjugate double bonds :

Some reactants of Diels-Alder reactions are :

dimerisation of cyclopentadiene

furan + maleic anhydride.

In this last reaction, the heat delivered for an equimolar mixture would be 94. kcal (0.06 Gcal/m^3). Unfortunately for such reactions, it is necessary to work in a solvent. The real volumetric storage capacity is thus lower.

3.3. Conclusion

Up to this point, reaction in liquid mixtures, such hydration and Diels-Alder reactions, have been discussed only in terms of their potentialities for energy storage. Before actually using a particular reaction, it will be necessary to determine the true heat storage capacity by calorimetry. The reactions must also be evaluated in terms of corrosion, toxicity, secondary reactions and initial cost of the chemicals.
The lack of available data justifies continued efforts in this area.

4. SOLID-SOLID REACTIONS

Reactions between four immiscible solids in four phases are considered briefly. Such a reaction may be written,

$$AB + CD \rightleftarrows AD + CD$$

The components AB and CD are inorganic salts where A and C are cations and B and D anions.

Well known examples are the conversion of Chili saltpetre with
potassium chloride :

$$NaNO_3 + KCl \xrightarrow{\leftarrow} KNO_3 + NaCl$$

and the ammonia-soda process :

$$NaCl + NH_4HCO_3 \xrightarrow{\leftarrow} NaHCO_3 + NH_4Cl$$

The number of degrees of freedom (W) for chemical reacting systems is
given by :

$$W = 2 + c - \varphi - r \qquad (15)$$

where c, φ and r are the numbers of components, phases and reactions.
For solid-solid reacting systems the number of phases is equal to the
number of components and thus

$$W = 2 - r \qquad (16)$$

For one chemical reaction occuring, these system have thus only one
degree of freedom and their behaviour is thermally equivalent to that of
phase transition or congruent melting point systems.
The affinity depending only of the temperature is equal to zero, when
the four phases coexist, at one temperature, called the transition temperature. Below this temperature the reaction goes in one direction until
at least one solid disappears completely. Above this temperature the
reaction goes in opposite direction until consumption of at least one
reactant.

Heat storage using this system imposes to maintain the storage medium
above the transition temperature and clearly the main advantage of the
chemical reacting systems of the two first classes,which are able to store
energy for a long term without insulation, is then lost.

THERMAL STORAGE SYSTEM BASED ON THE HEAT OF
ADSORPTION IN AIR-BASED SOLAR HEATING SYSTEMS

J.K.M. VERDONSCHOT

Institute of Applied Physics TNO-TH, Delft, The Netherlands

ABSTRACT

A thermal storage system based on the heat of adsorption of water in
hygroscopic materials has been studied as a component of a solar space
heating system. The aim of this project is to decrease the storage volume
in comparison to a rock-bed storage system by increasing the stored
energy density. The solar contribution of the system with an adsorbent
bed is compared with the solar contribution of the same system with a
rock-bed. The function of a storage system in an air-based solar heating
system is discussed. Furthermore, some experimental results are presented.

HEAT STORAGE IN AIR-BASED SOLAR HEATING SYSTEMS

Solar heating systems with air as heat transfer medium need a large
heat exchange area between the transfer medium and the storage medium.
Therefore solar air collectors are usually combined with a rock-bed
storage system. Such a system needs a large storage volume because of
the low energy density of rock-beds which is a big disadvantage.

The possibilities of using adsorbent materials seem to be very
promising. This is based on energy density comparisons which result in
a theoretical decrease of the storage volume by a factor 10.
A solar heating system with air collectors can be divided into three
air-flow circuits.
Each of them is given in figure 1.

I   The direct use circuit

The collected solar heat is transferred directly to the house. In
that case air from the house is passed through the collectors. If the
supply temperature of the air is still insufficient the auxiliary
unit is switched on.

II The heat storage circuit

If there is no heat demand of the house, the heat of the collectors is stored in the storage medium.

III The discharge unit

If there is no or insufficient heat collection by the solar collectors, heat is withdrawn from the storage medium.

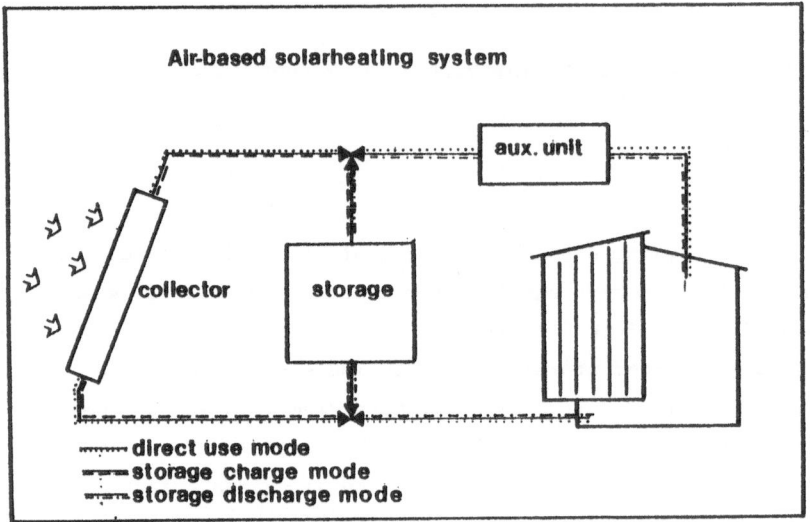

Figure 1

THE STORAGE PRINCIPLE

When solar energy is available air is heated by the collectors. This hot air has a small relative humidity which makes it possible to dry the adsorbent material. In fact heat is stored by drying an adsorbent material thus storing heat at a low temperature level.

Heat can be withdrawn from the storage system by cold air which has a high relative humidity. In that case the hygroscopic material will adsorb water.

During the process of water adsorption, heat is liberated. This means that cold, wet air is dried and heated, so it can be used to heat the house. It should be noted that the heat of adsorption is higher than the heat of evaporation.

A possible application in a solar heating system can be seen in figure 2. The heat removal cycle in this design is inefficient because the humidifier will decrease the air temperature. An improvement can be reached by the application of a heat exchanger.

The application of a solar energy installation with an adsorbent storage medium needs a careful study of humidity and temperature processes to provide an optimum solar contribution and a comfortable indoor climate.

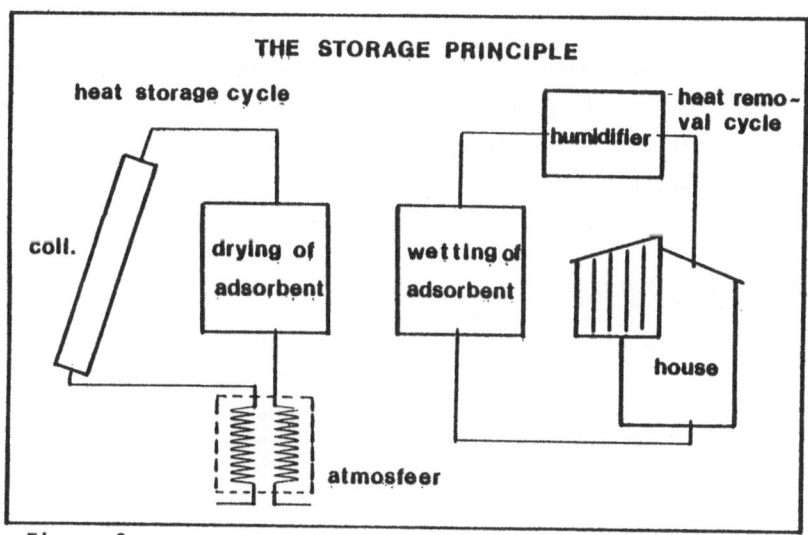

Figure 2

MEASURED PROPERTIES OF THE ADSORBENTS

In the adsorbent storage system combined heat and mass transfer takes place. The quantities of heat and mass transfer are determined by the temperature and the humidity of the airstream. The various physical properties and relations are measured for the following adsorbents: Silicagel, Sorbead-R, Sorbead-W, Activated Alumina and Molecular Sieves. The equilibrium states (Langmuir curves) of the adsorbent material Sorbead R are given in figure 3.

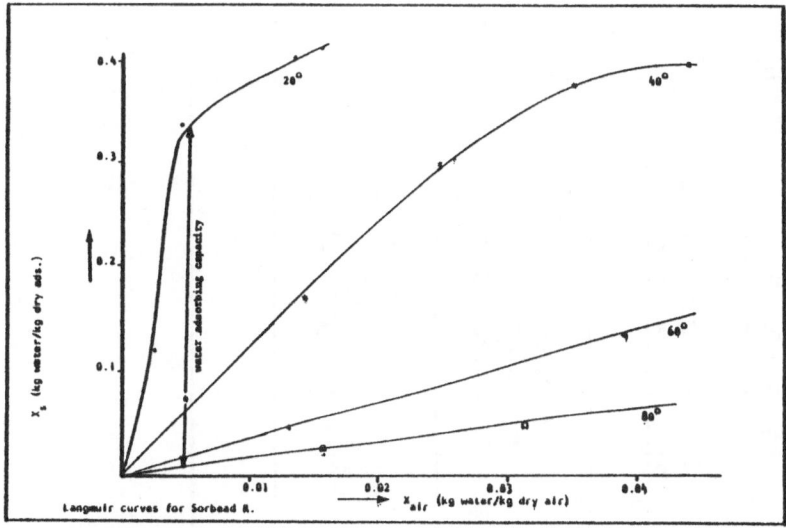

Figure 3

The stored energy density of an adsorbent storage system depends on the water adsorbing capacity and the heat of adsorption in the operating range of temperature and humidity. In an adsorbent storage system energy is stored both in the form of sensible heat and in the form of the heat of adsorption. The amount of heat stored in the form of heat of adsorption can be defined as:

$$Q_{ads} = \rho_s V \int r_a(X_s)\,dX_s$$

in which:

Q_{ads} = amount of heat stored in the form of heat of adsorption
ρ_s = density of the dry adsorbent
V = storage volume
r_a = heat of adsorption
X_s = water contents of the adsorbent

In order to make a clear comparison between the various adsorbents a reference level is defined as:

> The stored energy is zero when the adsorbent is in equilibrium with air of temperature 20 °C and absolute humidity 0.010 kg water/kg dry air.

This air condition is a realistic value for the air which is available for charging and discharging of the storage system in a solar space heating system. With the measured heat of adsorption and the equilibrium curves the stored energy density of five adsorbents are determined as can be seen in table I.

	sensible heat	heat of ads.	total MJ/m^3
silicagel	35	616	651
sorbead-R	38	768	806
sorbead-W	38	502	540
act. alumia	29	573	602
mol. sieves	32	119	151
rock bed	75	~	75

the stored energy density operation range 20-80 °C 0.010 kg water/air

Table I

It appears that Sorbead-R has the highest energy density.

THE COMPUTER MODEL

The computer model of a solar heating system with air-cooled collectors consists of three main parameters, viz.:
- the air-cooled collectors
- the adsorbent storage system
- the heat demand of the house

The aim of our calculations is to study the influence of several adsorbent storage systems. Therefore, we use a reference collector type. This flat plate collector type is spectral selective, single glazed and has an aperture area of 25 m^2. The total heat demand for space heating is determined by the insulation of the house and the ventilation rate. By changing these parameters the heat load is varied. The calculations are carried out for a heating season of nine months. So the calculation process starts at the end of the summer with a fully charged storage system. For the calculations hourly weather data from the Dutch reference year are used.

DESIGN OF THE SOLAR HEATING SYSTEM

The solar heating system in figure 4 is a result of calculations on several operation modes. The operation of this system shows some differences with respect to a solar space heating system with a rock-bed storage. The heat storage cycle operates with atmospheric air and a heat exchanger is applied to increase the cycle efficiency.

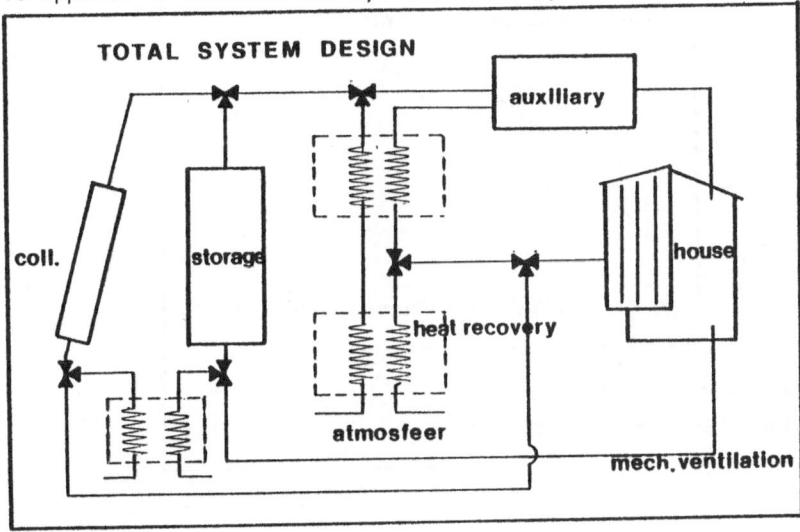

Figure 4

The heat removal cycle operates with mechanical ventilation air from the house. The outlet ventilation air which has a high relative humidity is used as inlet air for the storage system. This system therefore includes heat recovery from the exhaust ventilation air. From preliminary calculations it appeared that Sorbead-R gives the highest solar contribution. Therefore the reference system of figure 4 is equiped with a Sorbead-R storage system.

The various other parameters have the following values:

- total air flow through the direct use circuit   1200 m^3/hr
- air flow in the primary circuit   600 m^3/hr
- mechanical ventilation flow   200 m^3/hr
- heat exchanger efficiency   80%

RESULTS OF CALCULATIONS

For the above-described solar heating system the solar contribution to space heating has been calculated as a function of the heat load. The heat load for the solar space heating system is determined by the total heat demand of the house and the heat recovery. Thus the heat recovery is thought to be an energy saving method.

The result of the calculations is given in figure 5. The solar contribution is divided in two parts:

- the direct use contribution (dotted line)
- the storage system contribution

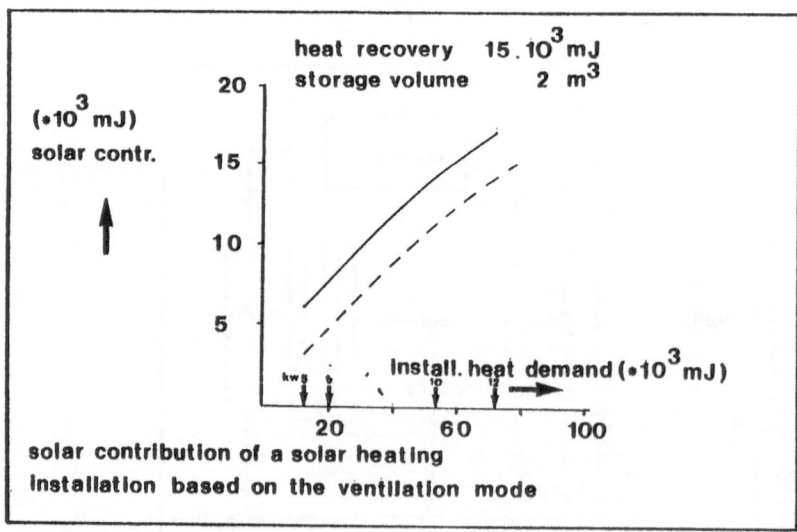

Figure 5

The direct contribution increases with increasing heat demand whereas the contribution of the storage system (volume of 2 m^3) is rather constant. Therefore, the function of the storage system is more important in better insulated houses because of the higher relative contribution. In better insulated houses more often during the heating season energy can be stored because the house does not need directly all the solar energy from the collector.

COMPARISON WITH A ROCK-BED STORAGE

The total solar contribution is strongly determined by the heat demand to the system. In order to make a clear comparison between a solar heating system with and adsorbent storage and a rock-bed storage, it is necessary for the heat demand to the solar system to be the same for both systems. In the system with an adsorbent storage the heat recovery is applied to improve the total system as such.

To adjust both heat demands, the system with a rock-bed storage has been extended with a heat recovery from the outlet ventilation air. The comparison is made for a house with an annual heat demand for space heating of 35.10^3 MJ. The contribution of the heat recovery system is 15.10^3 MJ. So the total heat load for the solar system is 20.10^3 MJ.
In figure 6 it can be seen that an adsorbent storage of about 1 m^3 gives the same solar contribution as a rock bed storage of 6 m^3. On the other hand a rock bed storage of 1 m^3 only gives 3% less solar contribution than the adsorbent storage of 1 m3. These 3% are related to the heat load of 20.10^3 MJ.

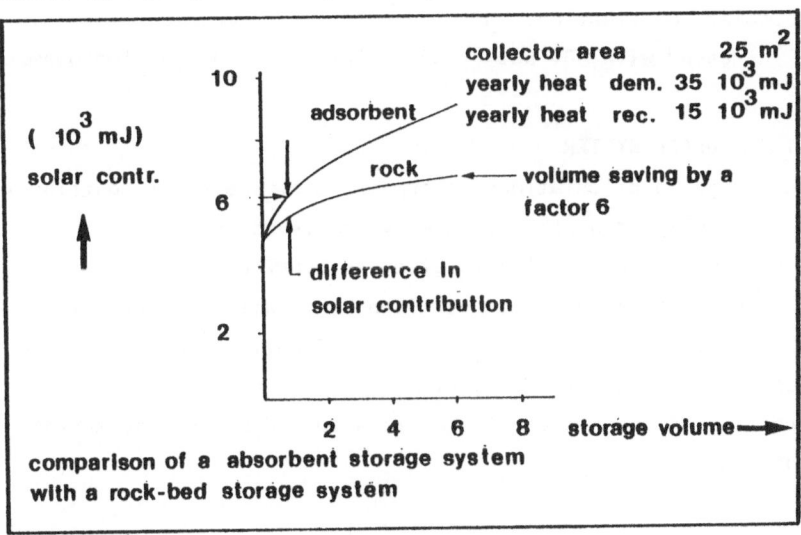

comparison of a absorbent storage system with a rock-bed storage system

Figure 6

This behaviour is not affected by the total heat demand of the system.
From figure 6 can also be seen that the application of a rock-bed storage
of more than 2 m^2 gives no system improvement. The higher solar contribution
of the adsorbent bed at 6 m^3 is due to the effect of seasonal storage.
The small improvement obtained with an adsorbent bed of 2 m^3 or less is
due to the small discharge flow of 200 m^3/hr (ventilation air). Therefore,
a system improvement was expected from the application of the so-called
recirculation mode. The discharge of the storage is arranged with air from
the atmosphere as is to be seen in figure 7.

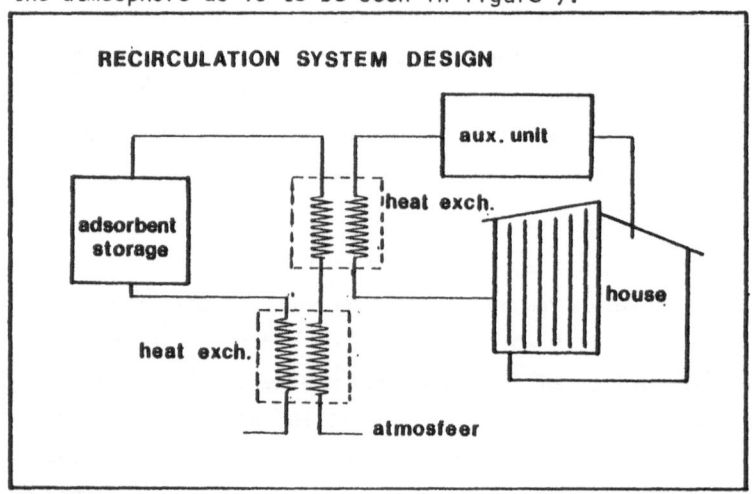

Figure 7

The recirculation mode was compared with the ventilation mode.
It appeared that there was a strong influence of the heat exchanger efficiency on the solar contribution. When this efficiency was 100% the solar
contribution increased with 30% but at an efficiency of 80% no improvement
was observed.

FUNCTION OF THE STORAGE SYSTEM

From the foregoing calculations it appears that the solar contribution
for space heating deliverd by the storage system is rather small.
In figure 8 this solar contribution is given as a percentage of the
heat load as a function of the same heat load. This solar contribution
is divided into a direct contribution and a contribution via the storage
system. It appears that the storage system is more important for small heat
loads. The function of the storage system is affected by the thermostat
setting of the house.

The results of figure 8 are based on a system with 25 m^2 of collector area
and a thermostat setting of 20 $^\circ$C during day and evening. From figure 8
can also be seen that an adsorbent storage system hardly improves the solar
contribution for yearly heat loads above 30.10^3 MJ. When it concerns
smaller heat loads an economical study will be necessary for the choice
between the application of an adsorbent or a rock-bed storage system.
When the thermostat setting is changed to 18 $^\circ$C during the day and 21 $^\circ$C
in the evening, the function of the storage system increases because the
heat load is mainly in the evening.

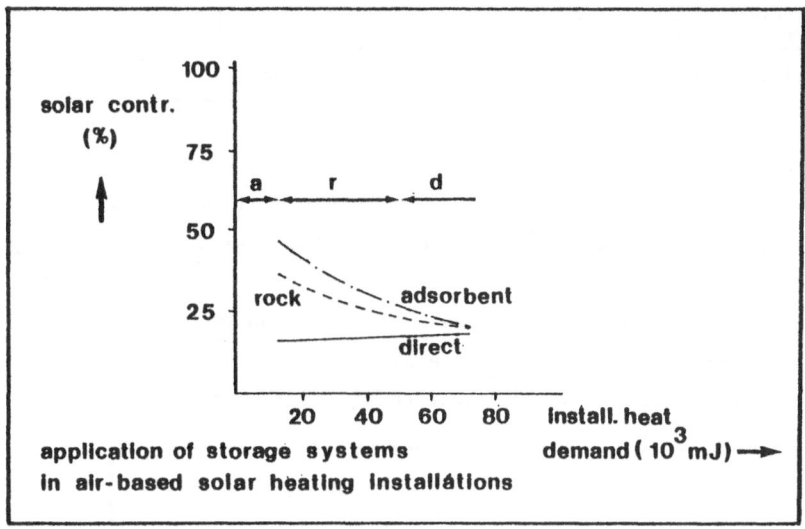

Figure 8

Also an increase of the collector area influences the solar contribution
through the storage system. The effect of collector area and thermostat
setting is shown in table II. The values are related to a storage volume
of 2 m^3. When the collector area is increased the difference between a
rock-bed and an adsorbent bed increases. The adsorbent storage results
in a higher solar contribution. More energy can be stored for a longer
period. A change of the thermostat setting results in a higher heat demand
for the evening. The direct use part of the solar contribution diminishes
while the storage system has to store the collected energy only for a
few hours. In that case the difference between the adsorbent and the
rock-bed is only 2%.

This is mainly due to small heat transfer rate of the adsorbent storage system which is determined by the ventilation airflow.

INSTALLATION HEAT DEMAND 20.10^3MJ/year	Solar contribution (%)		
DESIGN MODE	SYSTEM PARAMETERS		
	Ac = 25	Ac = 35	Ac = 25 thermostat
Direct use	18	21	7
Direct + rock	33	36	30
Direct + adsorbent	39	45	32
Table II: Solar contribution for different design modes.			

EXPERIMENTAL RESULTS

A test circuit capable of simulating the air conditions coming from the collectors and the air conditions to extract the heat from the storage has been realised. The design of the test circuit is based on a real storage system with a volume of 1 m^3, an airflow in the primary circuit of 600 m^3/h. These values are scaled down with a factor 2 for the realised prototype. Figure 9 gives a picture of the test facility.

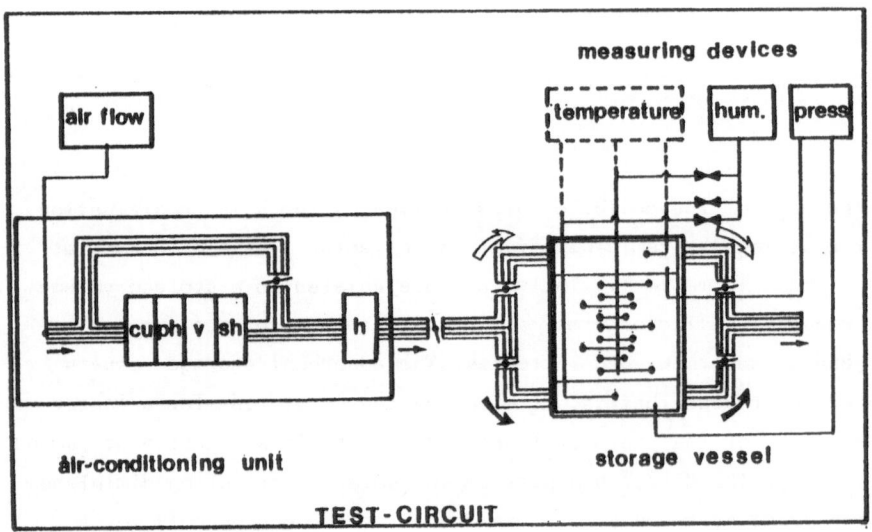

Figure 9

At the moment step-respons experiments are carried out to verify the
assumptions made in the computer model. The results of a step-change for
a charging and discharging mode are given in figure 10.
These testresults refer to a step change in inlet temperature of 20-60 °C
and air humidity of 0.010 kg water/kg adsorbent. The adsorbent material
was Sorbead-R.

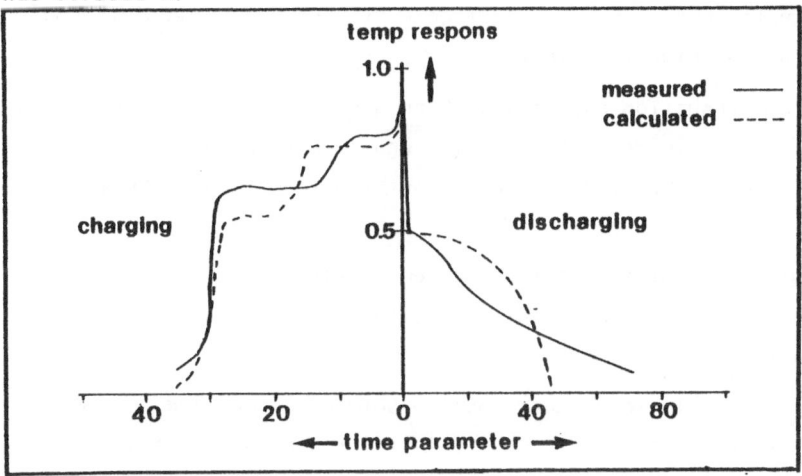

Figure 10

The dimensionless temperature response is defined as:

$$f = \frac{T_{out} - T_{in}}{\Delta T_o}$$

T_{out} = outlet temperature
T_{in} = inlet temperature
ΔT_o = temperature step change

The time parameter is defined as:

$$\theta = \frac{G \; t}{V}$$

G   = air mass flow
V   = volume of storage reservoir
t   = time

It appears that for the charging mode the calculated and measured
respons correspond quite well. For the discharge mode the measured
respons function is quite different from the calculated values. The
measured discharge time is longer than calculated.

The difference in the respons function can be translated into a heat transfer rate of the storage system. For the charging mode the mean heat transfer rate measured was 2.2 kW, calculated 2.1 kW whereas these values for the discharging mode are measure 1 kW, calculated 1.5 kW.

CONCLUSIONS

It apprears that a rock-bed storage system of more than 2 m^3 gives no improvement in solar contribution.

The volume reduction obtained with an adsorbent material will be in the range of 2 - 6. Therefor this factor should always be judged in combination with a comparison in solar contribution of the storage systems.

The application of the storage system in air based solar systems is only interesting for small heat loads. This is influenced by the thermostat setting of the house.

The difference in solar contribution between a rock-bed and an adsorbent storage is relatively small. Therefore an adsorbent storage system should be carefully designed, taking into account the extra investments such as heat exchangers and regulating valves. Special attention should be paid to the electrical energy consumption of ventilators.

Experiments are carried out to verify the results of the computer calculations.

CHEMICAL STORAGE AND PUMPING OF SOLAR ENERGY

A. VIALARON

INTRODUCTION

Chemical heat storage is familiar to us, in the form of carbon compounds, which are the basis of our present energy economy (wood-coal - natural gas - oil).

Storage implies renewal, and it is known that nature has performed this task using photosynthesis, almost isentropically, but at a low efficiency : nevertheless, new interest shown recently in the exploitation of biomass energy, reveals that this thechnique is viable according to technico-economic criteria.

Either photochemically, in imitation of nature, or by making use of the high temperatures created by concentrated solar radiation, chemists are now trying to prepare fuels, based principally on hydrogen (from water).

The narrowly limited use of high energy solar photons, or very high temperatures, cannot hope to give substantial yields : that is why, we have a right to question the necessity of building molecular structures able to deliver high quality energy, but at a low efficiency and so at great expense of occupied ground, if, for example, we need only to keep the temperature inside a house, at 20° to 30 °C above the external environment, and that only the deepest French winter.

In so far as a heat storage technique will provide an adequate retention time, it is obvious that the most economical solution will be to choose the device working with a thermal level at its inlet just above the thermal level of use.

On the other hand, when the storage device must be carried by a mobile vehicle, in which high thermal level ($\Rightarrow$CARNOT yield) and energy density ($\Rightarrow$ mission duration) are required, fuels such as we are already familiar, will represent the best chemical storage system.

These reasons, related or conflicting, lead us to the "a priori" conclusion, that chemical storage may be of value in cases where the following parameters are involved, separately or combined :

- thermal level
- energy density
- recovery time

We shall now make a brief survey of these aspects of chemical
storage, and afterwards try to define suitable areas of application.

THERMAL LEVEL AND ENERGY DENSITY

Figure 1 shows various reversible chemical systems which are at
present the object of numerous studies : we thought it useful to place
them on this diagram, taking into account the temperature at which the
internal energy variation in the reaction equals 0 ($\Delta G° = 0$). This is
what FUNK calls the "turning temperature"

$$T_{inv} = \frac{\Delta H°_R}{\Delta S°_R}$$

On the ordinate is shown, the $\Delta H°_R$ value for 1 g of substance, from
which one can calculate the theorical amount Q_{Kg} necessary for storing
1 KWht. We shall return later to this aspect of thermochemical storage,
dealing with the problem of retention time.

It is now possible to make the following remarks :

a. - Concerning the meaning of T_{inv} : it is a theoretical reference
point which permits the temperature range to be identified, in which
the reversibility of the chosen reaction will be located (ref. I,
WENTWORTH, p. 207). We can see, for example, that regeneration of hydrogen and/or carbon based fuels, needs thermically a very high temperature
level ; this is well known. Using thermochemical cycles, we can lower
this temperature to about 1000 °C or less, but with a disappointing
yield and a high investment.

Otherwise, if, with M. DUCARROIR (ref. II, p. 518), we compare
magnesium sulfate dissociation, with the dissociation of sulfur trioxide,

working conditions being the same, we note that the salt dissociation is completed below T_{inv}, whereas SO_3 dissociation is only partial at T_{inv}. Inversely, the reaction $MgO + SO_3 \rightarrow MgSO_4$ will occur at a recovery temperature close to the inlet temperature.

b. - Concerning reactions with high entropy variations : it is theoretically interesting to choose chemical reactions giving a large enthalpy variation associated with a large entropy variation. In this way one can find very high energy systems, still reversible at relatively moderate temperatures. The classical example (ref. I) is ammonium hydrogen sulfate which gives three gas molecules ($\overset{.}{SO_3}$, NH_3, H_2O) : unfortunately, because of the necessity of separating the three components, this chemical system looses a large part of its theoretical interest.

A well known paper by G. ERVIN (ref. 3, p. 60) shows that in practice the performances (energy density per m^3) of most of the systems at present studied are far from their theoretical possibilities.

c. - Concerning photochemically regenerated reactions : as an example we have mentioned the well known cycloaddition reaction of norbornadiene in quadricyclane which theoretically allows the liberation of 0.29 Kcal/g at a temperature higher than 100 °C. In this case the practical difficulties (catalyst - sensitizers - dilution) are also very great (ref. 4, H.D. SCHARF, p. 658).

d. - Concerning chemicals systems in which the condensation enthalpy of the gas phase resulting from the dissociation is very important compared with the enthalpy of the reaction itself : three examples are given, two of which are suitable for practical application (hydration of quick lime or of partially hydrated Mg chloride : in both cases the heat necessary to evaporate the water can be taken from a "free" cold source or from a source with an easily attainable thermal level (e.g. : a thermal loop using oil).

In this way we have a heat pump effect, coupled more or less effectively with a storage capacity.

It can be shown theoretically that the coefficient of performance (COP) of these systems can attain 2 (ratio : total released heat/high temperature heat), while the rise in temperature above the cold source is a function of the enthalpies and entropies of the two reaction systems according to the following equation :

$$\frac{T_u}{T_f} = \frac{\Delta H_I}{\Delta H_{II}} \cdot \left(1 + \frac{(\Delta S_I - \Delta S_{II})\, T_u}{\Delta H_{II}} \right)$$

These "chemical pumps" present a certain attraction for heating of dwellings : the principal difficulties in their use concern heat transfer (products are often in solid state) and material transfer (diffusion of a gaseous phase in a solid or in a non-agitated liquid) (ref. 5, WETTERMARK and al.).

B. LAURENT (ref. 6) recently proposed the use of coupled reactions between hydrides to carry out either the superheating of a steam thermodynamic cycle, or the low level heat contribution to a hot gas cycle, the heat requirement being completed by a solar concentrator at high temperature. In the first case, we have a thermochemical "booster" which allows the thermodynamic yield of the cycle to be improved by raising its temperature, while the basic amount of heat can be derived from a thermofluid loop, well known for its ease of use but also for its temperature limits.

Finally, without going into details, it is necessary to mention hybrid thermochemical systems, in which the regeneration reaction only partially makes use of heat ; the rest is provided either electrically or photonically. In the first case, one finds, for example, the hybrid electro-thermochemical cycles for water dissociation into hydrogen, and, in the second case, the thermochemical valorization of biomass (solar gasifier combined with photosynthesis).

As we mentioned before, the manufacture of hydrogen and/or carbon monoxide either from water or CO_2 may be considered as a chemical storage of solar energy.

C.N.R.S. THERMOCHEMICAL PROGRAM

The thermochemical program of the P.I.R.D.E.S. (Solar program of the Centre National de la Recherche Scientifique) at present includes :

A. - CHEMICAL TRANSFER AND STORAGE

    1. - Storing by reversible reactions using metal sulfates
        (M. DUCARROIR, L.U.R. ODEILLO et B. SPINNER, L.C.M.T. PERPIGNAN)

    2. - Storing by reversible reactions with lime
      2.a. - Hydration, dehydration
            (R. TORRENTI, L.S.E.E.M.P. SOPHIA-ANTIPOLIS & R. ROUANET, L.U.R. ODEILLO)

      2b. - Carbonation, decarbonation
            (G. FLAMANT & A. VIALARON, L.E.S. ODEILLO)

B. - CHEMICAL PUMPING

  - Use of ammoniacates in the solid phase
    (B. SPINNER, L.C.M.T. PERPIGNAN)
  - Use of ammoniacates in the dispersed phase (suspension or solution)
    (R. BUGAREL, I.G.C. TOULOUSE - collaboration with C. WETTERMARK STOCKHOLM).

C. - HYBRID SYSTEMS

  - Hybrid cycle based on metal sulfates. Chemical storing in the form of metal sulfite solutions, combined with water electrolysis
    (D. STEINMETZ & A. VIALARON, I.G.C. TOULOUSE)

  - Solar gasification of biomass for synthesis gas production
    (J. LEDE et J. VILLERMAUX, L.S.G.C. NANCY - X. DEGLISE University of NANCY I - C. ROYERE, L.E.S. ODEILLO)

  - Thermochemical dissociation of metal oxides and energy restitution either in the form of hydrogen from the decomposition of water or of electricity from batteries (Zn)

(M. COMTAT, L.G.C.E.C. University Paul-Sabatier TOULOUSE,

M. DUCARROIR, L.U.R. ODEILLO & B. SPINNER, L.C.M.T. PERPIGNAN).

Finally it is worth mentioning an association of three laboratories :
(J. VILLERMAUX, L.S.G.C. NANCY – A.M. ANTHONY, C.R.P.H.T. ORLEANS &
P. VALENTIN, L.P.T.M.R. ROUEN) in a joint effort to accomplish water
thermolysis at a very high temperature (1800 °C – 2500 °C).

All of these research projects are financed within the frame-work
of the "Actions Thématiques Programmées". The specialisation of certain
laboratories in chemical engineering (NANCY and TOULOUSE) will lead in
the near future to the construction of demonstration devices, which
will be financed at a later date. Without going into the details of the
chosen techniques, may we just mention, that we make considerable use
of fluidized systems for high temperatures studies.

RECOVERY TIME

Before taking up any position concerning the sectors in which
thermochemical storage might be applied, it seems necessary to consider
the economic aspect of the problem : a very simplified approach consists
in only considering the storage installation investments, neglecting
maintenance and exploitation costs. Figure II shows the cost of the
stored thermal KWh as a function of the investment, of refunding
conditions and of the number of cycles achieved by installation during
the reference period, i.e. one year.

At the present level of fossil energy costs, we can expect, depending
on the circumstances (individual house or centralized installation), a
storage cost going from 0.05 up to 0.20 FF per KWht (the cost of one
KWht of butane in a domestic installation is around 0.25 FF in 1980).

One can distinguish three ranges in the use of a storage system :
interseasonal, weekly, daily. We note that the interseasonal storing
implies an investment of about 1 FF per KWT of power, while in the
extreme case (daily storing) the investment can, of course, be as much

as 50 to 100 times higher.

If we look at the market prices of industrial chemical products (fig. III, for example) we can see that it will be difficult to find effective storage materials, with a price lower than 5 000 F.F. per ton and in which the storage capacity exceeds 200 KWh/Ton (fig. I).

Finally if we consider that the scale coefficient for estimating an installation cost, is not favourable to small units (ref. 7) and still represents (fig. IV) 80 % of the total investment for a capacity of 1000 KWt (in the chosen example) we realize the difficulty in finding at the present time a thermochemical system competitive with fossil fuels. We may hope for a more favourable situation for renewable systems in the future, when the scarceness of fossil energy sources will increase their price considerably.

In the particular case of chemical storage which makes use of products and techniques already industrialized (tanks, exchangers, pumps, ...) it is probable that installation costs will practically follow the cost of living. Only mass production could offer some hope of reducing prices.

CONCLUSION

What conclusions can be drawn from all that has been considered so far ?

― If there is an opening for chemical storage, it will be without a doubt, at the level of heat storage for one to two weeks, as far as housing is concerned. A recent study (J.L. PEUBF, ref.8) shown the advantage of a decadal storage which allows the use of auxiliary power (non solar) partially reduced in a moderate climate such as the Atlantic coast (La Rochelle).

J.J. IANNUCCI (ref. 9) came to similar conclusions, concerning chemical storage combined with thermoelectrical solar plants.

― Chemical pumping, if heat and material transfer problems can be solved, seems tempting for frequent use (daily) in homes, perhaps with an inversed operating mode in summer (total climatization) and also a complementary use of electrical fossil power, on sunless days.

- The thermochemical super heating of a thermodynamic cycle represents
a more distant goal but still a very attractive one : while reactions
at high restitution temperature in fluidized reactors, in which the
thermal losses are reduced during storage phases (sedimentation of poor
conducting solids ⇒ low conductive losses) may be competitive for
running hot gas cycles.

Finally the use of solar energy in the production of a storable
chemical component (methanol) seems to be the only solution expected
for a use of solar energy, independent of the factors of time and place
(transfer of energy from sunny countries to less favourable zones) :
in this case, the plant for converting solar energy to fuel is best
payed off with the annual production rate at a maximum, while the
chemical storage is reduced to its simplest form (fuel storage tanks).

REFERENCES

1. W.E. Wentworth, E. Chen. Simple thermal decomposition reactions
   for storage of solar thermal energy. Solar Energy, vol. 18, pp. 205-
   214
   Pergamon Press 1976

2. M. Ducarroir, M. Tmar, C. Bernard. Possibilités de stockage de
   l'énergie solaire à partir de sulfates. Revue Phys. Appl. 15 Mars
   1980, pp. 513-528

3. G. Ervin. Solar heat storage using chemical reactions. Journal of
   solid state chemistry 22, pp. 51-61. 1977

4. H.D. Scharf, J; Fleischlauer, H. Leinsmann, I. Ressler, W. Schleker,
   R. Weitz. Criteria for the efficiency, stability, and capacity of
   abiotic photochemical solar energy storage systems. Angew. Chem.
   Int. Ed. Engl. 18, pp. 652-662. 1979

5. G. Wettermark, B. Carlsson, H. Stymne. Storage of heat
   Document D2 : 1979. Swedish Council for Building Research

6. B. Laurent. Pompes à absorption à hydrures et cycles à vapeur
   Revue Phys. Appl. 15. 1980, pp. 417-422

7. A. Chauvel, P. Leprince. Manuel d'évaluation économique des
   procédés. Editions Technip. 1976

8. J.L. Peube, J. Pêcheux, E. Plancq. Sur les conditions optimales d'utilisation de l'énergie solaire pour le chauffage de l'habitat Revue Phys. Appl. 15 - 1980, pp. 553-557.

9. J.J. Iannucci, J.D. Fish, T.T. Bramlette. Review and assessment of thermal energy storage systems based upon reversible chemical reactions - SANDIA REPORT 79-8239 Août 1979.

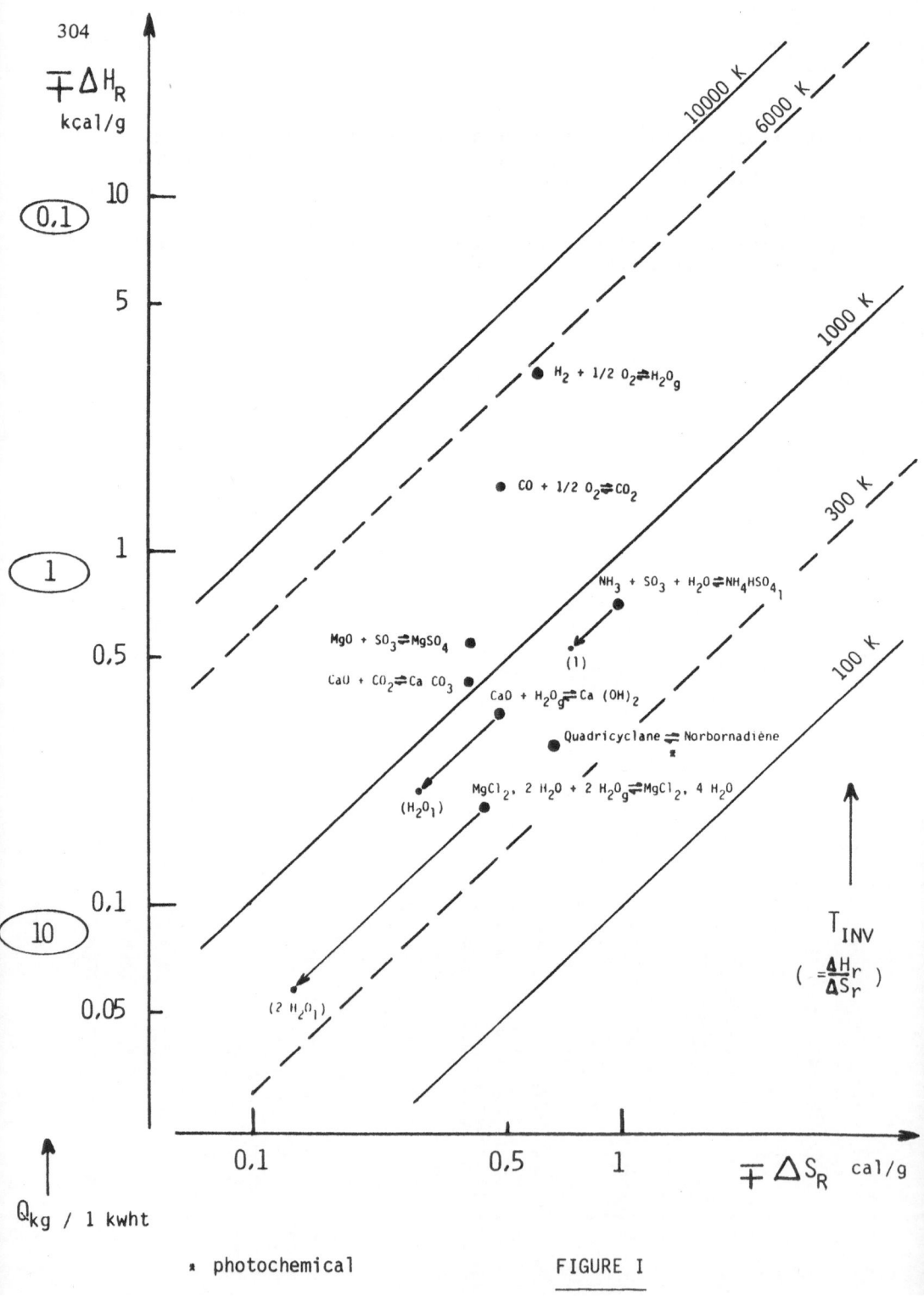

* photochemical

FIGURE I

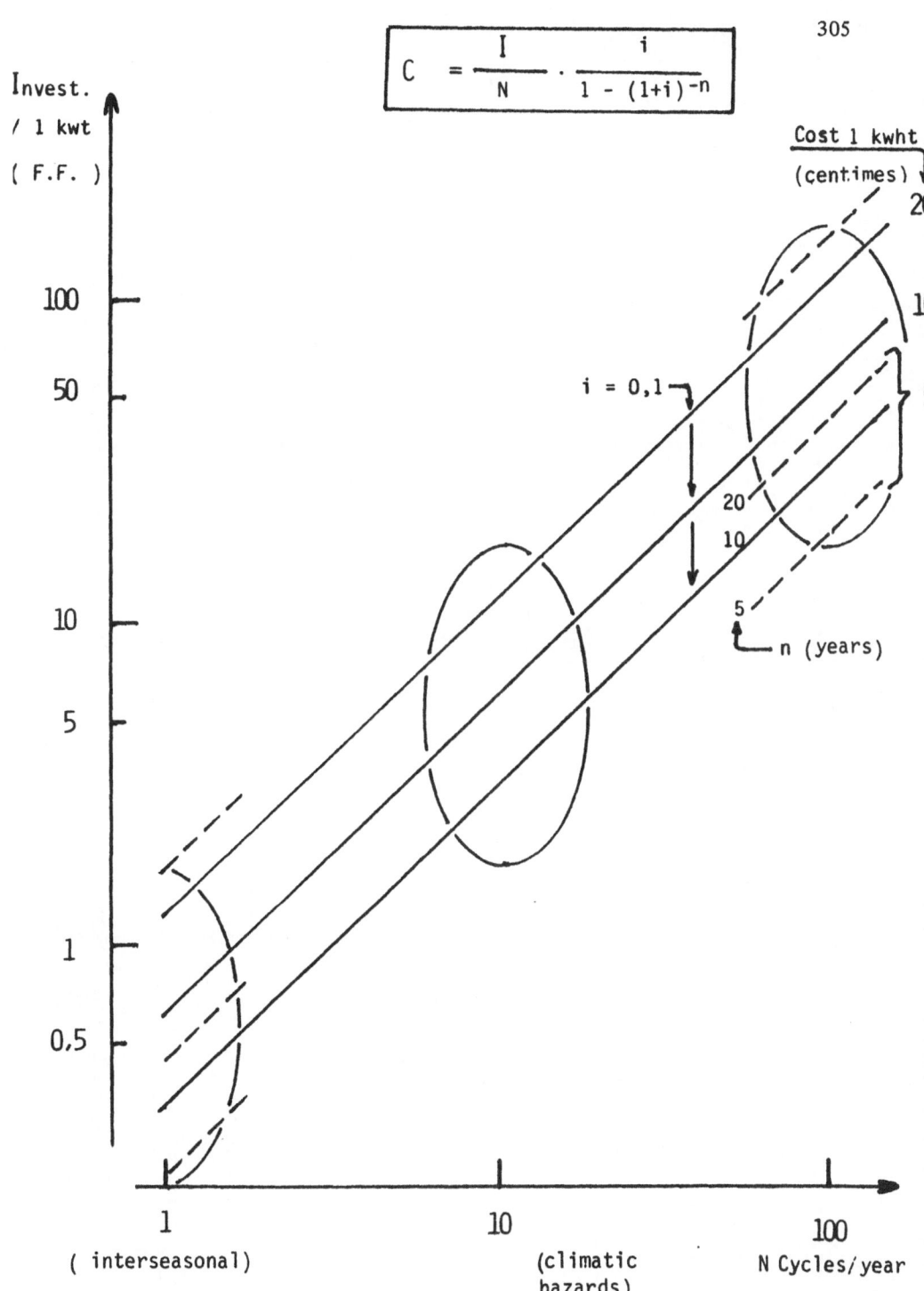

$$C = \frac{I}{N} \cdot \frac{i}{1 - (1+i)^{-n}}$$

FIGURE II

CHEMICAL PRICES

EN FF/TON
BULK/EX WORKS APRIL 1980

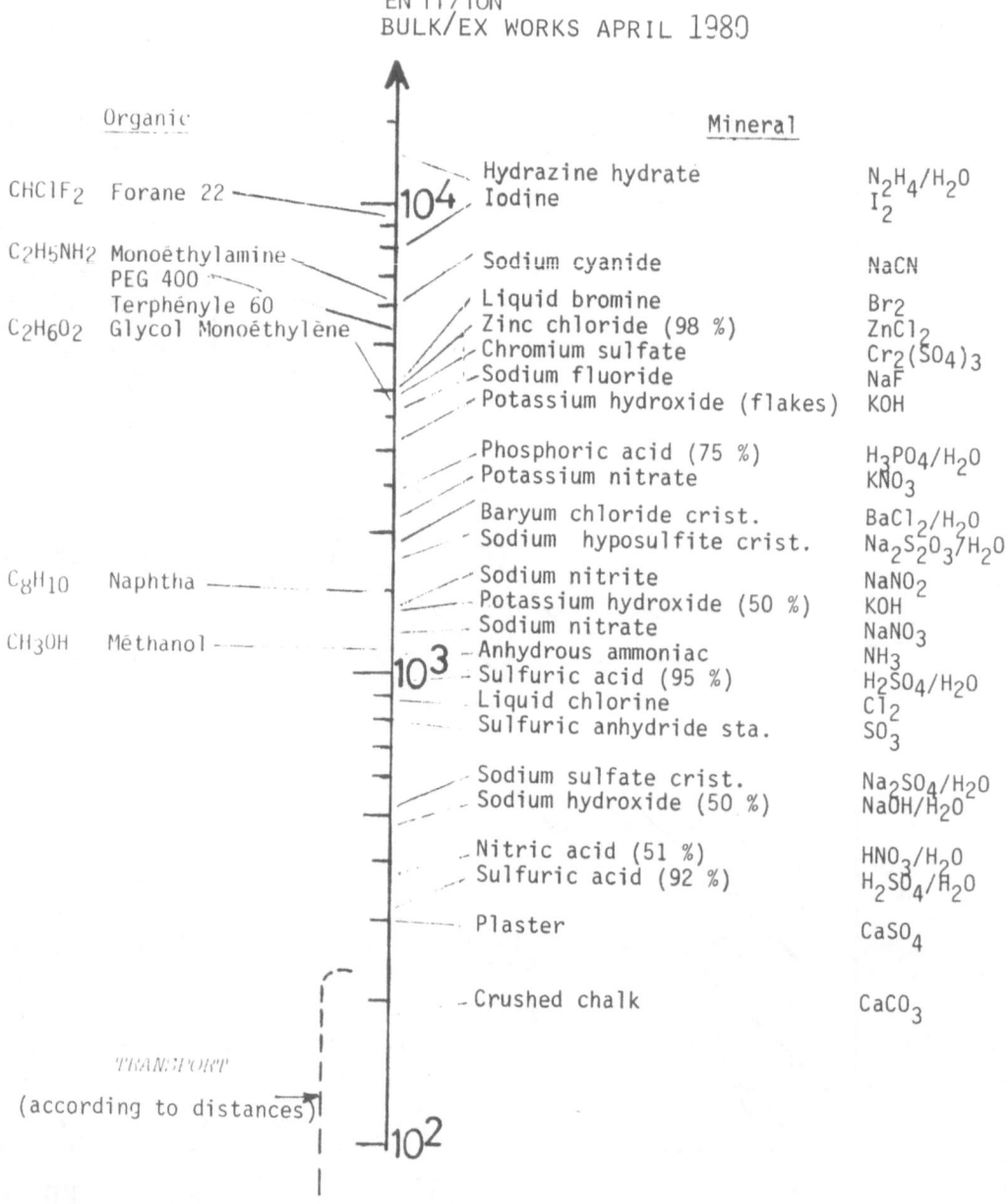

FIGURE III

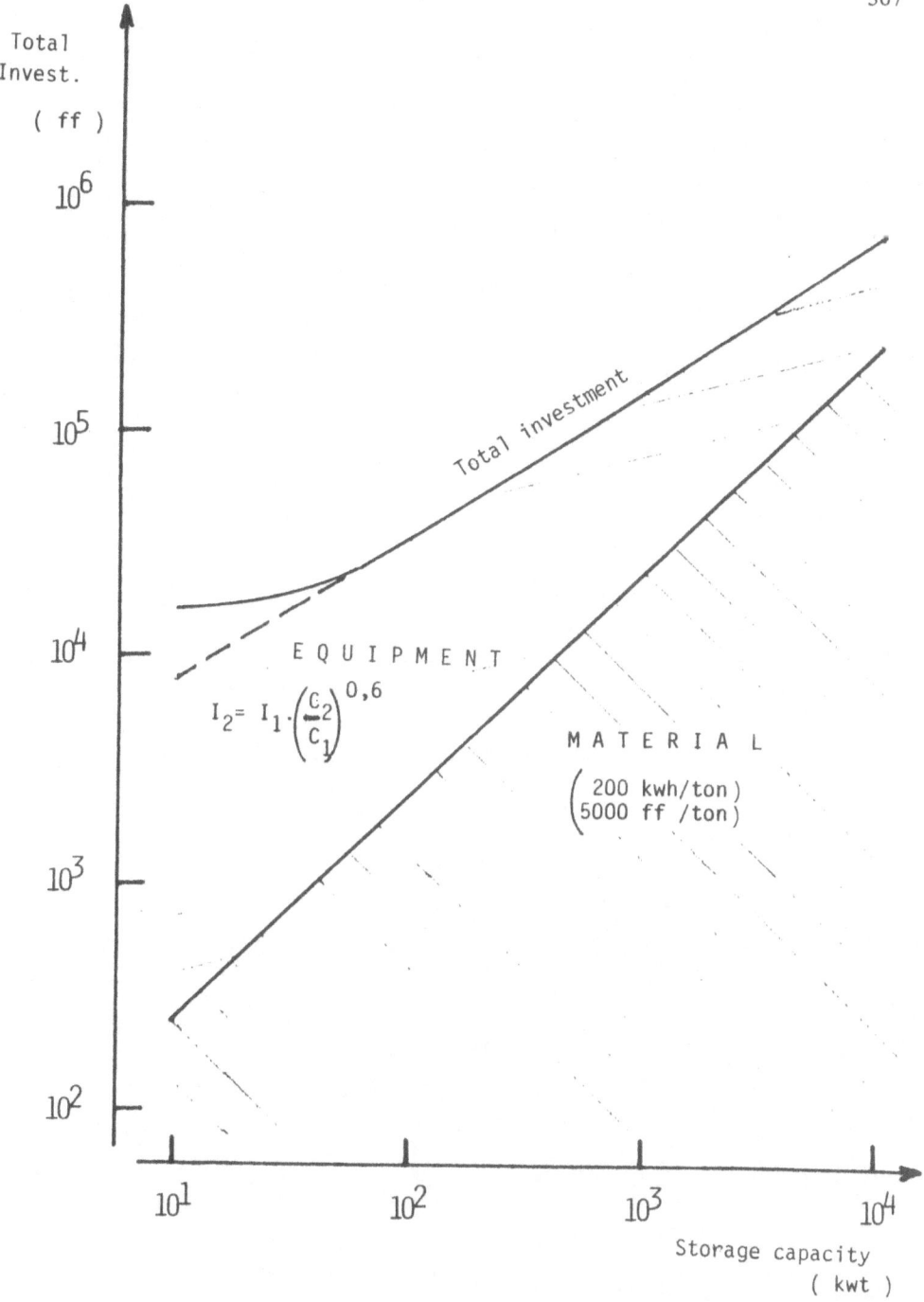

FIGURE IV

SOLID-SOLID PHASE TRANSITIONS FOR THERMAL ENERGY STORAGE

V. BUSICO[*], P. CORRADINI and M. VACATELLO
Istituto Chimico dell' Università, 80134 Naples (ITALY)

F. FITTIPALDI
Istituto di Fisica, Facoltà di Ingegneria, 80125 Naples (ITALY)

L. NICOLAIS
Istituto di Principi di Ingegneria Chimica, 80125 Naples (ITALY)

1. INTRODUCTION

It is well known that, among the thermal energy storage (TES) devices, those based upon latent heat effects are - at least in principle - the most promising ones, for their high heat capacity (usually 20-60 cal g^{-1}) and the fact that, at a given pressure and chemical composition, the storage proceeds at a fixed and invariant temperature.

On the other hand, despite the huge amount of research done in this field in recent years, the latent heat TES systems require a technology still far from being commercialized.

It is mainly for this reason that the solar plants which use latent heat TES can be counted on two hands fingers; of these, about 90% exploit the dehydration of hydrate salts and of their eutectics, and the remaining 10% ca. is based on the melting process of long chain aliphatic compounds, expecially mixtures of n-paraffins (1).

Only little attention has been given up to now to TES through phase transitions in the solid state, mainly because - excepting the structural rearrangements of some peculiar solids with globular molecules (for instance diaminopentaerithrol) - the enthalpic changes associated with the solid-solid phase transitions usually are quite low (1).

In the last few years, however, some consideration has been
paid to several organometallic compounds with the inclusive
denomination of "layer perovskites", undergoing high enthalpy
phase changes in the solid state, fully reversible and with a
wide choice of transition temperatures. We wish to report here
the results obtained from the study of these compounds, and
to discuss the possible prospects that such results introduce
in the field of latent heat TES devices.

2. THE LAYER PEROVSKITES: STRUCTURE - PROPERTIES RELATIONSHIPS

The bis (n-alkylammonium) tetrahalometallates (II) are organo
metallic compounds of the general formula $(\underline{n}-C_nH_{2n+1}NH_3)_2MX_4$,
where M is a divalent metal atom and X a halogen. Typical com
pounds studied have M = Mn, Cu, Hg, Fe, Co, Zn, X = Cl and n
varying between 8 and 18 (2-7). They have been termed "layer
perovskites" because, for M = Mn, Cu, Hg, Fe, their crystal
structure consists of layers similar to those present in the
mineral Perovskite, $CaTiO_3$.

Powder and single crystal X-ray diffraction studies (8,9)
have revealed that these compounds are typical "sandwich" sys
tems, their structure resulting from the regular alternation
of thin inorganic and thick hydrocarbon regions, so that each
inorganic layer is sandwiched between two hydrocarbon layers
and vice versa (figure 1).

For the compounds with M = Mn, Cu, Hg, Fe, the inorganic
regions consist of nearly two dimensional macroanions of com
position MX_4^{2-}, made up of sharing corners MX_6 octahedra (figure
2); isolated MX_4^{2-} tetrahedra have been found, on the other
hand, for M = Co, Zn (figure 3) (2-9).

The hydrocarbon regions are constituted (figures 1,3) by
the long paraffinic chains of the n-alkylammonium groups, ionic

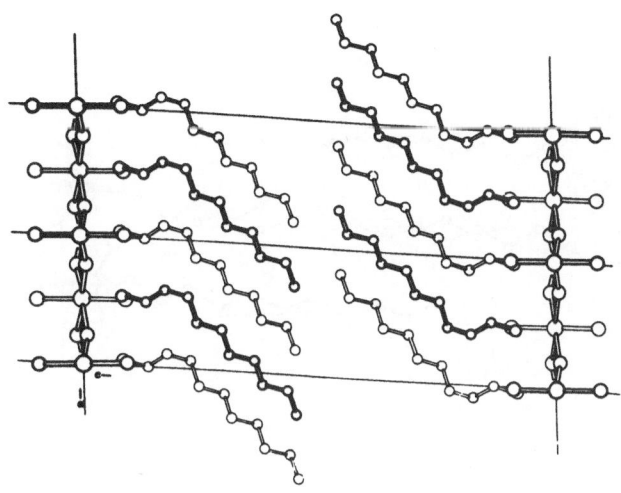

FIGURE 1. The structure of the layer perovskites: compound
($\underline{n}$-$C_{10}H_{21}NH_3$)$_2$MnCl$_4$ (8).

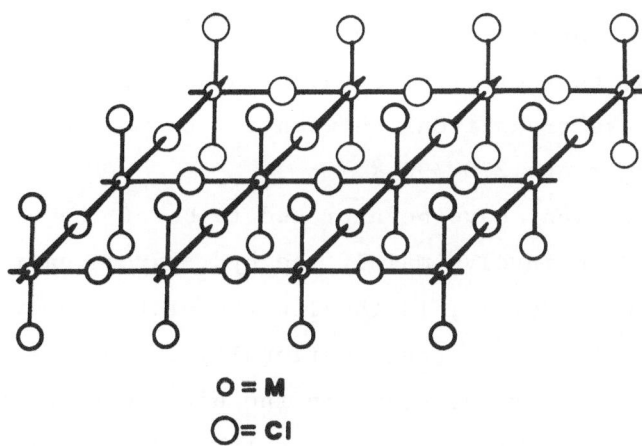

O = M

O = Cl

FIGURE 2. Schematic view of a portion of the macroanion in
the layer perovskites with M = Mn, Cu, Hg, Fe (8).

312

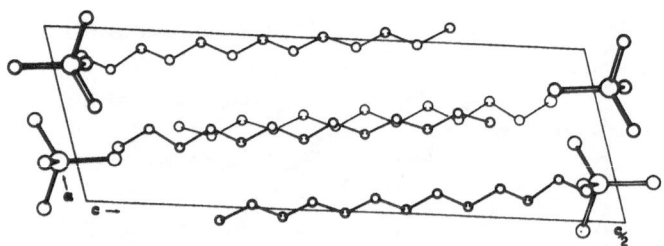

FIGURE 3. The structure of the layer perovskites: compound $(\underline{n}\text{-}C_{12}H_{25}NH_3)_2ZnCl_4$ (9).

ally linked to the inorganic layers. These linear alkyl chains are responsible for the peculiar thermal behaviour of the lay er perovskites, characterized by high enthalpy, reversible solid-solid phase transitions between two polymorphic forms in the temperature range 270-390 K.

These transitions have been proved to be of the order-dis order kind, and to involve mainly the hydrocarbon parts of the structure. In more detail, in the low temperature polymorphs the long alkyl chains are conformationally ordered and mainly in a planar zig-zag arrangement; in the high temperature poly morphs, on the other hand, the chains take on a disordered ar rangement, and gain a conformational freedom comparable with that found for the n-alkanes in the melt (10,11).

One of the most probing experimental evidences supporting such interpretation comes from the study of the IR spectra of

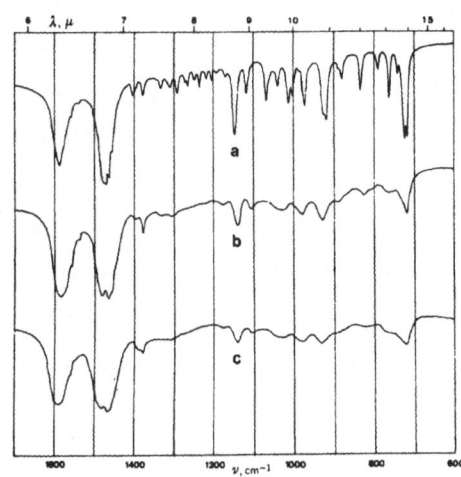

FIGURE 4. 1700-600 cm^{-1} region of the infrared spectrum of
$(\underline{n}\text{-}C_{12}H_{25}NH_3)_2CoCl_4$. (a) Room temperature ordered phase; (b)
High temperature disordered solid phase; (c) Molten phase (10).

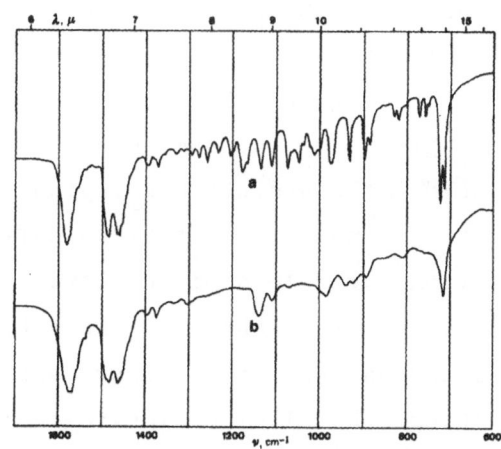

FIGURE 5. 1700-600 cm^{-1} region of the infrared spectrum of
$(\underline{n}\text{-}C_{13}H_{27}NH_3)_2MnCl_4$. (a) Room temperature ordered phase; (b)
High temperature disordered solid phase (10).

the compounds under investigation (10).

Figures 4,5 show that the low temperature IR spectra of
the compounds examined are characteristic of compounds contain
ing long straightchain alkyl groups mainly in the all-trans
configuration. This is unequivocally shown by the 725 cm^{-1} band
(methylene rocking fundamental) which is a doublet, because
there are two non equivalent chains in the crystallographic
unit cell, and by the methylene twisting absorption which is
split into several peaks in the 1200-1300 cm^{-1} region, due to
the regular interaction of CH$_2$ groups of the same chain.

Figure 4c shows the infrared spectrum of a molten Co salt.
The 725 cm^{-1} band is in this case broad and no obvious splitting
is observed. The methylene twisting absorption is, on the other
hand, very broad and unsplit, because of the lack of regular
interactions between CH$_2$ groups of the same chain.

Figures 4b, 5b show the infrared spectra of the high tem
perature solid polymorphs of the compounds investigated. The
725 cm^{-1} band and the 1200-1300 cm^{-1} region of these spectra
are practically identical to the corresponding regions of the
molten Co salt, showing the absence of correlations in these
forms among CH$_2$ groups of the same chain and of adjacent chains.

In other words, the hydrocarbon regions of these still crys
talline solid polymorphs are in a "liquid-like" state (10,11).

The transition parameters of the layer perovskites are
strongly dependent on the length of the alkyl chains and on the
specific metal and halogen in the inorganic regions (figure 6).
Anyway, even though the alkyl chains in the hydrocarbon layers
are fixed at one end, the transition enthalpies per mole of
alkyl chains for the layer perovskites are not much lower than
the molar melting enthalpies of the corresponding normal paraf
fins (of course, the total specific transition enthalpies are

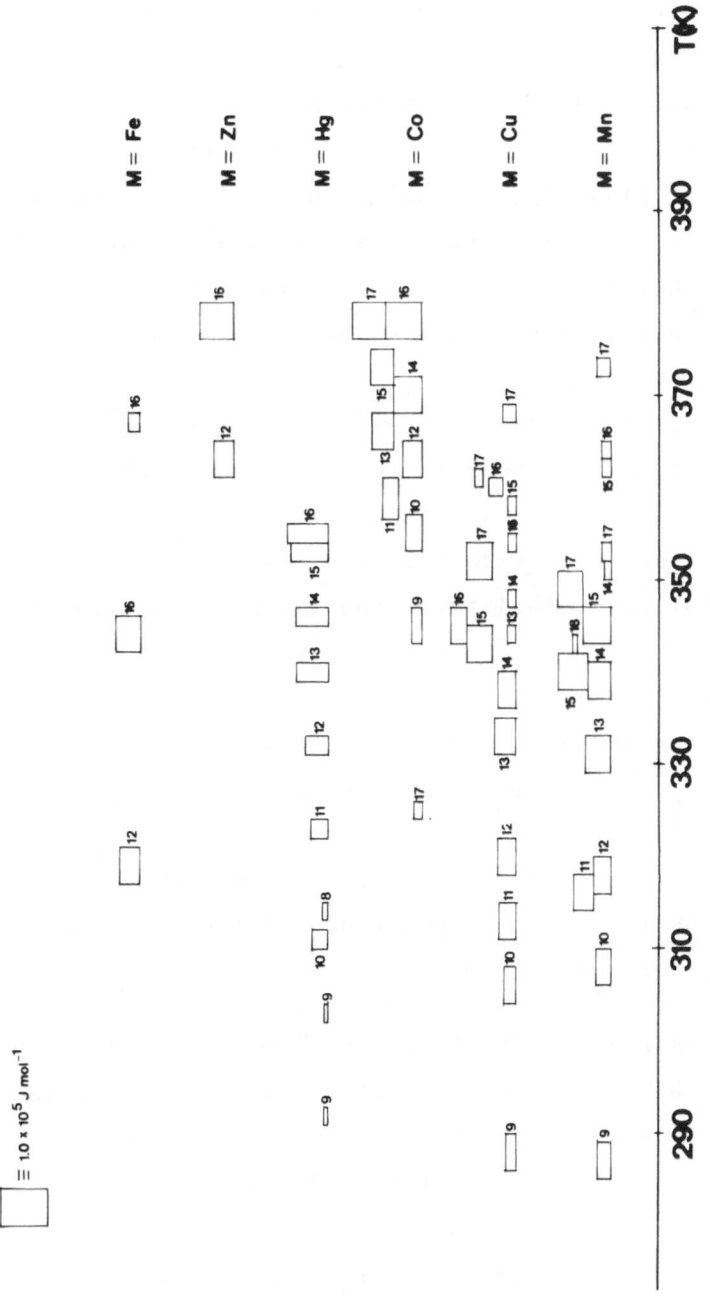

FIGURE 6. Transition temperatures and enthalpies of the layer perovskites with M = Mn,Cu,Hg,Fe,Co,Zn. The area of each rectangle is proportional to the transition enthalpy it represents.

lower because they take into account the weight of the "inert" inorganic regions).

It is also possible to obtain solid solutions of layer pero skites - at least for M = Mn - in which the alkyl chains have different lengths, by chemical methods or by mechanical hot mixing of finely powder pure compounds. These solid solutions exhibit a thermal behaviour not too far from the one consistent with the additivity predictions based upon the behaviour of the pure constituents (12,13).

The solid-solid phase transitions of the pure compounds, as well as those of their mixtures, are thus characterised by relatively high enthalpic changes (10-35 cal g^{-1}). The change of volume associated with the transitions is of the order of 5-10%. Total reversibility of the thermal behaviour is observ ed even after 1000 continuous thermal cycles through the trans ition point.

The layer perovskites show a specific heat at constant pressure, c_p, of the order of 0.4 cal g^{-1} K^{-1}, and their ther mal conductibility, λ, has a mean value of $2x10^{-2}$ W cm^{-1} K^{-1}.

They are chemically stable even at moderately high tempe rature, slow thermal decomposition intervening, in the presence of oxygen, only above 500 K.

3. THE LAYER PEROVSKITES AS TES SYSTEMS: RESULTS AND DISCUSSION

From the physico-chemical data reported in the last section it is possible to argue that the layer perovskites are possible candidate systems for latent heat TES devices (14), as are the normal paraffins. With respect to the latter, they have the ad vantage of remaining solid after the phase change, and of under going a more limited increment of volume with the transition; their thermal conductibility is also higher of about one order of magnitude. In table 1, some relevant thermodynamic data for the layer perovskites are summarized, and compared with the corresponding ones for the normal paraffins.

TABLE 1. Some relevant thermodynamic data for the layer perov skites and for the normal paraffins.

LAYER PEROVSKITES	NORMAL PARAFFINS
Solid-solid phase transition	Solid-liquid phase transition
Transition temperature range: 0 - 120 °C	Transition temperature range: 10 - 70 °C
Transition enthalpy: 10-35 cal g^{-1}	Transition enthalpy: 30-50 cal g^{-1}
Change of volume at the trans ition: 5-10%	Change of volume at the trans ition: 15%
Density (at 25°C): 1.1 - 1.5 $g\ cm^{-3}$	Density (molten): 0.8 - 0.9 $g\ cm^{-3}$
Specific heat at constant pressure: 0.4 cal $g^{-1}\ K^{-1}$	specific heat at constant pressure: 0.4 cal $g^{-1}\ K^{-1}$
Thermal conductibility (at 25°C): $2x10^{-2}$ W $cm^{-1}\ K^{-1}$	Thermal conductibility (at 25 °C): $0.5-2.0x10^{-3}$ W $cm^{-1}\ K^{-1}$

Anyway, over and above the obvious considerations related to physico-chemical properties, the practical applications of new materials depend on the attainment of acceptable production costs and economic storage and handling. In the case of the lay er perovskites, in an attempt to reduce the production costs, modifications of the normally used synthesis procedures have been successfully developed which eliminate the necessity of using highly purified amines and expensive solvents. The inherent difficulties associated with storage and handling have been over come by preparation of polymeric composites wherein the perov skites are incorporated as dispersed fillers, thereby protect ing the compounds from various forms of environmentally induc ed degradation (15).

As it is foreseable on the ground of the results for the solid solutions of the layer perovskites, compounds of the general formula $(n-C_nH_{2n+1}NH_3)_2MCl_4$, prepared from commercially available amines (Armeen, supplied by Akzo Chemie), exhibit a thermal behaviour very similar to the optimal behaviour observed in compounds prepared from the corresponding reagent grade chemicals. Large scale samples of such compounds for M = Mn have been prepared in the experimental apparatus shown schematically in figure 7. The process involves batch stirring of the

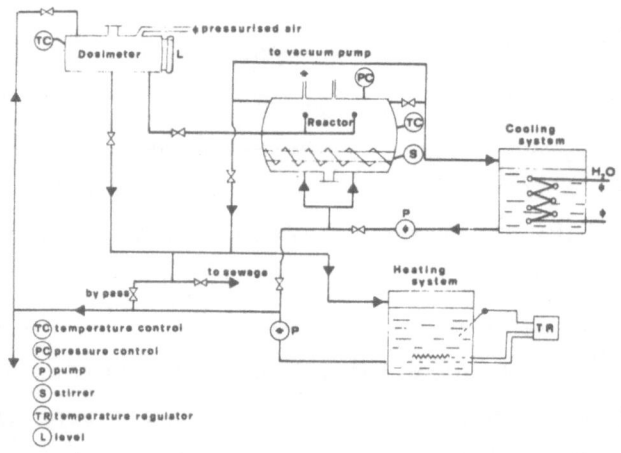

FIGURE 7. Schematic picture of the experimental apparatus used for the preparation of large scale samples of layer perovskites.

molten alkylamine in a reactor fitted with a heating mantle at a temperature of 70 °C. While maintaining continuous agitation, a stoichiometric quantity of concentrated HCl(aq) is added. At the end of the reaction, the resulting alkylammonium chlorhydrate is reacted stoichiometrically with $MnCl_2 \cdot 4H_2O$ in concentrated aqueous solution. The final product is then dried in the same reactor under vacuum at a temperature of 80 °C. A production cost of 1-2 U.S. dollars seems however to be the best achiev able one.

The resulting compounds, although characterised by excellent thermal behaviour, are recovered as friable powders which show poor mechanical properties even after prolonged syntherisation. These powders were therefore incorporated as "thermally active" fillers in polystyrene matrices by conventional compression moulding techniques. Systematic thermal and mechanical charac terisation of the composites derived from $(n-C_{12}H_{25}NH_3)_2MnCl_4$ dispersed in polystyrene (Montedison Edistir FA) have been per formed (15,16). This particular layer perovskite exhibits a so lid-solid phase transition at 332 K, with an enthalpy change of 19 cal g^{-1}.

The thermal behaviour of the perovskite remains unaltered after the inclusion into the polymeric matrix, as shown by ex tensive DSC characterisation. The flexural modulus of the com posites, obtained with an Instron 1112 testing machine at a strain rate of 5 cm min^{-1}, is shown in figure 8 at temperatures trav ersing the transition one. At lower temperatures the moduli are only slightly dependent upon the filler content, although at higher temperatures there is a discernible decrease of the me chanical properties with increasing filler content; also in this case, anyway, the materials remain fully self-carrying at least for filler loadings of less than 80%.

In figure 9, the time dependence of the temperature on the "cold" surface of a composite slab, heated continuously at a constant temperature from the opposite fase, is shown. The charac teristic plateau of temperature vs. time occurs near the trans ition temperature, since the latent heat effect associated with the solid-solid transition precludes additional temperature

320

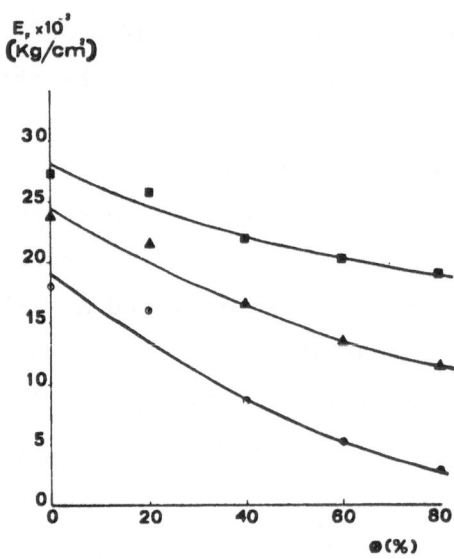

FIGURE 8. Flexural modulus, E_F vs. volumetric filler content, ϕ of the composites at various temperatures: ■ T = 20°C; ▲ T = = 40 °C; ⊙ T = 60 °C. Composites thickness: 5 mm.

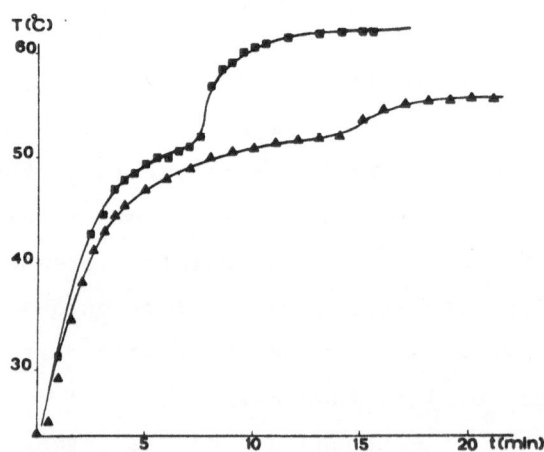

FIGURE 9. Time dependence of the temperature of the "cold" face of a 5 mm thick composite slab with a filler content of 60% heated at 65 °C (▲) and at 75 °C (■) from the opposite surface.

rises of the slab "cold" surface as long as the transition is taking place. The duration of the plateau effect is plotted vs. the filler content in figure 10 and, quite intriguingly, passes through a maximum at a filler loading of the order of 60%, presumably corresponding to incipient contact of the part

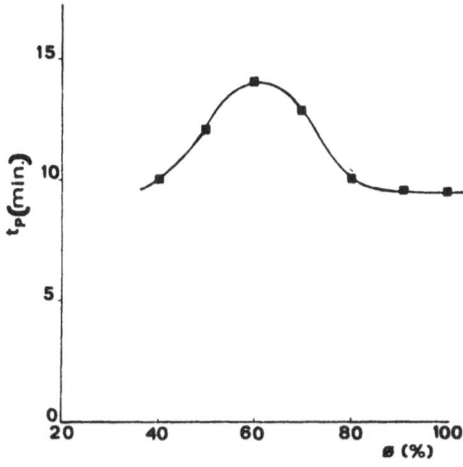

FIGURE 10. Duration of the plateau effect, t_p vs. the filler content, ϕ of 5 mm thick composite slabs heated continuously at a constant temperature of 65 °C.

icles constituting the dispersed phase. At higher filler cont ents, therefore, the dispersed phase serves as an intercomunicat ing locus for heat transfer through the composite slab (15,16).

Lastly, it could be worthwile to underline that nearly no changes of volume with the solid-solid phase transitions are observed for the layer perovskites/polystyrene composites with a filler content of less than 80%, presumably because the empty interstices between the filler particles and the polymeric ma trix are large enough to accomodate the filler volumic incre ments without deformations of the matrix itself.

4. CONCLUSIONS

A material which undergoes a latent heat change at a certain temperature must comply with the following well known criteria to be of potential interest for energy storage:

High enthalpy of transformation per unit volume of the material;
Perfectly reversible transformation;
Small volume change associated with the latent effect;
Chemical stability and compatibility;
Non toxic process effluents;
Adequate lifetime;
Cost-effective process performance.

At present there do not appear to be examples in the litera ture which document systems satisfaying all these requirements. As previously said, the most promising suggestions seem to be related to the exploitation of normal paraffins solid-liquid (physical) conversion or some eutectic salt hydrate mixtures which undergo chemical transformations (dehydration).

The thermal properties of the layer perovskites suggest various applications involving storage of thermal energy or, alternatively, special insulations at the respective transition temperatures. For such applications these materials offer several advantages, including the perfect reversibility of the thermal behaviour and the inherent convenience associated with a solid-solid, rather than a solid-liquid or solid-vapour, transition. Moreover, the available compounds offer a wide range of trans ition temperatures. Although the enthalpic changes associated with the phase changes are not very large ($10-35$ cal g^{-1}), they are of the same order of magnitude as the transformation enthal pies of other recently proposed latent heat TES systems - for example, a device based on a core of Glauber's salt, suggested by various American industrial groups and by researchers at

MIT (1,17).

When included into polymeric matrices as thermally active fillers, the layer perovskites present, as just said, nearly no change of volume at the phase transition; they are chemical ly stable and the polymeric matrix isolates them from the ex ternal environment.

Their production cost, on the other hand, is about ten times that of the normal paraffins. However, although a casual evalu ation of these systems suggests that at present there is a sig nificant economic advantage connected with the exploitation of less exotic devices, the technical problems associated with the confinement of the resulting liquid or vapour phases must be included into a comprehensive evaluation.

It would seem, therefore, that ultimately the convenience in handling and operating associated with thermal accumulation based upon solid-solid transitions will provide the requisite economic incentives for further development and - at least for special applications - ultimate market acceptance.

ACKNOWLEDGEMENTS

We thank Drs. A. Addeo, C. Carfagna and C. Migliaresi for their collaboration.

Financial support for this study was provided by the Italian National Research Council (CNR) (Energy Research Programme - Subproject: Solar Energy - Topic A1: Materials and Thermophy sical Properties).

REFERENCES

1. V. Rosselli. Stato dell' arte di sistemi di accumulo dell' energia termica per impianti ad energia solare. 1978. Rome, CTIP Solar Ed.

2. M. Vacatello and P. Corradini, Gazz.Chim. Ital. 103 (1973), 1027

3. M. Vacatello and P. Corradini, Gazz. Chim. Ital. 104 (1974), 773

4. M. Vacatello, Ann. Chim. (Rome) 64 (1974), 13

5. V. Busico, V. Salerno and M. Vacatello, Gazz. Chim. Ital. 109 (1979), 581

6. E. Landi and M. Vacatello, Thermoc. Acta 13 (1975), 441

7. E. Landi, V. Salerno and M. Vacatello, Gazz. Chim. Ital. 107 (1977), 27

8. M.R. Ciajolo, P. Corradini and V. Pavone, Gazz. Chim. Ital. 106 (1976), 807

9. M.R. Ciajolo, P. Corradini and V. Pavone, Acta Cryst. B33 (1977), 553

10. C. Carfagna, M. Vacatello and P. Corradini, Gazz. Chim. Ital. 107 (1977), 131

11. M. Vacatello and P. Corradini, Rend. Accad. Sci. Fis. Mat. (Naples) IV, XLIV (1977), 505

12. V. Salerno, A. Grieco and M. Vacatello, J. Phys. Chem. 80 (1976), 2444

13. V. Busico, V. Salerno and M. Vacatello, Gazz. Chim. Ital. 109 (1979), 577

14. V. Busico, C. Carfagna, V. Salerno, M. Vacatello and F. Fittipaldi, Solar Energy 24 (1980), 575

15. A. Addeo, L. Nicolais, V. Busico and C. Migliaresi, Appl. Energy 6 (1980), 353

16. V. Busico, P. Corradini, C. Migliaresi and L. Nicolais, J. Appl. Pol. Sci., in press

17. Anon. Chemical Week (1 March, 1978), 34

THE ABSORPTION PROCESS FOR STORAGE OF LOW-TEMPERATURE HEAT

WIKTOR RALDOW
Swedish Council for Building Research
Sankt Göransgatan 66, S-112 33 Stockholm, Sweden

ABSTRACT

The absorption principle is presented and illustrated by a description of an advanced Swedish chemical heat pump project for seasonal storage of solar thermal energy. A short overview of the usable working pairs and heat reservoirs is given. In addition, the economic feasibility of some applications of thermochemical energy stores is briefly discussed.

INTRODUCTION

The energy system of today consists of a centralized large producer network and decentralized small consumer sites. The producers store, transform and distribute energy. To make these processes feasible on a large scale, the energy carriers employed, fossil fuels and electricity for example, are of a very high thermodynamic quality. The high quality of the energy distributed to the consumers makes its utilization very convenient. On the other hand, a large part of the supply is used for the heating and cooling of space, i.e. low quality needs. This mismatch in the thermodynamic quality between supply and demand clearly demonstrates that the potential for saving energy is high. A radical step towards accomplishing this goal would be to link the consumer price to the exergy content of the energy supplied.

The future restructuring of the energy system will proceed towards better utilization of the conventional energy sources and the use of intermittent and dispersed ones, e.g. solar and wind. For both paths a need for decoupling of the energy supply and demand in time and space will have to be met, this being accomplished by efficient storage and transport of different forms of energy.

Since a very large part of the demand constitutes heat, efficient stores for this form of energy will have to be developed. Thermochemical energy storage offers many advantages: the energy quality can be preserved and the storage densities are high. In addition, storage and transport can occur at ambient temperatures without losses, and conversion from heat to other forms of energy is possible.

THE ABSORPTION PRINCIPLE

One of the possible designs of a thermochemical store is the so called chemical heat pump, which works according to the absorption principle. (Absorption machines, known for more than a hundred years, are suitable for decentralized systems as well as energy storage.)

In order to describe this principle, let us start with a vessel, H, in which a complex, C, between an inorganic salt, S, and a volatile compound, V, is contained. The vapour pressure of V in H is described by

$$\ln P_H = -\frac{\Delta H_H}{RT} + \frac{\Delta S_H}{R} \qquad (1)$$

if we assume that ΔH_H and ΔS_H, the enthalpy and the entropy change for the reaction

$$C_{(s)} \rightleftharpoons S_{(s)} + V_{(g)} \qquad (2)$$

are not dependent on temperature. In order to separate V from C at a certain temperature T_m, we can (reversibly) evaporate V by providing heat of a reaction $|\Delta H_H|$ to the system. Then, for storage, V can be liquified in a separate vessel, L, at T_m. The condensation pressure, P_h, can be determined from

$$\ln P_L = -\frac{\Delta H_L}{RT} + \frac{\Delta S_L}{R} \qquad (3)$$

which equation describes the vapour pressure above liquid V

$$V_{(l)} \rightleftharpoons V_{(g)} \qquad (4)$$

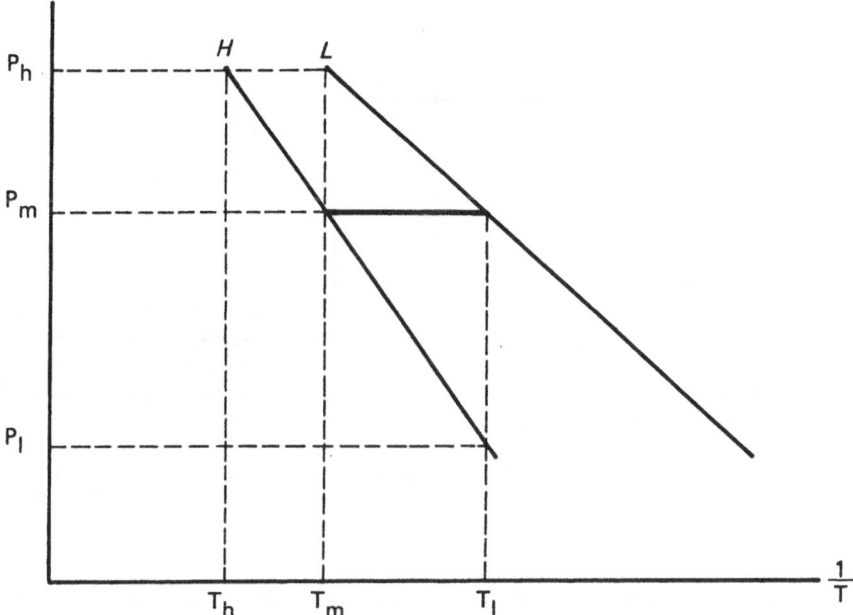

Fig 1. An absorption system in a pressure - temperature diagram.

At this point different equilibrium vapour pressures prevail in the two vessels and this is demonstrated by a chemical potential difference: if the containers were linked together, V would flow from L to H. In order to prevent this, the temperature of L can be decreased from T_m to a temperature T_1 at which, see fig 1:

$$P_H(T_m) = P_L(T_1) = P_m \tag{5}$$

Under these conditions, a steady state of no mass-transfer between the two tanks, one at T_m (high-temperature tank) and one at T_1 (low-temperature tank), will prevail.

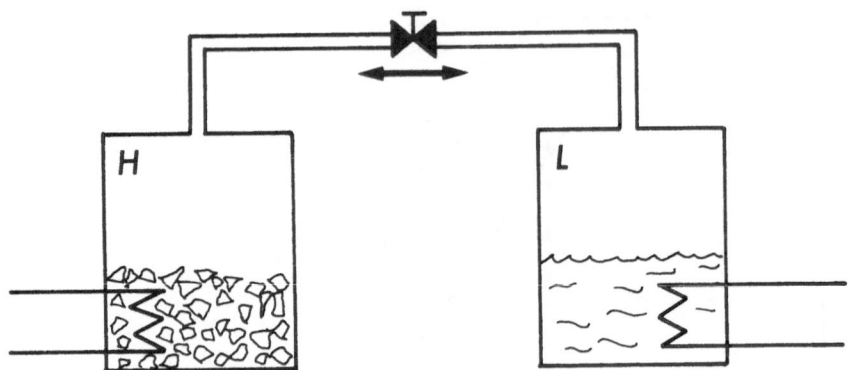

Fig 2. A simple periodic absorption machine. Here it is assumed that in tank L the working fluid is condensed whereas it is bound by a salt to a complex in tank H.

A simple absorption machine can be built of two such vessels, H and L, in interaction with two heat reservoirs: a warm at T_M and a cold at T_L, see fig 2.

The overall reaction which takes place in the system is

$$C_{(s)} \underset{\longleftarrow}{\overset{\longrightarrow}{}} S_{(s)} + V_{(1)} \tag{6}$$

and for this reaction

$$\Delta H = \Delta H_H - \Delta H_L \tag{7}$$

and

$$\Delta S = \Delta S_H - \Delta S_L \tag{8}$$

The products of (6) are separated (on a slightly different path than the one described above) during the regeneration mode, which proceeds as follows:

Heat $|\Delta H_H|$ from the warm reservoir is supplied to H, and the complex decomposes. The vapour pressure in H will rise (infinitesimally), V will distill over to L and condense at T_L. The heat of condensation $|\Delta H_L|$ will have to be dissipated into the cold reservoir.

In this charging mode, energy is stored in the machine. The stored energy

$$\Delta H = \Delta H_H - \Delta H_L = \Delta G - T \Delta S \tag{9}$$

can be devided into a work (of separation) term

$$w_m = \Delta G(T_m) = RT_m \ln \frac{P_h}{P_m} \tag{10}$$

and a heat term

$$q_m = T_m \cdot \Delta S \tag{11}$$

The energy stored (9) in the thermochemical reaction (7) is, under the above specified conditions, not dependent on the temperature. However, the work (10) and heat (11) terms are temperature dependent.

The work part of the stored energy can be utilized for pumping heat from the cold (T_L) into the warm (T_M) reservoir. In this working mode, i.e. in the reversed process to the one described above:

$$w_m = RT_m \ln \frac{P_h}{P_m} = \Delta H_L \frac{T_m - T_1}{T_1} \tag{12}$$

Another possible utilization of the absorption machine is the production of work e.g. in a device such as the one presented in figure 3. Ideally, in this device the working fluid will be expanded from P_m to P_1 at a constant temperature T_1. (The regeneration still requires the temperature T_m.) The work obtainable from the process is

$$w_1 = RT_1 \ln \frac{P_m}{P_1} = |\Delta H_H| \frac{T_m - T_1}{T_m} \tag{13}$$

Note that in this case the whole of the exergy provided to the system during the regeneration is converted to work.

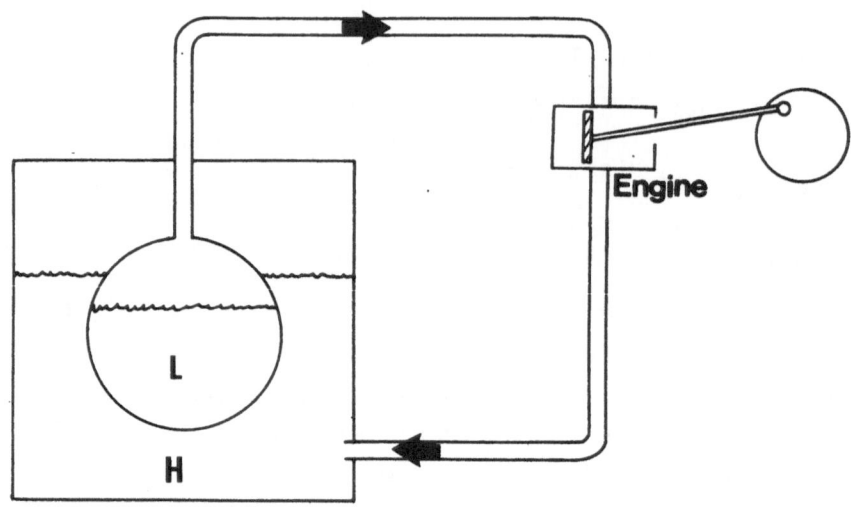

Fig 3. An absorption engine.

POSSIBLE APPLICATIONS

From the discussion so far, it is apparent that the absorption process can be used for heating (heat of reaction in H is utilized), for cooling (when heat withdrawn into L is pumped from a cooled space) and for production of high-quality (mechanical) energy. The time gap between regeneration and working modes means energy storage, during which period the working pair, S and V, can be transported from a regeneration to a consumer site and there recombined. Furthermore, if after regeneration, the vessel L is linked into the warm reservoir, heat can be upgraded from T_m to a higher temperature T_h, see figure 1. In summary, the following applications are possible:

- energy storage
- energy transportation
- heating
- cooling
- upgrading of heat
- transformation of heat to high-quality energy

Some of the above can be combined within a single machine.

The following section will describe one application: the storage of heat for heating purposes. The machines designed for this purpose are usually called chemical heat pumps.

THE TEPIDUS PROJECT*

An advanced Swedish project aiming at seasonal storage of solar thermal energy for space heating and hot water production in a single-family house is taken as an illustration of the present development. Its goal is to prove the feasibility of seasonal storage and to verify the technique.

The actual installation is shown in figure 4. The main features can be presented as follows:

The working fluid and the absorber in the system are H_2O and H_2S, respectively. The volume of tank L is very small since water is not stored but added to or withdrawn from the unit upon demand. This greatly reduces the volume of the system. A vacuum pump is permanently attached to the system to degas new portions of water and remove residual gases. The ground is used as the low-temperature heat source/sink to prevent the freezing of water; its temperature varies from 0 to 15^0 C. Heat is extracted from and dissipated to the soil through a ground coil - a technique widespread in Sweden. 40 m^2 flat plate solar collectors provide heat for charging. The collectors can also be used for direct heating or evaporation of water.

*Further information can be obtained from the project manager, Mr K Bakken, Tepidus AB, P O Box 5607, S-114 86, Stockholm, Sweden. The work is jointly sponsored by the Swedish Council for Building Research and the Swedish Board for Technical Development.

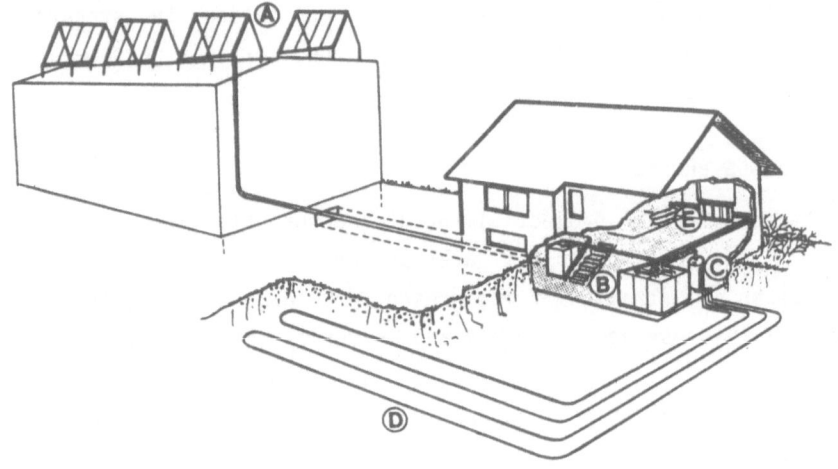

Fig 4.   Prototype of the Tepidus chemical heat pump in a single-family
house

    a) solar collectors
    b) modules with Na_2S (H-tank)
    c) condensor/evaporator (L-tank)
    d) soil heat exchanger
    e) radiator/tap water

(from E Å Brunberg: The Tepidus System for Seasonal Heat Storage and for
Cooling)

The main data for this 100 % solar system can be summarized as follows:

- storage capacity - approximately 7 000 kWh
- 7 000 kg of Na_2S is stored in 8 modules
- 2/3 of the energy is pumped from ground through
- 3 x 300 m polyethylene pipes
- operating pressure range 5-20 torr
- charging temperature $75 < t < 100^0$ C
- discharging temperature 50-60^0 C
- maximal effect: 20 kW upon charging, 10 when discharged

CHEMICAL HEAT PUMP CLASSIFICATION

The system outlined above is just one of many under development. The others cannot be described here. However, some characteristics of the main types are sketched out in this article in a very condensed form, see Tables 1-4. Accordingly, the Tepidus prototype, fig 4, can be described as an intermittent, semi-closed, dry absorption machine in which

working fluid is	H_2O
absorbent	Na_2S
heat source	solar thermal
heat sink	ground
low-temperature heat source	ground

and which is connected to a hydronic heating system in a single-family house.

Table 1

Some types of absorption machines

Criterium	Type	Comments
Periodicity	Intermittent Continuous	machines working between three temperature levels. Sensible heat recovery improves thermal efficiency
Mass transport from and to the system	Closed Open	the working fluid (H_2O, O_2 is not condensed but stored in the atmosphere, e.g. dryers, dehumidifiers
State of sorbent material	Wet	liquid absorbents. Moving temperatures facilitate adaptation to the temperatures of heat reservoirs. Pumping of solutions easily accomplished. Solvent may boil away with the working fluid. Considerably better heat transfer as compared to dry systems
	Dry	solid absorbents or adsorbents. Charging and discharging at nearly constant temperatures in systems with absorbents. Easy regulation. High energy densities. Solid phase stability is a problem in achieving long life-times. Low thermal conductivity complicates heat exchanging and limits the heat flows from and to the system
Others		e.g. resorption systems in which working fluid is bound by sorbents in both tanks; multistage machines

Table 2

Some working fluids utilized in the absorption process

Fluid	Comments
H_2O	cheap. Environmentally safe. High melting point limits the working temperatures. Ambient air not suitable as low-temperature heat source. Open systems feasible. High energy density achievable in semi-closed systems. Very careful design necessary due to low operating pressures. Carrier gas increases parasitic power requirements.
NH_3	toxic (Swedish standards allow up to 3 kg of NH_3 inside a house.) Relatively high condensation pressure. Ambient air can be used as low-temperature heat sink/source
CH_3OH	low storage energy densities. Fire and toxicity hazards. Low melting point, medium pressure range
H_2	resorption systems. Expensive metal alloy adsorbents
freons	organic solvents as absorbents
others	e.g. O_2, SO_2, CO_2, CH_3NH_2, $C_2H_5NH_2$, C_2H_5OH, CH_3COCH_3

Table 3

Some sorbent materials

Working fluid	Sorbent	
	wet	dry
H_2O	$LiBr^{1)}$, H_2SO_4, NaOH, $CaCl_2$, glycols	Na_2S, $MgCl_2$, CaO, zeolites, silica gel
NH_3	H_2O, $LiBr^{1)}$	$CaCl_2^{2)}$, $SrCl_2$, $FeCl_2$, $MnCl_2$, $NH_4Cl^{3)}$
CH_3OH	$LiBr^{1)}$	$CaCl_2$
H_2		metal alloys e.g. of $LaNi_{5-x}$ Al_x type
freons	organic solvents	

1) also salt mixtures
2) also systems with inert solvent
3) liquid ammoniates formed

Table 4

Examples of heat reservoirs for chemical heat pumps for space heating[1,2]

Type	Source
High-temperature source	solar thermal energy
	industrial or utility waste heat
	fossil fuels
	wind energy
	others, e.g. resistance heating
Heat sink on charging	ambient air
	ground
	water reservoirs, e.g. aquifers, lakes
	heated space
Low-temperature source	ambient air
	ground
	water reservoirs
	cooled space
	exhaust air

1) An interesting alternative is to charge a chemical heat pump by using a small conventional compressor heat pump.

2) In the reversed absorption process heat at an insufficient temperature can be upgraded (through a linkage to a heat sink) to temperatures suitable for space heating.

SUMMARIZING REMARKS

Several difficult problems, e.g. concerning heat transfer, mass transfer, solid phase stability, environmental safety, materials, parasitic power requirements, will have to be solved before a reliable absorption storage machine is constructed. Another challenge to meet is making the systems economical at present or in the near future.

Economic assessment of the field of thermochemical energy storage is rather difficult due to the novelty of the applications and technology. A comprehensive project, sponsored by the Council and aiming at a detailed analysis of the field is presently being undertaken. However, some preliminary conclusions can be presented here. Once more, the Tepidus single-family installation described above will serve as a starting point for the discussion.

An economic assessment of this installation reveals that the energy price is, at present, not competitive. The main costs in the system are:
- solar collectors
- storage unit
- ground coil

The last one, the cost associated with the low-temperature heat source, can be reduced if lower melting fluids as NH_3 or CH_3OH were chosen. On the other hand, these choices necessitate larger storage volumes and more extensive safety precautions. As far as the costs associated with the high-temperature heat source are concerned, it seems that a substantial reduction in the price of solar collectors is needed to significantly improve the economy of solar driven units. However, charging the system with a small vapour compression heat pump which works under nearly optimal conditions during the summer, appears promising. Another very interesting possibility is to utilize waste heat.

Finally, the investment in the storage unit is very high due to the fact that the unit cycles only once a year. To lower the capital cost the units have to be charged and discharged more often. This can be accomplished by e.g. transporting smaller movable units to and from an industrial (waste heat) source. In this respect one should remember that a certain application demands a certain effect. In the case of a single-family house, much larger effects per mass unit will have to be reached than those presently achieved in the Tepidus system.

A large number of cycles means also that the storage time decreases. The competition from other, presumably cheaper techniques such as sensible and latent heat storage, will then stiffen. In other words, the unique properties of thermochemical storage, i.e. the potentiality of long term energy storage, of energy transportation and flexibility in application, have to be utilized to design a system profitable under present conditions.

LITERATURE

1. W Niebergall, Sorptions-Kältemaschinen, Springer-Verlag, Berlin, Göttingen, Heidelberg, 1959.
2. F Daniels, Direct Use of the Sun's Energy, Ballentine Books, New York, 1974.
3. G Wettermark, B Carlsson and H Stymne, Storage of Heat. A Survey of Efforts and Possibilities, Swedish Council for Building Research, Document D2:1979. Available from Svensk Byggtjänst, Box 1403, S-111 84 Stockholm, Sweden.
4. H Bjurström and W Raldow, The Absorption Process for Heating, Cooling and Energy Storage - An Historical Survey, International Journal of Energy Research, in press.
5. G Wettermark (editor), International Seminar on Thermochemical Energy Storage, Swedish Council for Building Research, Document D25:1980. Available from Svensk Byggtjänst, Box 1403, S-111 84 Stockholm, Sweden.

AUTHORS INDEX

AUTHORS INDEX PAGE

344

HOOGENDOORN, Prof. ir. C.J.   75
Technische Hogeschool Delft
Afdeling der Technische Natuurkunde
Vakgroep Warmtetransport
Lorentzweg 1
2628 CJ DELFT
THE NETHERLANDS

IRIS, P.   197
Ecole Nationale Supérieure des
Mines de Paris
Centre d'Informatique Geologique
35, Rue Saint Honoré
77305 FONTAINEBLEAU
FRANCE

JADOT, R.   265
Faculté Polytechnique de Mons
9, Rue de Houdain
B-7000 MONS
BELGIUM

JONG, Ir. A.G. de   123
Grontmij N.V.
P.O. Box 513
3700 AD ZEIST
THE NETHERLANDS

JONG, Prof. Ir. W.A. de   3
Chairman of the Board of Directors of the
Netherlands Organization for Applied Scientific
Research
P.O. Box 297
2501 BD DEN HAAG
THE NETHERLANDS

JOUANNA, Prof. P.   233
Laboratoire de Génie Civil
Université de Montpellier II
Place Eugène Bataillon
34060 MONTPELLIER CEDEX
FRANCE

KOPPEN, Prof. ir. C.W.J. van   11
Technische Hogeschool Eindhoven
Afdeling Werktuigbouwkunde
Vakgroep Warmte- en Stromingstechniek 4.06
P.O. Box 513
5600 MB EINDHOVEN

346

MALATIDIS, N.A.   157
Institut für Kernenergetik und
Energiesysteme (IKE) der
Universität Stuttgart
Postfach 801140
7000 STUTTGART-80 (Vaihingen)
FEDERAL REPUBLIC OF GERMANY

MANCINI, Prof. N.A.   99
Instituto di Fisica dell'
Università
57, Corso Italia
I 95129 CATANIA
ITALY

MARSHALL, Dr. R.H.   111
University College Cardiff
Newport Road
Cardiff CF2 ITA
GREAT BRITAIN

MICHAELS, Dr. Allan I.   79
Solar Thermal Storage Program
Argonne National Laboratory
9700 South Cass Avenue
Building 362
ARGONNE, ILLINOIS 60439
UNITED STATES OF AMERICA

NICOLAIS, L.   309
Instituto di Principi di Ingegeneria Chmica
80125 Naples
ITALY

OFVERHOLM, E.   91
Byggforskningsradet
St. Göransgatan 66
112 33 STOCKHOLM
SWEDEN

OUDEN, Ir. C. den   prefa
Technisch Physische Dienst TNO-TH
P.O. Box 155
2600 AD DELFT
THE NETHERLANDS

RADEMAKER, Prof. O.   61
Technische Hogeschool Eindhoven
Gebouw Warmte en Stroming
P.O. Box 513
5600 MB. EINDHOVEN
THE NETHERLANDS

RALDOW, Dr. W.   325
Department of Physical Chemistry
The Royal Institute of Technology
100 44 STOCKHOLM
SWEDEN

SCHOLZ, Dr. Ing. F.   221
Institut für Reaktorbauelemente
Kernforschungsanlage Jülich GmbH
P.O. Box 19 12
D-5170 JULICH
FEDERAL REPUBLIC OF GERMANY

TABOR, Dr. H.   17
The Scientific Research Foundation
Hebrew University Campus
P.O. Box 3745
JERUSALEM
ISRAEL

TORRENTI, R.   179
Ecole Nationale Supérieure des
Mines de Paris
Centre d'Energétique
Stockage de Chaleur
Sophia Antipolis
06560 VALBONNE
FRANCE

TSANG, CHIN FU   185
Lawrence Berkeley Laboratory
University of California
Berkeley, California 94720
UNITED STATES OF AMERICA

VACATELLO, M.   309
Instituto Chimico dell' Università
80134 Naples
ITALY

VANHEELEN, J.   33
Katholieke Universiteit Leuven
Departement Werktuigbouwkunde
Afd. Toegepaste Mechanica en
Energiekonversie
Celestijnenlaan 300 A
B-3030 HEVERLEE
BELGIUM

VELTKAMP, Ir. W.B.   47
Technische Hogeschool Eindhoven
Gebouw Warmte en Stroming
Kamer 01.32
P.O. Box 513
5600 MB EINDHOVEN

348

VERDONSCHOT, Ir. J.K.M.   283
Technisch Physische Dienst TNO-TH
P.O. Box 155
2600 AD DELFT
THE NETHERLANDS

VIALARON, A.   295
Centre National de la Recherche Scientifique
Laboratoire d'Energetique Solaire
B.P. 5, Odeillo
66120 FONT-ROMEU
FRANCE

WIJSMAN, Ir. A.J.Th.M.   209
Technisch Physische Dienst TNO-TH
P.O. Box 155
2600 AD DELFT

LIST OF PARTICIPANTS

BELGIUM

ARNOULD, L.
Programmation de la Politique
Scientifique
8 Ave de la Science
B-1040 Brussels

BLONDEEL, C.
CEN/SCK
Boeretang 200
B-2400 Mol

BOUGARD, J.
Faculté Polytechnique de Mons
31 Boulevard Dolez
B-7000 Mons

DUTRÉ, W.L.
K.U.L.
300 A, Celestijnenlaan
B-3030 Heverlee

JADOT, R.
Faculte Polytechnique de Mons
31 Bd Dolez
B-7000 Mons

KINNAER, L.
CEN/SCK
Boeretang 200
B-2400 Mol

KLEPFISCH, G.
Centre Scientifique et Technique
de la Construction
Rue du Lombard 41
B-1000 Bruxelles

PONCELET, J.P.R.M.
Fondation Universitaire
Luxembourgeoise,
Rue des Déportés 140
B-6700 Arlon

STEEMERS, T.C.
Commission of the European Communities
Directorate General XII
Rue de la Loi 200
B-1049 Brussels

STEENBERGHE, T. van
Belgonucleaire,25, Rue du Champ de Mars
1050 Brussels

STRUB, A.
Commission of the European Communities
Directorate General XII
Rue de la Loi 200
B-1049 Brussels

THEUNISSEN, P.H.
Von Karman Institute for Fluid Dynamics
Chaussee de Waterloo 72
1640 Rhode St. Genase

VERLATEN, J.
Solvay et Cie S.A.
310 Rue de Ransbeek
B-1120 Brussels

DENMARK

CHRISTENSEN, P.
Techn. Univ. Den. Thermal Insulation
Laboratory
Bygn 118 DTH
DK-2800 Lyngby

FURBO, S.
Thermal Insulation
Technical University of Denmark
Bldg 118
DK-2800 Lyngby

HELSHOJ, E.
Effex Innovation A/S
Reventlowsgade 8
DK-1651 Copenhagen V

FINLAND

HUHTINEN, M.
Valmet Oy Pansio Works
SF-20240 TURKU 24

KOPONEN, V.
Valmet Oy Pansio Works
SF-20240 TURKU 24

MATILAINEN, V.
Osy Oy
Box 15, SF 26101
Rauma 10

FRANCE

ALLIER, M.
Gilles Olive Ingenieur Conseil
16, Rue Nansouty
74014 Paris

Aubert-Dassé, C.
Centre de Recherches sur les très
basses Températures
25 Ave des Martyrs
38042 Grenoble Cedex

BENET, J.C.
Laboratoire Genie Civil
USTL Montpellier II
Place Eugene Bataillon
34060 Montpellier Cedex

BLAY, D.
Laboratoire d'Energetique
Solaire C.N.R.S.
40, Av. du Recteur Pineau
86022 Poitiers Cedex

BRUN, M.J.
Institut Francais de l'Energie
3, Rue Henri-Heine
75016 Paris

CALZIA, J.
Rhone Poulenc
Division Chimique Fine
21, Rue Jean Goujon
75008 Paris

CONSEIL, B.
Centre de Recherche ELF de Solaize
BP 22
69360 Saint Symphorien d'Ozon

DELCAMBRE,
Centre Scientifique et Technique
du Batiment (CSTB)
Boite Postale 21
06562 Valbonne Cedex

DESPOIS, J.
Cen/Saclay - Demt/Seen,
B.P. no. 2
91190 Gif sur Yvette

HOCHON,
Desmousseaux Information
Head Laboratoires de Marcoussis
Route de Nozay
91460 Marcoussis

IRIS, P.
Ecole des Mins de Paris
35 Rue Saint Honore
77305 Fontainebleau

JOUANNA, P.
Laboratoire Genie Civil
Place Eugene Bataillon
34060 Montpellier Cedex

LARUE, J.
Institut Francais du Petrole
1 et 4 Avenue de Bois-Práu
92500 Rueil Malmaison

LAURENT, P.A.
Consultant Solar Energy
25, Rue de la Sourdière
75001 Paris

LECOMTE, D.
Ecole Nat. Supérieure des Mines de Paris
Sophia Antipolis
06560 Valbonne

LIEVRE, P. C.E.A.
31 Tue de la Fédération
75015 Paris

ROMIJN, M.M.
Energy Materials S.A.
7, Rue Chambiges
75008 Paris

TORRENTI, R.
Ecole Nationale Supérieure des Mines
de Paris
Centre d'Energétique ENSMP
Sophia Antipolis
F-06560 Valbonne

GERMANY (F.R.)   ABHAT, A.
IKE, Universität Stuttgart,
Pfaffenwaldring 31
7000 Stuttgart 80

BOESE, F.K.
Interaton
Fr. Ebertstrasse
5060 Berg-Gladbach

HAHNE, E.
Institut für Thermodynamik und Wärmetechnik
Universität Stuttgart
Pfaffenwaldring 6
D-7000 Stuttgart

JANSEN, K.H.
Interaton
Internationale Atomreaktorbau GmbH
Friedrich Ebert-str.
5060 Berg Gladbach

JENSEN, H.W.
Fachhochschule Flensburg
Kanzleistr. 91-93
D-2390 Flensburg

LOCHOWITS, G.
Installa
Heizungs- und Installations GmbH
1, Eltenerstrasse
433 Muhlheim-Ruhr

LOTHAR, S.
Stiebel Eltron GmbH & Co.
Dr. Stiebelstrasse
3450 Holzminden

LOTTNER, V.
Kernforschungsanlage Jülich GmbH
Postfach 1913
5170 Julich

MÄRTIN, H.
Senator für Wissenschaft und Forschung
Bredtschneiderstrasse 5
D-1008 Berlin 8

MEINECKE, W.
Interatom, Internationale Atomreaktorbau GmbH
Postfach 5060
Bergisch Fladbach 1

MENKE, Ch.
Interatom
Oberstrasse 1
3300 Braunschweig

MROTZEK, H.
B.P. A.G.
Hamburg

PACK, M.
Happel GmbH & Co.
Südstrasse
4690 Herne 2

POSORSKI, R.
IKP-SOL KFA Julich Gmbh
Postfach 1913
5170 Julich

SCHOLZ, F.
Kernforschungsanlage Julich GmbH
Institut für Reaktorbauelemente
Postfach 1913
D-5170 Julich

SOYKA, M.
Rutgerswerke AG
Mainzer Landstrasse 217
D-6000 Frankfurt am Main

STOCKMEYER, R.
Institut für Festkörperforschung
der Kernforschungsanlage
Postfach 1913
D-5170 Julich

STORK, A.
Ingenieursburo Stork München
Weinbergerstrasse 39
D-8430 Neumarkt/Opf.

THEYSE, F.H.
Theyse Energieberatung
Habichtweg 1
D-5060 Bergisch Gladbach 1

WEIBLEN, R.D.
Installa
Heizungs- und Installations GmbH
1, Eltenerstrasse
433 Muhlheim Ruhr

WEIK, H.
Fachhochschule Lübeck
Stephansonstrasse 3
D-2400 Lubeck

ISRAEL   TABOR, H.
The Scientific Research Foundation
P.O. Box 3745
Jerusalem

ITALY   ARANOVITCH, E.
European Commission
Joint Research Centre ISPRA
21020 VARESE

ASTEGIANO, L.
University of Rome
Via Taddeo da Sessa 12
00165 Rome

BELLA   C. 1a
Rome

BENANNO ASSUNTE
Universita della
Calabria
Via Nationale
Calabria

FITTIPALDI, F.
Instituto di Fisica
Facolta di Ingegneria
Piazzale Techio
80125 Napoli

GAETANO, C.
CNR "Transformazione ed accumulo di energia"
Salita S. Lucia sopra Contesse 39
Pistunina - Messina

MANCINI, N.A.
Instituto di Fisica Dell'Universita
57, Corso Italia
Catania

DE MARIA, G.
Chimica Fisica
Università di Roma
P. le A.Moro 5
Rome

MASSIMO CONTI
C.I.R.A.E.S.
Universita della Calabria
Rende (Cosenza)

MORONI, E.
C.I.S.E. S.p.A.
Via Reggio Emillia 39
Segrate (Milano)

SIMONE, F.
Instituto di Fisica Dell'Universita
57, Corso Italia
95129 Catania

SORTA, E.
Assoreni
S.Donato Milanes
Milan

TURCO, E.
Italeco S.p.A.
Gruppo IRI Italstat
Via Arno 9A
00198 Rome

VELLONE, R.
Comitato Nazionale per l'Energia Nucleare (C.N.E.N.)
Via Anguillarese
00100 Rome

THE NETHERLANDS   BAARDMAN, M.
Energie Besparende Systemen EBS B.V.
Postbus 95
5360 AB GRAVE

BLOCK, H.G. de
Gemeentewerken Schiedam
Postbus 61
3100 AB SCHIEDAM

BOER, G. de
Gemeentewerken Schiedam
Postbus 61
3100 AB SCHIEDAM

BOGAARD, A.W.M. van den
Van den Bogaard B.V.
Postbus 131
4870 AC ETTEN LEUR

BOO, A. de
Psychologisch Laboratorium
Universiteit van Amsterdam
Weesperplein 8
1018 XA AMSTERDAM

BOTTRAM, A.M.M.
Technische Hogeschool
Vakgroep Systeem & Regeltechniek
TH-Gebouw W & S 01.30
EINDHOVEN

BRAAMS, P.
Grabowsky & Poort B.V.
Postbus 84319
2508 AH 's-GRAVENHAGE

BRINK, G.J. van den
Technical University Delft
Esdoornlaan 19
2923 EG KRIMPEN A/D IJSSEL

BRONSEMA, B.
Adviesgroep Ketel en van Scheindelen
Postbus 2886
2601 CW DELFT

BROUG, N.
Tebodin
Laan van N.O. Indië 25
's-GRAVENHAGE

BROUWER, G.
Raadg. Techn. Bureau van Heugten
St. Annastraat 145
6524 EP NIJMEGEN

BUSTRAAN, M.
Voorlopige Raad voor het Energieonderzoek
Postbus 20101
2500 EC 's-GRAVENHAGE

CROOCKEWIT, P.
Calcol B.V.
Westerlaan 1
3016 CK ROTTERDAM

CRIJNS, H.M.H.J.
N.V. Philips' Gloeilampenfabrieken
Li. D.E.C., EM3
Emmasingel
5600 MD EINDHOVEN

DARMADJI,
Wolter & Dros B.V.
Amsterdamseweg 53
3812 RP AMERSFOORT

DEKKERS, C.
Ubbink Nederland B.V.
Postbus 26
6988 AA DOESBURG

DEKKERS, F.J.M.
Technical University
Hoogstraat 71
EINDHOVEN

DELIL, A.A.M.
National Aerospace Laboratory N.L.R.
Postbus 153
8300 AD EMMELOORD

DUIKERS, S.
Verhaar Uniad
Technische Adviesbureaus B.V.
Stationsstraat 9
9711 AR GRONINGEN

ELPRAMA, R.
TH-Delft
Vakgroep Geotechniek
Stevinweg 1
DELFT

FRANCKEN, J.C.
Rijksuniversiteit Groningen
Vakgroep Technisch Fysica
GRONINGEN

GALEN, E. van
Institute of Applied Physics TNO-TH
Stieltjesweg 1
2628 CK DELFT

GEURDEN, J.M.G.
NEOM
Postbus 17
6130 AA SITTARD

GLIMMERVEEN, J.
TH-Delft
A. Fokkerstraat 66
3331 KB ZWIJNDRECHT

GOED, J.
Gemeentewerken Schiedam
Postbus 61
3100 AB SCHIEDAM

GOOTJES, J.H.
Ned. Energie Ontwikkelings Mij. Sittard
c/o Verloren Engl. 5
1261 CP BLARICUM

GRIJS, J.C. de
N.V. Philips Gloeilampenfabrieken
Afdeling Ontwikkeling Zonnekollectoren
Gebouw EB5
Emmasingel
EINDHOVEN

HAVINGA, J.
Laboratorium voor Technische Natuurkunde
University of Groningen
Nijenborgh 18
9747 AG GRONINGEN

HEEL, J.M. van
Bouwcentrum
Binnenmilieutechniek
Postbus 299
ROTTERDAM

HEIL, J.A.
ECN
Westerduinweg 3
1755 LE PETTEN

HEK, R.R. van
Architectenbureau BHB
Verrijn Stuartlaan 27
2248 EL RIJSWIJK

HENDAL,
Steenfabriek "De Hoogewaard"
Hoogewaard 21
6624 KP HEEREWAARDEN

HENDRIKS L.W.J.L.
Architect AB
Postbus 298
3300 AG DORDRECHT

HEIJNEN, W.J.
Delft Soil Mechanics Laboratory
Postbus 69
2600 AB DELFT

HOOGENDOORN, C.J.
Delft University of Technology
Lorentzweg 1
2628 CJ DELFT

HORSMAN, J.W.
Stork L.P.I.
Staringstraat 46
2150 AC NIEUW VENNEP

HULST L.P.D.M. van
Kema N.V.
Utrechtseweg 310
ARNHEM

JONG, A.C. de
Rijksuniversiteit Groningen
Chemische Laboratoria
Afd. VSS, kA312
Nijenborgh 16
9747 AG GRONINGEN

JONG, A.G. de
Grontmij. N.V.
Postbus 153
3700 AD ZEIST

JONG, W.A. de
Chairman of the Board of Directors
of the Netherlands Organization for
Applied Scientific Research TNO
Postbus 297
2501 BD DEN HAAG

JOON, K.
Netherlands Energy Research Foundation ECN
Postbus 1
1755 ZG PETTEN

KAMMINGA, W.
Laboratorium voor Technische
Natuurkunde, R.U. Groningen
Th. v.d. Dijkweg 14
9801 EL ZUIDHORN

KEMP, J.W.
TH-Eindhoven
Memlinestraat 9
EINDHOVEN

KETELAAR, J.A.A.
Markeloseweg 91
7491 PB RIJSSEN

KOPPEN, C.W.J. van
Eindhoven University of Technology
5600 MB EINDHOVEN

KORSTANJE, H.P.
Enka Research Institute Arnhem
Velperweg 76
6824 BM ARNHEM

KOUFFELD, R.W.J.
Installatietechniek Bredero B.V.
Afd. Ontwerp
Postbus 2419
3500 GK UTRECHT

KRISTINSSON, J.
Architect & Ingenieurs Bureau Kristinsson
Noordenbergsingel 10
7411 SE DEVENTER

LAMPHEN, H.
Gemeente Schiedam
Postbus 1501
3100 EA SCHIEDAM

LANGE, J.M.
IMAG
Dr. S.L. Mansholtlaan 12
6708 DA WAGENINGEN

LANGE, F. de
N.V. Philips' Gloeilampenfabrieken
Energy Systems, EH2, room 3
5600 MD EINDHOVEN

LENTZ, A.
Bronswerk Technische Import
Postbus 28
3800 AC AMERSFOORT

LOOGMAN, J.D.
Openbare Werken Amsterdam
Chopinstraat 37
2162 VR LISSE

LOON, P.M. van
N.V. Kema
Afd. WPB
Postbus 9035
6800 ET ARNHEM

LUINSTRA, J.
LGM
Groene Wetering
GOUDERAK

MEURS, G.A.M. van
Laboratorium voor Technische Natuurkunde
Technical University
Lorentzweg 1
2600 GA DELFT

MEYER VIOL, E.
Technical University
Ganzebloemstraat 33
5643 JM EINDHOVEN

MISCHGOFSKY, F.H.
LGM
Postbus 69
2600 AB DELFT

MULDER, G.
Laboratorium voor Technische Natuurkunde
University of Groningen
Nijenborgh 18
9747 AG GRONINGEN

OUDEN, C. den
Institute of Applied Physics TNO-TH
Stieltjesweg 1
2628 CK DELFT

PASSCHIER, G.
Passchier Vandensteen
Studio voor architektuur B.V.
Schiedamsedijk 77
3011 EM ROTTERDAM

POLLAERT, P.
Ing. Bureau Jongen-Laura B.V.
Scharnerweg 129
6224 JC MAASTRICHT

RADEMAKER, O.
Technische Hogeschool Eindhoven
Gebouw W & S
Kamer 01.33
Postbus 513
5600 MB EINDHOVEN

RENZEN, T.J.
Stichting Milieuwoningen
Zonnewende 10
2317 VW LEIDEN

RIENKS, C.G.
Verhaar Uniad
Technische Adviesbureaus B.V.
Stationsstraat 9
9711 AR GRONINGEN

ROEST, R.A.
Isolatie Dienst Nederland B.V.
Vaalserberg 86
2905 PS CAPELLE A/D IJSSEL

RON, A.J. de
Van Swaay Installaties B.V.
Postbus 220
2700 SE ZOETERMEER

RUITER, A.L. de
Stichting Milieuwoningen
Zonnewende 10
2317 VW LEIDEN

RIJN, A.H.J. van
N.V. Philips' Gloeilampenfabrieken
C.D. Light - Solar Collectors
Emmasingel - Bldg.EKp
EINDHOVEN

RIJNEN, A.F.
N.V. Noordelijke Ontwikkelingsmij.
Postbus 424
9700 AK GRONINGEN

ROMIJN, A.H.
Hollandsche Beton Groep N.V.
Postbus 81
2280 AB RIJSWIJK

SCHAAP,
Steenfabriek "De Hoogewaard"
Hoogewaard 21
6624 KP HEEREWAARDEN

SCHELLEMAN, F.
Ministerie van Economische Zaken
Postbus 20101
2500 EC DEN HAAG

SMEENK, G.
Energiebesparende Systemen
Postbus 95
5360 AB GRAVE

SMIDT, E.H.
Free University Amsterdam
Osdorperweg 489
1067 SR AMSTERDAM

SNIJDERS, A.
Bredasingel 54
ARNHEM

STRAATEN, J.A.B. van
Cetema B.V.
Postbus 19
5340 AA OSS

TEST, F.L.
Delft University of Technology
Dept. of Applied Physics
Lorentzweg 1
2628 CJ DELFT

THOLEN J.P.P.
Esmil International B.V.
Postbus 7811
1008 AA AMSTERDAM

THOMASSEN, J.
Energiebesparende Systemen
Postbus 95
5360 AB GRAVE

VELTKAMP, W.B.
Technische Hogeschool Eindhoven
Gebouw W & S 1.11
Postbus 513
5600 MB EINDHOVEN

VENHEELEN, J.
Katholieke Universiteit Leuven
Celestijnenlaan 300a
3030 HEVERLEE

VERDONSCHOT, J.K.M.
Institute of Applied Physics TNO-TH
Stieltjesweg 1
2628 CK DELFT

VERLOOP, J.
S.I.P.M.
Postbus 162
2501 AN DEN HAAG

VIEZEE, D.J.
Enka B.V.
Velperweg 76
6824 BM ARNHEM

VLESSING, L.
B.V. Groen Technisch Installatie
Kabelweg 25
1014 BA AMSTERDAM

VOLLMAR, R.
Goudenregenstraat 28
EINDHOVEN

VOORTER, P.H.C.
Red. Tijdschrift 'Energiebesparing'
Uitgeverij Ten Hagen
Postbus 34
2501 AG DEN HAAG

WESDORP, J.
Heineken Brouwerijen B.V.
Heineken Technisch Beheer B.V.
Postbus 510
2380 BB ZOETERWOUDE

WESSELINGH, A.I.
Shell Internationale Chemie Maatschappij
Carel van Bylandtlaan 30
2596 HR   DEN HAAG

WESTERBEEK VAN EERTEN, L.F.A.
Westburg Trading B.V.
Arnhemseweg 52
3811 NN   AMERSFOORT

WIT, M.H. de
TH Eindhoven
H.G. 10.75
Postbus 513
5600 MB   EINDHOVEN

WIJCKERHELD BISDOM, W.H.
Calcol B.V.
Westerlaan 1
3016 CK   ROTTERDAM

WIJSMAN, A.J.Th.M.
Institute of Applied Physics TNO-TH
Stieltjesweg 1
2628 CK   DELFT

ZUIDERWIJK, A.P.
Solargie B.V.
Hollewatering 1
2295 LV   KWINTSHEUL

ZIJDEVELD, C.
Chairman of the Municipality of Schiedam
Emmastraat 1
3111 GA   SCHIEDAM

NORWAY

GRØNVOLD, F.
University of Oslo
Department of Chemistry
Blindern, OSLO 3

MEISINGSET, K.K.
University of Oslo
Department of Chemistry
P.O. Box 1033
Blindern, OSLO 3

MONSEN, B.E.
Institute for Inorganic Chemistry
7034 NTH   TRONDHEIM

SWEDEN

ANDERSSON, Eng. S.
AIB - Allmänna Ingenjörsbyran A.B.
P.O. Box 5511
S-11485 STOCKHOLM

EFTRING, B.
Department of Mathematical Physics
Lunds Institute of Technology
P.O. Box 725
S-220 07 LUND

JOSEPHSON, T.
State Power Board
Jämlandgs. 99
16287 VALLINGBY

LAGERKVIST, K.O.
Statens Provningsanstalt
National Testing Institute
P.O. Box 857
S-501 15 BORAS

NILSSON, T.
Statens Provningsanstalt
National Testing Institute
P.O. Box 857
S-501 15 BORAS

OFVERHOLM, E.
Swedish Council for Buildings Research
Sankt Gorangsgaten 66
S-112 33 STOCKHOLM

PERMAN, G.
Sydkraft
21701 MALMO

RALDOW, W.M.
Department of Physical Chemistry
The Royal Institute of Technology
10044 STOCKHOLM

SJÖBLOM, C.A.
Physics Department
Chalmers University of Technology
S-412 96 GOTHENBURG

STYMNE, H.
Department of Physical Chemistry
Royal Institute of Technology
S-100 44 STOCKHOLM

WENNERHOLM, H.
Statens Provningsanstalt
(National Testing Institute)
P.O. Box 857
S-501 15 BORAS

SWITZERLAND

BURG, H.P. von
Scheller Ltd. Zürich
P.O. Box 688
Ch-8021 ZÜRICH

GILST, J. van
SORANE S.A.
52, Route du Chatelard 52
1018 LAUSSANNE

GUÉNAT, R.
Bureau Auto Routes (B.A.R.)
1772 Nierlet-les-Baiss
FRIBOURG

HEFEL, W.
Meyhall Chemical AG
Sonnenwiesenstrasse 18
CH-8280 KREUZLINGEN

NITTNER, E.
Meyhall Chemical AG
P.O. Box 862
CH-8280 KREUZLINGEN

UNITED KINGDOM

ALPER, B.
E.T.S.U.
Building 156
A.E.R.E. Harwell
Oxon OX11 ORA

BELL, M.A.
Cranfield Institute of Technology
Cranfield BEDFORD, MK43 OAL

BRINKWORTH, B.J.
University College Cardiff
Newport Road
Cardiff CF2 iTA

BURNS, A.P.
BHRA Fluid Engineering
Cranfield Bedford MK43 OAJ

COHEN, R.R.
Cranfield Institute of Technology
C.T.T. Cranfield, Beds

CREPAUX, A.
The British Petroleum Company Limited
BP Research Centre
Chertsey Road
Sunburry-on-Thames
Middlesex TW16 7LN

LEWIS, D.R.
Miller Homes Northern Ltd.
Miller House,
18 South Groathill Avenue
Graigleith, Edinburgh

MARSCHALL, R.
Solar Energy Unit of the Dept.
Mech. Engr.
U.C. Cardiff
Newport Road
Cardiff CF2 1TA

MORTIMER, J.V.
BP Trading Limited
BP Research Centre
Chertsey Road
Sunbury-on-Thames
Middlesex TW16 7LN

PAGE, J.K.R.
Calor Group Ltd.
Calor House, Windsor Road
Slough, Berkshire SL1 2EQ

PYE, D.B.
Shell Research Ltd.
Thornton Research Centre
P.O. Box 1
Chester CH1 3SH

STAMBOLIS, C.
Heliotechnic Assoc. Int.
5, Dryden Street
London WC2

SWAYNE, R.E.H.
Calor Group Ltd.
Calor House, Windsor Road
Slough, Berkshire SL1 2EQ

USA

CHIN FU TSANG
Lawrence Berkely Laboratory
University of California
1, Cyclotron Road
Berkely
California 94720

MICHAELS, A.I.
Argonne National Laboratory
9700, South Cass Avenue
Building 362
Argonne, Ill. 60439

YUGOSLAVIA   POPOVSKI, K.
SOZT ZZPK "MAKEDONIJA"
P.O. Box 326
91000 SKOPJE